AF594952

Obtaining and Characterization of New Materials

Obtaining and Characterization of New Materials

Editor

Andrei Victor Sandu

MDPI • Basel • Beijing • Wuhan • Barcelona • Belgrade • Manchester • Tokyo • Cluj • Tianjin

Editor
Andrei Victor Sandu
Faculty of Materials Science and
Engineering, "Gheorghe Asachi"
Technical University of Iasi
Romania

Editorial Office
MDPI
St. Alban-Anlage 66
4052 Basel, Switzerland

This is a reprint of articles from the Special Issue published online in the open access journal *Materials* (ISSN 1996-1944) (available at: https://www.mdpi.com/journal/materials/special_issues/Obtain_Charact_Mater).

For citation purposes, cite each article independently as indicated on the article page online and as indicated below:

LastName, A.A.; LastName, B.B.; LastName, C.C. Article Title. *Journal Name* **Year**, *Volume Number*, Page Range.

ISBN 978-3-0365-3092-5 (Hbk)
ISBN 978-3-0365-3093-2 (PDF)

Contents

About the Editor

Andrei Victor Sandu is an associate professor at the Faculty of Materials Science and Engineering, Technical University "Gheorghe Asachi" of Iaşi. He obtained his PhD in materials engineering in 2012 summa cum laude. He has published over 350 scientific articles, with over 350 indexed by SCOPUS and more than 300 indexed by ISI Web of Science. His H-index is 24. He is the co-author of 31 patents and other 9 patent applications (Romania, R. Moldova and Malaysia), and he has published 9 books, with 2 of them in the USA. He is the publishing editor for the International Journal of Conservation Science (Web of Science and Scopus indexed) and the European Journal of Materials Science and Engineering, and a reviewer for more than 15 Web of Science indexed journals. He is a visiting professor at Universiti Malaysia Perlis and the president of the Romanian Inventors Forum. Based on his expertise, he is also a senior researcher for the National Institute for Research and Development for Environmental Protection INCDPM and a representative for Romania at IFIA (International Federation of Inventors' Associations) and WIIPA (World Invention Intellectual Property Associations. http://afir.org.ro/sav/.

Editorial

Obtaining and Characterization of New Materials

Andrei Victor Sandu [1,2,3]

1 Faculty of Material Science and Engineering, Gheorghe Asachi Technical University of Iasi, 41 D. Mangeron St., 700050 Iasi, Romania; sav@tuiasi.ro
2 Romanian Inventors Forum, 3 Sf. Petru Movila St., L11, III/3, 700089 Iasi, Romania
3 National Institute for Research and Development for Environmental Protection INCDPM, 294 Splaiul Independentei, 060031 Bucharest, Romania

Citation: Sandu, A.V. Obtaining and Characterization of New Materials. *Materials* **2021**, *14*, 6606. https://doi.org/10.3390/ma14216606

Received: 16 October 2021
Accepted: 26 October 2021
Published: 3 November 2021

Publisher's Note: MDPI stays neutral with regard to jurisdictional claims in published maps and institutional affiliations.

The main objective of this Special Issue was to publish outstanding papers presenting cutting-edge research in the field of new materials and their understanding.

At present, more and more obtaining procedures and technologies are available next to advanced characterization techniques. The Special Issue managed to gather several outstanding articles in a broad field, from obtaining new materials to their understanding using the latest characterization techniques.

In this way, the focus was on new materials used in the water treatments: one study allowed the selection, on the basis of the caustic module, of ceramics with a high capacity for ionic exchange [1]; and another focused on new ecological materials for replacing cement (geopolymers and cementitious materials), highlighting the thermal stability of geopolymers in the 25–1000 °C temperature range through the use of thermogravimetric analysis, differential thermal analysis, and XRD [2]. Other studies on geopolymer composite showed that temperatures exceeding 400 °C accelerated the strength development, thus increasing the strength of the DFA composites [3], respectively. The use of embedded biofilm-resistant photoactivated TiO_2 nanoparticles at low concentrations in the cementitious composite matrix is an effective method to increase material durability and reduce maintenance costs [4]. All these published articles focused on materials which reduce the CO_2 footprint [2–4].

A very interesting field of research is that of insulating materials, concluding that the sheep wool has a comparable sound absorption performance to mineral wool or recycled polyurethane foam [5]. For the first time, three new parameters of integration efficiency of the thermal insert, thermal insulation efficiency parameters, and efficiency parameters of the integration of the textile material integrated into the clothing system were introduced; based on these parameters, it was possible to perform an effective and accurate comparative analysis of the thermal insulation of multi-layer thermal inserts in clothing [6].

An interesting approach looked at biomaterials and medical applications, concluding that structural material flaws account for a high percentage of the observed causes of failure, and identifying the breaking mechanisms macroscopic and microscopic investigations, including stereomicroscopy and scanning electron microscopy, is required—this can help us identify the causes that lead to the failure of implants [7]. A study on magnesium biocompatible alloys showed that the addition of 2–3 wt.%Y in the Mg-0.5Ca alloy improved both the biodegradability rate and cytocompatibility behavior [8]. Duceac et al. demonstrated that ceftriaxone-loaded chitosan nanoparticles can be used as a carrier in antibiotic delivery [9].

Seifert et al. presented research, which furthered the understanding of some mechanisms: co-sputtered or multilayered Ti-Al films with a thickness of 200 nm were deposited on thermally oxidized Si substrates. It was concluded that in order to realize a high temperature stability of γ-TiAl thin films, a contact to SiO_2 needs to be avoided by substituting the barrier with another material, e.g., AlN, or by using an additional protection layer. Also, a combined barrier layer consisting of 20 nm AlN and 20 nm SiO_2 was able to prevent the oxidation of the RuAl film on CTGS up to 800 °C in air and 900 °C in HV. On LGS, stable films are realized up to 600 °C in air and 900 °C in HV [10,11].

Zinc oxide films were produced by means of the electron beam evaporation method; the ones produced at room temperature consisted of ZnO and Zn crystallites, and their optical constants exhibit a metallic-like behaviour [12].

Hashim et al. [13] demonstrated the effect of Ni on the suppression of Sn whisker formation in a Sn-0.7Cu solder joint, concluding that a small amount of Ni addition (~500 ppm) was able to alter the microstructure of Cu_6Sn_5 to form a $(Cu,Ni)_6Sn_5$ IMC intermetallic layer, and it is very significant to the nucleation and growth of Sn whiskers.

The performance of Sn-3.0Ag-0.5Cu composite solder with kaolin geopolymer ceramic reinforcement on microstructure and mechanical properties under isothermal ageing, resulting in the morphology of interfacial IMC layer of non-reinforced SAC305 and SAC305-KGC composite solder joints, showed a duplex IMC structure comprises of scallop-type Cu_6Sn_5 and layer-type Cu_3Sn [14].

A study by Titu et al. [15] concluded that there is an economic advantage obtained when cutting semi-finished products of alloy steels by EDMCB, using the metal band as TO (tool).

A different approach was discussed in order to evaluate some medieval documents, a very important piece of our history [16]; the study revealed the morphological changes of parchment that occurred at various levels in the collagen fibrous mesh and established the state of conservation of the support, writing, and decorations, as well as the pigments involved.

All this published research will offer a new approach for further studies in order to create a sustainable society based on knowledge.

This issue is in memoriam Henriette Szilagyi, Cluj-Napoca Branch Director of the URBAN INCERC Institute, Romania.

Funding: This research received no external funding.

Data Availability Statement: Not applicable.

Acknowledgments: The Guest Editor of this Special Issue would like to thank all the authors from all over the world (Romania, Malaysia, Ivory Coast, Croatia, Poland, Germany), who contributed their valuable works to the accomplishment of the Special Issue. Special thanks are due to the reviewers for their constructive comments and thoughtful suggestions. Finally, the editor is grateful to the Materials Editorial Office, particularly Fannie Xu, for their kind assistance.

Conflicts of Interest: The author declares no conflict of interest.

References

1. Sandu, A.; Vasilache, V.; Sandu, I.; Sieliechi, J.; Kouame, I.; Matasaru, P.; Sandu, I. Characterization of the Acid-Base Character of Burned Clay Ceramics Used for Water Decontamination. *Materials* **2019**, *12*, 3836. [CrossRef] [PubMed]
2. Nergis, D.B.; Abdullah, M.; Sandu, A.; Vizureanu, P. XRD and TG-DTA Study of New Alkali Activated Materials Based on Fly Ash with Sand and Glass Powder. *Materials* **2020**, *13*, 343. [CrossRef] [PubMed]
3. Azimi, E.; Abdullah, M.; Vizureanu, P.; Salleh, M.; Sandu, A.; Chaiprapa, J.; Yoriya, S.; Hussin, K.; Aziz, I. Strength Development and Elemental Distribution of Dolomite/Fly Ash Geopolymer Composite under Elevated Temperature. *Materials* **2020**, *13*, 1015. [CrossRef] [PubMed]
4. Hegyi, A.; Lăzărescu, A.; Szilagyi, H.; Grebenişan, E.; Goia, J.; Mircea, A. Influence of TiO_2 Nanoparticles on the Resistance of Cementitious Composite Materials to the Action of Bacteria. *Materials* **2021**, *14*, 1074. [CrossRef] [PubMed]
5. Borlea (Mureşan), S.; Tiuc, A.; Nemeş, O.; Vermeşan, H.; Vasile, O. Innovative Use of Sheep Wool for Obtaining Materials with Improved Sound-Absorbing Properties. *Materials* **2020**, *13*, 694. [CrossRef] [PubMed]
6. Rogale, D.; Majstorović, G.; Rogale, S.F. Comparative Analysis of the Thermal Insulation of Multi-Layer Thermal Inserts in a Protective Jacket. *Materials* **2020**, *13*, 2672. [CrossRef] [PubMed]
7. Nica, M.; Cretu, B.; Ene, D.; Antoniac, I.; Gheorghita, D.; Ene, R. Failure Analysis of Retrieved Osteosynthesis Implants. *Materials* **2020**, *13*, 1201. [CrossRef] [PubMed]
8. Istrate, B.; Munteanu, C.; Lupescu, S.; Chelariu, R.; Vlad, M.; Vizureanu, P. Electrochemical Analysis and In Vitro Assay of Mg-0.5Ca-xY Biodegradable Alloys. *Materials* **2020**, *13*, 3082. [CrossRef] [PubMed]

9. Duceac, L.; Calin, G.; Eva, L.; Marcu, C.; Bogdan Goroftei, E.; Dabija, M.; Mitrea, G.; Luca, A.; Hanganu, E.; Gutu, C.; et al. Third-Generation Cephalosporin-Loaded Chitosan Used to Limit Microorganisms Resistance. *Materials* **2020**, *13*, 4792. [CrossRef] [PubMed]
10. Seifert, M. High Temperature Behavior of RuAl Thin Films on Piezoelectric CTGS and LGS Substrates. *Materials* **2020**, *13*, 1605. [CrossRef] [PubMed]
11. Seifert, M.; Lattner, E.; Menzel, S.; Oswald, S.; Gemming, T. Phase Formation and High-Temperature Stability of Very Thin Co-Sputtered Ti-Al and Multilayered Ti/Al Films on Thermally Oxidized Si Substrates. *Materials* **2020**, *13*, 2039. [CrossRef] [PubMed]
12. Skowronski, L.; Ciesielski, A.; Olszewska, A.; Szczesny, R.; Naparty, M.; Trzcinski, M.; Bukaluk, A. Microstructure and Optical Properties of E-Beam Evaporated Zinc Oxide Films—Effects of Decomposition and Surface Desorption. *Materials* **2020**, *13*, 3510. [CrossRef] [PubMed]
13. Hashim, A.; Salleh, M.; Sandu, A.; Ramli, M.; Yee, K.; Mohd Mokhtar, N.; Chaiprapa, J. Effect of Ni on the Suppression of Sn Whisker Formation in Sn-0.7Cu Solder Joint. *Materials* **2021**, *14*, 738. [CrossRef] [PubMed]
14. Zaimi, N.; Salleh, M.; Sandu, A.; Abdullah, M.; Saud, N.; Rahim, S.; Vizureanu, P.; Said, R.; Ramli, M. Performance of Sn-3.0Ag-0.5Cu Composite Solder with Kaolin Geopolymer Ceramic Reinforcement on Microstructure and Mechanical Properties under Isothermal Ageing. *Materials* **2021**, *14*, 776. [CrossRef] [PubMed]
15. Țîțu, A.; Vizureanu, P.; Țîțu, Ș.; Sandu, A.; Pop, A.; Bucur, V.; Ceocea, C.; Boroiu, A. Experimental Research on the Cutting of Metal Materials by Electrical Discharge Machining with Contact Breaking with Metal Band as Transfer Object. *Materials* **2020**, *13*, 5257. [CrossRef] [PubMed]
16. Haulică, M.; Florescu, O.; Vasilache, V.; Sandu, I. The Comparative Study of the State of Conservation of Two Medieval Documents on Parchment from Different Historical Periods. *Materials* **2020**, *13*, 4766. [CrossRef]

Article

Characterization of the Acid-Base Character of Burned Clay Ceramics Used for Water Decontamination

Andrei Victor Sandu [1,2,*], Viorica Vasilache [3], Ioan Gabriel Sandu [1,*], Joseph M. Sieliechi [4], Innocent Kouassi Kouame [5], Petre Daniel Matasaru [6] and Ion Sandu [3,7]

1 Faculty of Materials Science and Engineering, Gheorghe Asachi Technical University of lasi, Blvd. D. Mangeron 71, 700050 lasi, Romania
2 Center of Excellence Geopolymer & Green Technology (CeGeoGTech), School of Material Engineering, Universiti Malaysia Perlis (UniMAP), P. O. Box 77, d/a Pejabat Pos Besar, 01000 Kangar, Perlis Malaysia
3 Arheoinvest Centre, Institute of Interdisciplinary Research – Department of Science, Alexandru Ioan Cuza University, 11 Carol I Blvd, 700506 Iasi, Romania; viorica_18v@yahoo.com (V.V.); ion.sandu@uaic.ro (I.S.)
4 National School of Agro Industrial Sciences (ENSAI), University of Ngaoundere-Ngaoundere, PO.Box 455 00237 Ngaoundere, Cameroon; jsieliechi@yahoo.fr
5 Laboratory of Geosciences and Environment, UFR Science and Environmental Management, Nangui Abrogoua University, 28 BP 847 28 Abidjan, Ivory Coast; innocent_kouassi@yahoo.fr
6 Telecommunications and Information Technology, Faculty of Electronics, Gheorghe Asachi Technical University of lasi, Carol I 11A, 700506 lasi, Romania; dmatasaru@etti.tuiasi.ro
7 Romanian Inventors Forum, Str. Sf. P. Movila 3, 700089 Iasi, Romania
* Correspondence: sav@tuiasi.ro (A.V.S.); gisandu@yahoo.com (I.G.S.); Tel.: +40-745-438604 (A.V.S.)

Received: 28 October 2019; Accepted: 19 November 2019; Published: 21 November 2019

Abstract: The paper presents the results of ample investigations performed on industrial and traditional ceramics of fired clay used in processes of water potabilization in the last stage of filtration, after that of active charcoal. Using the data obtained through the scanning electron microscope coupled with energy dispersive X-ray analysis (SEM-EDX) and pH analyses, on the basis of the atomic composition and free concentration of hydronium ions, the normal caustic (Si/Al) and summative [(Si+Ti+FeIII+Cl)/(Al+Ca+Mg+Na+K)] modules were assessed, which were correlated with the free acidity and, respectively, the capacity of absorption and ionic exchange of the Fe^{3+} and Al^{3+} ions. The study allowed the selection, on the basis of the caustic module, of the ceramics with high capacity for ionic exchange.

Keywords: ceramics; acid-base character; caustic module; water treatment; SEM-EDX

---

1. Introduction

The capacity for ionic exchange is due to the acidic marginal structures of the type Si(IV)–O–H+ and of the hydroxide ones Al(III)–OH. The base-acidic activity of these groups from the structure of the fired-clay ceramics is due to the Si:Al stoichiometric ratio (known in practice as the caustic module, which varies from 1:1 to 4:1), but also to the position of the two coordination centers of the basic elementary cell. It is known that the basic units of fired-clay ceramics are tetrahedrons of silicate anions and octahedrons of aluminate anions, connected by piro-links in various stratified or coplanar two-dimensional and, respectively, three-dimensional structural forms, which at the surface, present hydroxide (HO–) groups, amphoteric, with double-linked oxygen and acidic structures (–O–H+). Generally, because of the different mineralogical nature of the base clay (composed of kaolinite, montmorillonite, chlorite, halloysite, illite, vermiculite, smectite, pyrophyllite, etc.) and of the firing technology, but also the foundations, plasticizers and other additives employed in the manufacturing, the acid-basic character, given by the marginal groups, varies widely, from weakly acidic (pH ≤ 4,5) to weakly alkaline (pH ≤ 9,5) [1–6]. An important role in the capacity for ionic exchange is that of the porosity and, respectively, the active surface area of the granulite's. The capacity for ionic exchange is, thus, directly proportional to the capacity of absorption and the active surface.

The number and distribution of groups capable of ionic exchange is given both by the mineralogical composition of the raw materials, the composition before the plastic and thermal processing, as well as the conditions of calcination, vitrification, and glazing, which have variable parameters: heating speed, firing temperature, firing duration, firing system (closed, open or semi-open), with a direct or hidden source, oxidative, or reducing, etc. [6–9].

The capacity for ionic exchange is due only to the acidic structures Si(IV)–$O^{-}H^{+}$, Ti(IV)–$O^{-}H^{+}$, and Fe(III)–$O^{-}H^{+}$. The ceramics with high concentrations of Al(III), Ca(II), and Mg(II) have a character that varies from amphoteric to weakly basic, while those with Si(IV), Ti(IV), and Fe(III) vary from amphoteric to acidic [10–16].

The main advantage of ceramic products used in treating and purifying water is due to the mechanic resistance, the acid-basic and redox stability, but also to the thermal and photochemical stability, as well as to an extensive series of very important characteristics, such as the apparent density or specific weight, under 1500 kg/m^3, and compression resistance, which varies between 50 and 200 daN/cm^2; the maximum quantity of absorbed water, which varies between 8% and 20%; the gelifraction or the phenomenon of mechanical deterioration (breakup) of the products saturated with water due to freezing and thawing, then their biodegradation and biodeterioration through biochemical and chemical processes, due to the specific loads of the treated waters. Likewise, another advantage of the industrial ceramics is given by the minimum concentration of soluble components in aqueous systems [17–29].

For this reason, the granules used in water treatment are beforehand subjected to cleaning processes by dispersion in containers with deionized—weakly acidic water in which they are lightly agitated for several tens of minutes, after which the water is decanted and the ceramics are dried with warm air [6].

The aim of this study is to assess the acid-basic character in correlation with the scanning electron microscopy coupled with energy dispersive X-Ray Spectroscopy (SEM-EDX), the free acidity, and, respectively, to assess the capacity for absorption and ionic exchange of the cations Fe^{3+} and Al^{3+}, of the fired-clay ceramics originating from industrial bricks and traditional pots. The study allowed the selection of the ceramics optimal for eliminating the traces of Fe^{3+} and Al^{3+}, of the taste and smell of the treated waters. The ceramics will be employed in the final filtration stage of the treatment of underground and surface waters in order to make them drinkable, producing quality water with impressive organoleptic characteristics.

2. Experimental

2.1. Ceramics Under Consideration

The research focused on two types of ceramics:

- *Industrial ceramics*, labelled CI, in the form of granules from freshly-fired bricks that were crushed;
- *Traditional ceramics*, labelled CT, in the form of granules obtained by crushing the neck of glazed pottery, freshly fired.

Three groups of samples were collected from the two sample groups, separated using a granulometric sieve with openings of 1.5–2.5 mm, 2.5–3.5 mm, and, respectively, 3.5–6.5 mm. The resulting samples were stored in plastic containers with screwed caps.

2.2. Processing and Analyzing the Samples

The composition of each lot was determined by collecting small chips weighing ca. 10–20g, with two parallel surfaces and indexed as following:

- CI_a, acidic industrial ceramics;

- CI_{sa}, weakly-acidic industrial ceramics;
- CI_{am}, amphoteric industrial ceramics;
- CI_{saHCl}, weakly-acidic industrial ceramics treated with HCl 3M;
- CI_{amHCl}, amphoteric industrial ceramics treated with HCl 3M;
- CT_a, acidic traditional ceramics;
- CT_{sa}, weakly-acidic traditional ceramics;
- CT_{amsa}, amphoteric to weakly-acidic traditional ceramics;
- CT_{am}, amphoteric traditional ceramics;
- CT_{amsb}, weakly-basic amphoteric traditional ceramics;
- CT_{amHCl}, amphoteric traditional ceramics treated with HCl 3M;
- $CT_{amsbHCl}$, weakly acidic amphoteric traditional ceramics treated with HCl 3M.

These samples were analyaed from the point of view of elemental chemical composition and internal structure by means of scanning electron microscopy coupled with X-ray spectrometry (SEM-EDX). The equipment used for this stage consisted of an SEM VEGA II LSH scanning electron microscope produced by Tescan (Brno, Czech Republic) and a Quantax QX2 EDX detector produced by Bruker/Roentec (Berlin, Germany).

For each sample, we determined the free acidity and, respectively, the pH of the solution resulting from dispersing 90 g of finely grounded (1.5–2.5 mm fraction separated using the granulometric sieve) material from each sample into 100 mL of twice-distilled water. The resulting data were correlated with the elemental composition provided by the EDX.

The free acidity was determined from the value of the pH when the granules with a diameter of 1.5–2.5 mm were dispersed in the twice-distilled water, namely:

$$[H^+] = 10^{-pH}, \text{ moles/L}.$$

2.3. *Evaluation of retention capacity*

To evaluate the capacity of retention of ions of Fe^{3+} and Al^{3+} for those two groups of ceramics three sets of samples (1.5–2.5mm; 2.5–3.5mm and 3.5–6.5mm granulometry) were involved. The granules were indexed as follow:

- CI_m, control industrial ceramic (untreated with aqueous solution of chloride acid for solubilization of labile components) having granulometry: CI_m(1.5–2.5); CI_m(2.5–3.5); CI_m(3.5–6.5);
- CI_{HCl}, industrial ceramic treated with aqueous solution of chloride acid 5M, having granulometry: CI_{HCl}(1.5–2.5); CI_{HCl}(2.5–3.5); CI_{HCl}(3.5–6.5);
- CT_m, control traditional ceramic (untreated), having granulometry: CT_m(1.5–2.5); CT_m(2.5–3.5); CT_m(3.5–6.5);
- CT_{HCl}, treated traditional ceramic (aqueous solution of HCl 5M), having granulometry CT_{HCl}(1.5–2.5); CT_{HCl}(2.5–3.5); CT_{HCl}(3.5–6.5).

The capacity of retention of ions of Fe^{3+} and Al^{3+} from aqueous solutions of 0.5 M was performed by dispersion in 100 mL water of 90 g ceramic granules, under weak agitation at room temperature for 20 min. Before dispersion, ceramic granules were washed in bi-distilled water, dried for 4 h at 105 °C in an oven. After 20 min of dynamic dispersion, residual iron quantity was determined by atomic absorption spectroscopy. The retaining capacity was calculated by the formula: $CR = (C_r/C_i)\times 100\%$, where CR is the retaining capacity (%), C_r—is the final concentration (mol/L), and C_i—initial concentration (mol/L).

3. Results and Discussions

3.1. Determining the Chemical Nature and the Physical Microstructures of the Ceramics

Figures 1 and 2 presents the SEM microphotographs of the 12 samples analyzed (five from industrial bricks and seven from traditional pots).

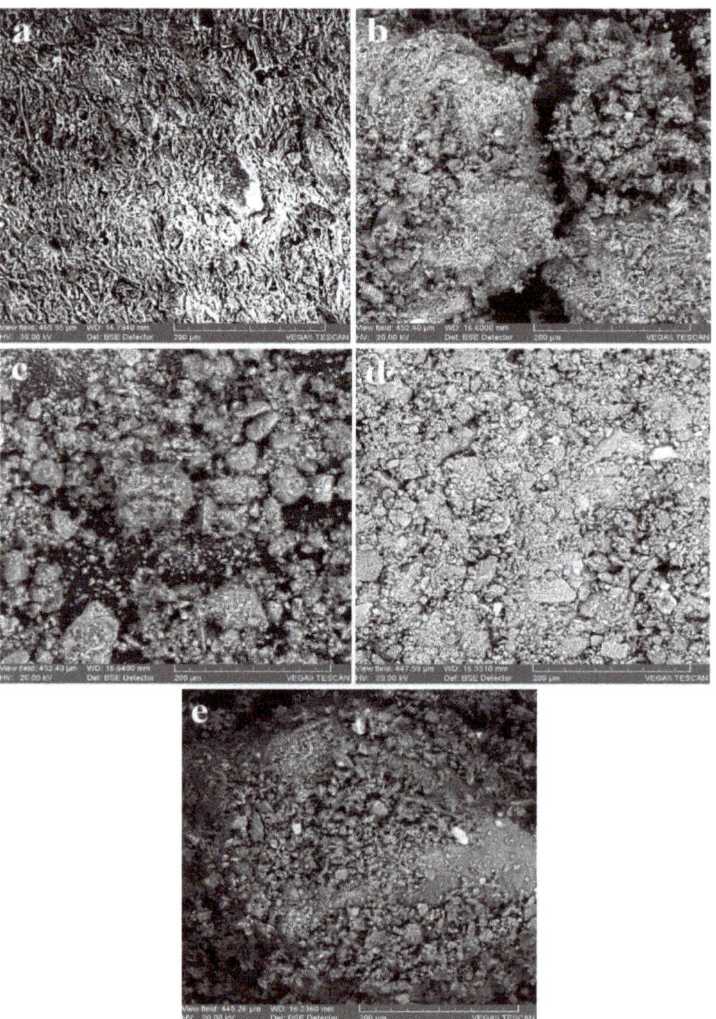

Figure 1. Scanning electron microscope images of the industrial ceramic samples: (**a**) CI_a acidic ceramic; (**b**) CI_{sa} weakly-acidic ceramic; (**c**) CI_{am} amphoteric ceramic; (**d**) CI_{saHCl} weakly-acidic ceramic treated with HCl 3M; (**e**) CI_{amHCl} amphoteric ceramic treated with HCl 3M

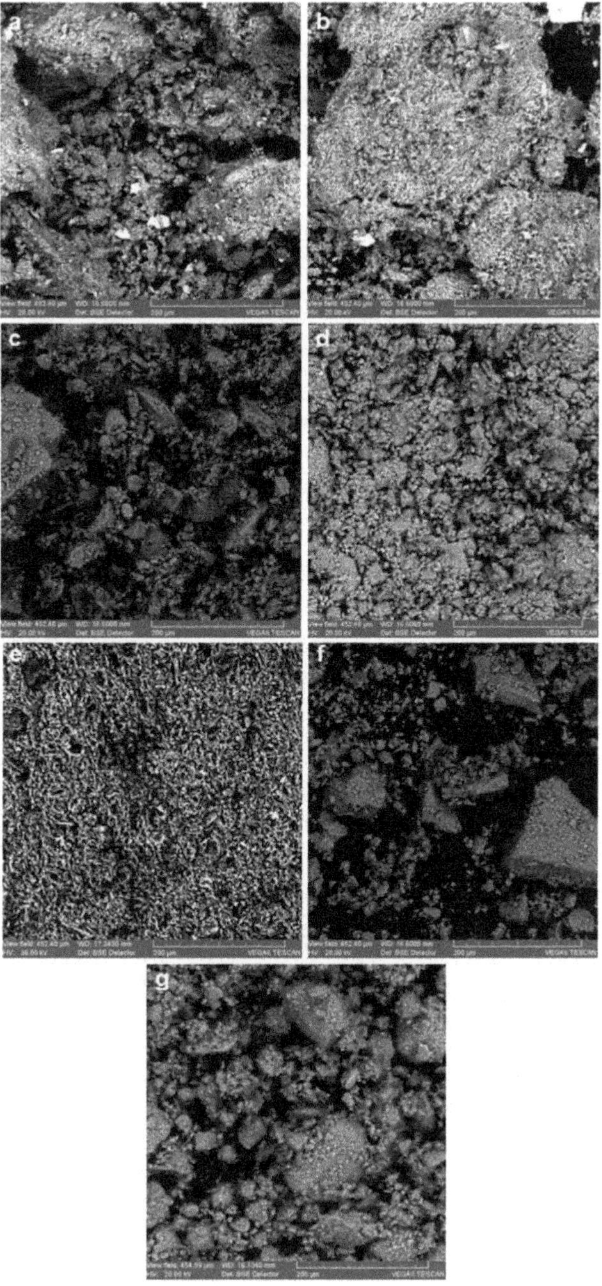

Figure 2. Scanning electron microscope images of the traditional ceramic samples (CT): (**a**) CT_a acidic ceramic; (**b**) CT_{sa} weakly-acidic ceramic; (**c**) CT_{amsa} amphoteric low-acidic ceramic; (**d**) CT_{amsb} amphoteric low-alkaline ceramic, (**e**) CT_{am} amphoteric ceramic, **f**) CT_{amHCl} amphoteric ceramic treated with HCl 3M, (**g**) $CT_{amsbHCl}$ amphoteric low-alkaline ceramic treated with HCl 3M.

The SEM analyzes allow, on the basis of the microphotographs from figures 1 and 2, assessing the structure and morphology of the sintering granules of the ceramic material.

The industrial ceramics display in the volume phase a finer granulation and the somewhat uniform distribution of the acidic alumino-silicate granulite's, whereas the traditional ceramics have a coarse and nonhomogeneous granulation, in which besides the acidic base structures there are often fragments of glazing and basic glasses.

Table 1 presents the elemental composition of the 12 ceramics under analysis.

Table 1. Chemical elemental composition of the studied ceramics.

Sample	Elemental Composition, Atomic Percentages (at %)									
	Si	Al	Fe	Ca	Mg	K	Na	Cl	Ti	O
CI_a	23.685	6.085	3.948	3.555	2.320	1.998	1.238	–	0.560	56.611
CI_{sa}	16.604	6.541	2.741	3.519	1.472	1.380	0.561	–	0.377	66.805
CI_{am}	14.183	13.123	0.895	0.165	0.411	0.517	0.298	–	0.610	69.798
CI_{saHCl}	15.058	5.688	2.190	1.724	1.092	1.207	0.433	0.230	0.347	72.031
CI_{amHCl}	14.720	10.703	1.162	0.110	0.244	0.817	0.406	0.107	0.494	71.237
CT_a	15.451	5.510	1.871	3.809	1.279	1.226	0.467	–	0.627	69.760
CT_{sa}	13.976	5.088	2.262	3.830	1.186	1.260	0.244	–	0.605	71.549
CT_{amsa}	17.354	10.667	2.299	0.569	0.721	0.780	0.521	–	0.496	66.593
CT_{am}	8.602	10.015	2.060	0.125	0.037	0.082	0.071	–	0.456	78.552
CT_{amsb}	20.780	8.424	2.187	5.755	1.700	2.394	0.505	–	0.522	57.733
CT_{amHCl}	17.477	11.084	2.266	–	0.564	0.766	0.094	0.189	0.529	67.031
$CT_{amsbHCl}$	14.909	13.180	0.744	0.170	0.279	0.473	0.066	0.178	0.590	69.410

The data from the energy dispersive X-ray (EDX) elemental analysis from Table 1 confirms the high compositional similarity of the industrial ceramics, explainable by the rigorously controlled dosing of the bisque or paste ingredients, whereas in the case of the traditional ceramics, the compositions vary widely for the same and between potters, who moreover used different clays and compositions for producing the bisque.

It must be recalled that initially, a much higher number of samples were advanced for these analyses, both from industrial ceramics (20) and traditional wares (30), from which only the most representative, in terms of chemical composition as ascertained from the SEM-EDX analyses, were selected.

The data from the EDX elementary analysis (Table 1) allow correlations between the value of the normal caustic module and that of the summative module and the free acidity. It must be stressed that in the case of the free acidity, an important role is played by the Cl^- ion, native, or induced from the HCl treatment.

The basic module of a ceramic structure that determines its acid-base behavior is given by the Si(IV)/Al(III) ratio. With respect to the three-dimensional structure of a ceramic, we speak of structures with *acidic functions* (titanium $[TiO_4]$ octahedrons or tetrahedrons), interspersing those of silicate $[SiO_4]$ and aluminate $[AlO_3]$, which are joined by the groups with *basic functions* (Ca–OH, Mg–OH, etc.) that co-act in the acid-base behavior. The exception is the marginal ions Cl^- and S^{2-}, which have dominant acidic function.

Table 2 presents the values of the normal caustic Si/Al modules, and the summative Si(Ti/Fe^{III}+Cl)/Al(Ca/Mg+Na/K) of the 12 types of ceramics, which are correlated with their free acidity.

3.2. The Free Acidity of the Ceramics

The free acidity of the samples listed above and, respectively, the pH of the solution resulting from dispersing 90 g of finely grounded (1.5–2.5 mm fraction separated using the granulometric sieve) material from each sample into 100 mL of twice-distilled water. The resulting data were correlated with the elemental composition provided by the EDX.

Table 2 presents the correlation between the normal caustic and summative modules, and the free acidity, expressed through the value of the pH obtained by dispersing the 1.5–2.5 mm granules into the twice-distilled water and, respectively, the $[H^+]$ concentration calculated by the formula:

$$[H^+] = 10^{-pH}, \text{ moles/L}.$$

Table 2. Correlation between the normal caustic and summative modules, and the free acidity.

Sample	Normal Caustic Module Si/Al	Summative Module Si(Ti/Fe^{III}+Cl)/ Al(Ca/Mg+Na/K)	pH	$[H^+]$ ($\times 10^{-5}$ mol/L)
CI_a	3.892	1.855	5.0	1.000
CI_{sa}	2.538	1.464	6.0	0.1000
CI_{am}	1.081	1.081	7.0	0.0100
CI_{saHCl}	2.647	1.757	4.7	1.9953
CI_{amHCl}	1.375	1.342	5.8	0.1585
CT_a	2.804	1.460	6.5	0.3162
CT_{sa}	2.747	1.451	6.2	0.0631
CT_{amsa}	1.627	1.520	5.2	0.3162
CT_{am}	0.859	1.076	7.0	0.0100
CT_{amsb}	2.467	1.251	7.1	0.0080
CT_{amHCl}	1.577	1.636	4.5	3.1623
$CT_{amsbHCl}$	1.131	1.159	7.5	0.0031

The resulting data show a high correlation between the normal caustic module and the summative one, obtained on the basis of the EDX atomic composition of the 12 types of ceramics. The data allowed sorting the 12 types of untreated and treated ceramics according to the decrease in the free acidity.

The present study allowed selecting the optimal ceramics for use in processes of chemo-absorption from the final stage of filtering ground and surface waters in order to make them drinkable, respectively, to obtain pure water with high organoleptic characteristics. In this regard, of interest are the ceramics with the caustic module between 1.1 and 2.8.

These ceramics have been featured in a new procedure for water potabilization using current city wastewater plants [6].

3.3. Retention Capacity of Fe^{3+} and Al^{3+} Ions

To evaluate the capacity of retention of ions Fe^{3+} and Al^{3+} of two types of considered granules, using only acid ceramics, CI_a and CT_a, after fine crushing the samples were separated in sieves having stitches of: 1.5–2.5mm; 2.5–3.5mm and 3.5–6.5mm, and indexed as follow:

- *Control industrial ceramic*, granulometry: CI_m(1.5–2.5); CI_m(2.5–3.5); CI_m(3.5–6.5);
- *Industrial ceramic treated* with solution of HCl 5M, granulometry CI_{HCl}(1.5–2.5); CI_{HCl}(2.5–3.5); CI_{HCl}(3.5–6.5);
- *Control traditional ceramic*, granulometry: CT_m(1.5–2.5); CT_m(2.5–3.5); CT_m(3.5–6.5);
- *Traditional ceramic treated* with solution of HCl 5M, granulometry CT_{HCl}(1.5–2.5); CT_{HCl}(2.5–3.5); CT_{HCl}(3.5–6.5).

Figures 3 and 4 present the retaining capacity of ions Fe^{3+} and Al^{3+} by percent of iron and aluminum changed by chemo-sorption for analyzed ceramics, function of three granulometric groups (for industrial ceramics CI and traditional ceramics CT), as such or treated with an aqueous solution of HCl 3 M.

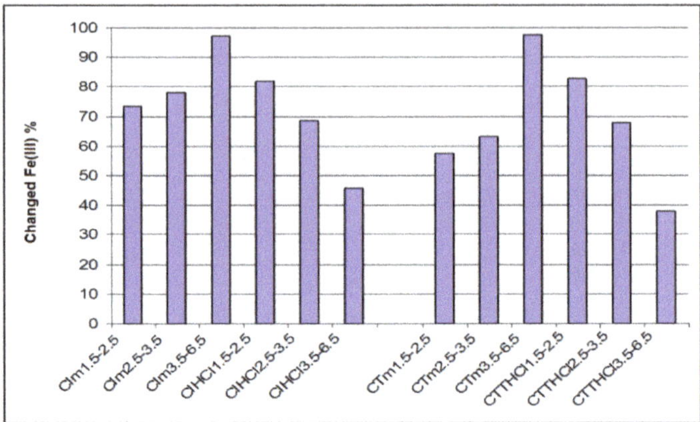

Figure 3. Iron percent changed through chemo-sorption in industrial and traditional ceramics treated with HCl and untreated.

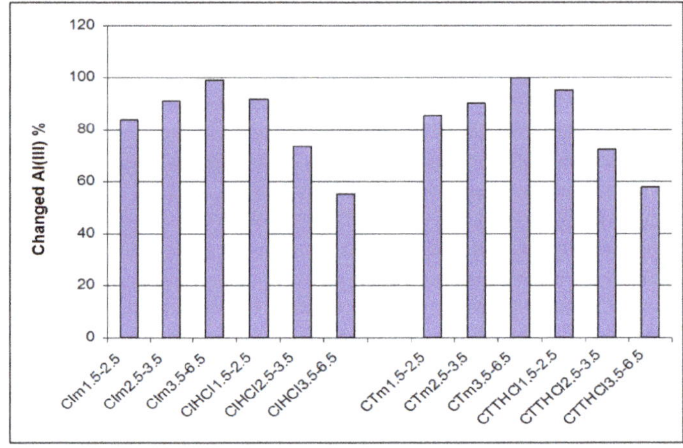

Figure 4. Aluminum percent changed through chemo-sorption in industrial and traditional ceramics treated with HCl and untreated.

Conforming of data from Figure 3, the change in iron percent of traditional ceramics treated with a solution of HCl 3 M decreased with increasing granulometry, so for granulometry of 1.5–2.5 mm, compared to 3.5–6.5 mm, the change in iron percent decreased from 81.9% to 45.6%. Tendency is inverted for the untreated traditional ceramics, where increases in iron percent were directly proportional with granulometry. So, for granulometry of 1.5–2.5 mm, comparing to 3.5–6.5mm, percent of the changed iron increase was proportional from 73.5% to 97.2%.

The same tendency was observed for industrial ceramic. Treated industrial ceramic presented for granulometry of 1.5–3.6 mm comparing to 3.5–6.5 mm a percent of changed iron in decreasing from 82.7% to 37.9%, while for the untreated one, the effect was inverted, the values of the change in iron percent increased from 57.5% to 97.5%.

Similar, the data from Figure 4 prove that for traditional ceramics treated with HCl 3 M, the percent of changed aluminum decreased when granulometry increased, so for granulometry of 1.5–2.5 mm, compared with 3.5–6.5 mm, the percent of changed aluminum decreased from 92.1% to 55.3%. The tendency is inverted for untreated traditional ceramic because the percent of changed aluminum increase was directly proportional with granulometry. So, for granulometry of 1.5–2.5 mm, compared to 3.5–6.5 mm, the percent of changed aluminum increased proportionally from 83.8% to 99.2%.

The same tendency was noted for industrial ceramic. So, treated industrial ceramic presented a granulometry of 1.5–3.6 mm compared to 3.5–6.5 mm, a decreasing percent of changed iron from 95.2% to 57.8%, while for those untreated, the effect was inverted, the percentage change in aluminum increased from 85.7% to 99.8%.

Initially, the capacity of ionic change for five industrial and seve traditional ceramics was studied, and it was observed that only those acids (CI_a and CT_a) presented a good reproducibility of the results. These ceramics were used for ionic change of other cations, considered as dangerous to surpass the limit levels of toxicity, like Hg^{2+}, Pb^{4+}, Pb^{3+}, Ni^{2+}, As^{3+}, etc., but of which results did not present reproducibility similar to Fe^{3+} and Al^{3+}. This fact is caused by the acid-base and redox action of ceramics in the caustic module, with polyhedron structures Si(IV), Ti(IV), and Fe(III) which confer a redoxite behavior with detoxification of some molecular structures and retention in potabilization process, as molecular chlorine reduced to chloride ion and a serial of cations (Hg^{2+}, Pb^{3+}, Pb^{4+}, As^{5+}, etc.), which are reduced to hardly soluble structures (Hg(I), Pb(II), As(III), etc.) retained in the ceramic. A cumulative effect was not be observed.

This domain is very interesting for science and technology; therefore, deeper research must be conducted. This study allowed the selection of optimum ceramics which could be used for chemiorption processes in the final stage of filtering and treating of ground and surface waters. These ceramics were the base to elaborate a new method of potabilization of waters [6].

The resulted data were re-validated using statistical, probabilistic, and numerical methods for analysis, keeping in mind future research involving IoT-based methods for the analysis.

4. Conclusions

As stated by the specialized literature, the capacity for ionic exchange of the normal fired-clay ceramics is due to the acidic structures of the type Si(Ti)–O^-H^+. The capacity for ionic exchange is influenced by the caustic module (Si/Al ratio), the porosity, and active surface of the ceramic granules. The number and distribution of grouping capable of ionic exchange is given, on the one hand, by the mineralogical composition of the raw materials, and the composition of the mixture before the plastic processing and the firing, and, on the other hand, by the conditions of calcination and vitrification, with their variable parameters—heating speed, firing temperature, firing duration, firing system (closed, open, semi-open), source type (direct or hidden), firing type (oxidative or reducing), etc.

The present study was carried out on two lots of freshly-produced ceramics, namely industrial bricks and traditional pots, in order to select the optimal type, from the acid-base point of view, for use in chemo-absorption processes during the final filtration stage of ground and surface water potabilization treatment, allowing the following conclusions.

The normal ceramic materials used in treating and purifying waters have, compared to other filtration materials, a series of advantages, including good mechanical resistance, acid-basic, and redox chemical stability, thermal and photochemical stability, but also an extensive series of very important characteristics for the purpose at hand, such as the apparent density or specific weight under 1500 Kg/m^3 and resistance to compression, which varies between 50 and 200 daN/cm^2; the maximum quantity of water absorbed (between 8% and 20%); gelifraction, or mechanical deterioration (breaking), and degradation of products saturated with water under the action of freezing and thawing and the biochemical and chemical processes, under the influence of the specific composition of the waters undergoing treatment, and last but not least, the low price.

Conforming with data from ionic change for ion Fe^{3+}, it was noted that for traditional ceramics treated with HCl 3 M, the percent of change in iron decreased when granulometry increased; This tendency is inverted for untreated traditional ceramic where the percentage of changed iron increased when granulometry increased.

The same tendency was noted for industrial ceramic. Treated industrial ceramic exhibit an decreasing of percent of changed iron when granulometry increase, while for the untreated ceramic, an increasing of percent of changed iron when granulometry increase;

Regarding the percent of changed aluminum, for traditional ceramics treated with HCl 3M, this decrease when granulometry increase, and the behavior inverted for untreated ceramics, when the percent of changed aluminum increase when granulometry increase; The same tendency was deduced for treated industrial ceramic, when the percent of changed aluminum decreased with increased granulometry, while for untreated industrial ceramics the percentage of changed aluminum increased when granulometry increased.

The study allowed selecting ceramics with optimum properties regarding the ionic change, able to be used in the final stage of filtering and treating the process of ground and surface waters to obtain fresh potable water. The better results exhibit acid granules notes as CI_a and CT_a;

The ceramics with the caustic module between 1.1 and 2.8 can be used for the ionic exchange of other cations that risk exceeding the threshold for toxicity.

As a general conclusion, these materials are very good to be used in water treatment due to their low cost, being waste materials or scraps from the technological flow of construction bricks/tiles or traditional ceramics. The resulting water has good organoleptic properties.

Author Contributions: writing and editing, formal analysis, A.V.S.; investigation, writing original draft, V.V.; investigation, reviewing I.G.S.; validation, J.M.S. and I.K.K.; data curation and visualization, P.D.M.; Conceptualization and supervision, I.S.

Funding: This research received no external funding.

Conflicts of Interest: The authors declare no conflict of interest.

References

1. Murray, H.H. *Applied Clay Mineralogy*; Elsevier: Amsterdam, The Netherlands, 2007.
2. Niculica, B.P.; Vasilache, V.; Boghian, D.; Sandu, I. An archaeometric Study of Several Ceramic Fragments from the Komariv (Komarow) Settlement of Adancata—Sub Padure, Suceva Conty, Fragments and Mineral Pigments from the Cucutenian Site of Tacuta—Dealul Miclea/Paic, Vaslui Conty. In *Third Arheoinvest Congress Programme and Abstracts, Proceedings of Third Arheoinvest Congress, Interdisciplinary Research in Archaeology, Iasi, Romania, 6–8 June 2013*; Cotiuga, V., Ed.; Alexandru Ioan Cuza University: Iasi, Romania, 2013; pp. 67–68.
3. Vasilache, V.; Filote, C.; Cretu, M.A.; Sandu, I.; Coisin, V.; Vasilache, T. Monitoring of groundwater quality in some vulnerable areas in Botosani County—Nitrates and nitrites based pollutants. In Proceedings of the 6th International Conference on Environmental Engineering and Management ICEEM 06, Balatonalmádi, Hungary, 1–4 September 2011.
4. Vasilache, V.; Sandu, I.; Vasilache, T. Studies Regarding the Quality of Waters for Some Rivers from Suceava County, Annals of "Dunarea de Jos" University of Galati, fasc. *Ix Met. Mater. Sci.* **2011**, *29*, 169–174.
5. Sandu, I.; ENitescu, E.; Calu, N.; Berdan, I.; Smocot, R.; Bialus, A.; Popa, G.; Vascu, V.; Stanila, A. Procedeu Şi Instalaţie Pentru Obţinerea Apei Potabile. Patent RO114318(B1), 30 March 1999.
6. Sandu, I.; Cretu, M.A.; Lupascu, T.; Vasilache, V.; Sandu, A.V.; Sandu, I.G.; Sieliechi, J.-M.; Kouame, I.K.; Kayem, J.G.; Sandu, A.V.; et al. Process for Water Treatment of Ground and Surface Waters. Patent MD 4298(B1), 31 October 2014.
7. Sentana, I.; Puche, R.D.S.; Sentana, E.; Prats, D. Reduction of chlorination byproducts in surface water using ceramic nanofiltration membranes. *Desalination* **2011**, *277*, 147–155. [CrossRef]
8. Koster van Gross, A.F.; Guggenheim, S. Dehydroxylation of Ca- and Mg-exchanged montmorillonite. *Am. Miner.* **1989**, *74*, 627–636.
9. Grim, R.E. *Clay Mineralogy*; McGraw-Hill Book Comp. Inc.: New York, NY, USA; Toronto, ON, Canada; London, UK; Sydney, Australia, 1968.
10. Martin, D.; Aparicio, P.; Galan, E. Time evolution of the mineral carbonation of ceramic bricks in a simulated pilot plant using a common clay as sealing material at superficial conditions. *Appl. Clay Sci.* **2019**, *180*, 105191. [CrossRef]

11. Beni, A.A.; Esmaeili, A. Design and optimization of a new reactor based on biofilm-ceramic for industrial wastewater treatment. *Environ. Pollut.* **2019**, *15*, 113298. [CrossRef]
12. Yasui, K.; Goto, S.; Kinoshita, H.; Kamiunten, S.; Yuji, T.; Okamura, Y.; Mungkung, N.; Sezaki, M. Ceramic waste glass fiber-reinforced plastic-containing filtering materials for turbid water treatment. *Environ. Earth Sci.* **2016**, *75*, 1135. [CrossRef]
13. Moga, I.C.; Ardelean, I.; Donu, O.G.; Moisescu, C.; Băran, N.; Petrescu, G.; Voicea, I. Materials and Technologies Used in Wastewater Treatment. *IOP Conf. Ser. Mater. Sci. Eng.* **2018**, *374*, 012079. [CrossRef]
14. Zheng, Y.; Yuan, Q.; Yan, Z.; Liu, J.; Liang, J.; He, L.; He, J.; Yan, M.; He, L. Effect of waste ceramic adsorbent on wastewater treatment. *MATEC Web Conf.* **2018**, *175*, 01010. [CrossRef]
15. Tociu, C.; Deák, G.; Maria, C.; Ivanov, A.A.; Ciobotaru, I.E.; Marcu, E.; Marinescu, F.; Cimpoeru, C.; Savin, I.; Constandache, A.C. Advanced treatment solutions intended for the reuse of livestock wastewater in agricultural applications. *IOP Conf. Ser. Mater. Sci. Eng.* **2019**, *572*, 012109. [CrossRef]
16. Olteanu, M.; Baraitaru, A.; Panait, A.-M.; Dumitru, D.; Boboc, M.; Deák, G. Advanced SiO2 Composite Materials for Heavy Metal Removal from Wastewater. *Water Air Soil Pollut.* **2019**, *230*, 179. [CrossRef]
17. Peterson, S. *The Craft and Art of Clay*; Harry N. Abrams: London, UK, 2003.
18. Aredes, S.; Klein, B.; Pawlik, M. The removal of arsenic from water using natural iron oxide minerals. *J. Clean. Prod.* **2012**, *29–30*, 208–213. [CrossRef]
19. Bociort, D.; Gherasimescu, C.; Berariu, R.; Butnaru, R.; Branzila, M.; Sandu, I. Comparative Studies on Making the Underground Raw Water Drinkable, by Coagulation-Flocculation and Adsorption on Granular Ferric Hydroxide Processes. *Rev. Chim.* **2012**, *63*, 1243–1248.
20. Bociort, D.; Gherasimescu, C.; Berariu, R.; Butnaru, R.; Branzila, M.; Sandu, I. Research on the Degree of Contamination of Surface and Groundwater used as Sources for Drinking Water. *Rev. Chim.* **2012**, *63*, 1152–1157.
21. Goldan, E.; Nedeff, V.; Barsan, N.; Mosnegutu, E.; Sandu, A.V.; Panainte, M. The effect of biochar mixed with compost on heavy metal concentrations in a greenhouse experiment and on Folsomia candida and Eisenia andrei in laboratory conditions. *Rev. Chim.* **2019**, *70*, 809–813.
22. Besnea, D.; Gheorghe, G.I.; Dontu, O.; Moraru, E.; Constantin, V.; Moga, I.C. Experimental researches regarding realization of wastewater treatment elements by means of modern technologies. *Int. J. Mechatron. Appl. Mech.* **2018**, *4*, 61–65.
23. Tataru, L.; Nedeff, V.; Barsan, N.; Sandu, A.V.; Mosnegutu, E.; Panainte-Lehadus, M.; Sandu, I. Applications of polymeric membranes ultrafiltration process on the retention of bentonite suspension. *Mater. Plast.* **2019**, *56*, 97–102.
24. Vu, D.H.; Wang, K.S.; Bac, B.H. Humidity control porous ceramics prepared from waste and porous materials. *Mater. Lett.* **2011**, *65*, 940–943. [CrossRef]
25. Chen, R.Z.; Lei, Z.F.; Yang, S.J.; Zhang, Z.Y.; Yang, Y.N.; Sugiura, N. Characterization and modification of porous ceramic sorbent for arsenate removal. *Colloids Surf. A Physicochem. Eng. Asp.* **2012**, *414*, 393–399. [CrossRef]
26. Jain, A.; Sharma, V.K.; Mbuya, O.S. Removal of arsenite by Fe(VI), Fe(VI)/Fe(III), and Fe(VI)/Al(III) salts: Effect of pH and anions. *J. Hazard. Mater.* **2009**, *169*, 339–344. [CrossRef]
27. Jeong, Y.; Fan, M.; Slingh, S. Evaluation of iron oxide and aluminium oxide as potential arsenic(V) adsorbents. *Chem. Eng. Process.* **2007**, *46*, 1030–1039. [CrossRef]
28. Keely, J.; Smith, A.D.; Judd, S.J.; Jarvis, P. Reuse of recovered coagulants in water treatments: An investigation on the effect coagulants purity has on treatment performance. *Sep. Purif. Technol.* **2014**, *131*, 69–78. [CrossRef]
29. Lee, S.J.; Kim, J.H. Differential natural organic matter fouling of ceramic versus polymeric ultrafiltration membranes. *Water Res.* **2014**, *48*, 43–51. [CrossRef] [PubMed]

Article

XRD and TG-DTA Study of New Alkali Activated Materials Based on Fly Ash with Sand and Glass Powder

Dumitru Doru Burduhos Nergis [1], Mohd Mustafa Al Bakri Abdullah [1], Andrei Victor Sandu [1,2,3,*] and Petrică Vizureanu [1,*]

1 Faculty of Materials Science and Engineering, "Gheorghe Asachi" Technical University, Blvd. D. Mangeron 71, 700050 Iasi, Romania; bunduc.doru@yahoo.com (D.D.B.N.); mustafaalbakri79@gmail.com (M.M.A.B.A.)

2 Romanian Inventors Forum, Str. Sf. P. Movila 3, 700089 Iasi, Romania

3 National Institute for Research and Development in Environmental Protection, 294 Splaiul Independenței Blv, 060031 Bucharest, Romania

* Correspondence: sav@tuiasi.ro (A.V.S.); peviz2002@yahoo.com (P.V.)

Received: 18 November 2019; Accepted: 8 January 2020; Published: 11 January 2020

Abstract: In this paper, the effect on thermal behavior and compounds mineralogy of replacing different percentages of fly ash with compact particles was studied. A total of 30% of fly ash was replaced with mass powder glass (PG), 70% with mass natural aggregates (S), and 85% with mass PG and S. According to this study, the obtained fly ash based geopolymer exhibits a 20% mass loss in the 25–300 °C temperature range due to the free or physically bound water removal. However, the mass loss is closely related to the particle percentage. Multiple endothermic peaks exhibit the dihydroxylation of β-FeOOH (goethite) at close to 320 °C, the $Ca(OH)_2$ (Portlandite) transformation to $CaCO_3$ (calcite) occurs at close to 490 °C, and $Al(OH)_3$ decomposition occurs at close to 570 °C. Moreover, above 600 °C, the curves show only very small peaks which may correspond to Ti or Mg hydroxides decomposition. Also, the X-ray diffraction (XRD) pattern confirms the presence of sodalite after fly ash alkaline activation, whose content highly depends on the compact particles percentage. These results highlight the thermal stability of geopolymers in the 25–1000 °C temperature range through the use of thermogravimetric analysis, differential thermal analysis, and XRD.

Keywords: geopolymers; fly ash; thermal behavior; Thermogravimetry-Differential Thermal Analysis (TG-DTA); XRD

---

1. Introduction

In recent years, strong technological development, the population increase, and the rapid development of the house-building industry in particular have led not only to a large lack of housing areas but also to high demand for building materials. The use of waste resulting from coal combustion in power plants offers two major advantages for this purpose: first, large tailings areas can be liberated by utilizing the waste, and second, a soil contaminant material can be converted into an advanced material with appropriate chemical and mechanical properties for engineering application through a geopolymerization process [1]. Geopolymers are inorganic materials, based on silica-alumina, which are chemically balanced by Group I oxides [2]. These are rigid gels, created under normal conditions of temperature and pressure, which can then be transformed into crystalline or glass-ceramic materials that are similar to zeolite materials [3]. A geopolymer, resulting from the exothermic process involving oligomers, is a very long reticular polymer with silicon groups (SiO_4) and a specific tetragonal network of aluminum oxide (AlO_4) [4]. The bonds between these tetragons are balanced by alkaline ions of K^+, Na^+, or Li^+ [4]. Any geopolymer can be divided into two main constituents: the base material and the activator (an alkaline liquid) [5]. The major constituent is the base material, which must be rich in silicon and aluminum and can be a natural mineral, such as clays, kaolin, etc. or alternatively can be a form of waste, such as fly ash, red mud, slag, etc. [1].

Due to their physical [6], mechanical [7], and chemical properties [8,9], geopolymers show high usefulness in multiple civil engineering applications [10,11] as a replacement material for conventional cement or ceramics [12–15]. Therefore, the thermal behavior and phase transition of fly ash based geopolymers during the exposed temperature range must be analyzed in order to evaluate the stability of their structure.

An additional advantage is the fact that the geopolymer microstructure contains multiple unreacted particles, which are continuously reacting with the extra-gel remained in the micropores [16,17]. As a result, some harmful cracks and pores could be repaired through the self-healing mechanism [18,19]. Obviously, this self-healing feature positively influences the time depending behavior of the composites due to its high durability. Despite the fact that geopolymers possess many chemical and mechanical properties and can be obtained through simple methods, most of them are obtained from natural minerals instead [20]. Therefore, it is essential to design, create, and characterize new geopolymers that use mineral waste as a source of raw materials, especially indigenous waste, and recyclized reinforcing particles. This is encouraged for both economic and environmental reasons, because through the geopolymerization reaction, we can obtain useful materials using "free" wastes that have negative effects on dumping areas [21].

There are multiple studies regarding the influence of different types of particles on the mechanical properties of geopolymers [17,22–26]. However, the presence of these particles will influence all the characteristics of the geopolymers, including their thermal behavior. The aim of this study is to evaluate the thermal behavior changes and the phase transition due to the introduction of different types of particles in new geopolymers based on indigenous fly ash.

2. Materials and Methods

Geopolymerization is a multiple-stage chemical reaction which occurs when a raw material rich in aluminum and silicon oxides is mixed with an alkaline solution of sodium silicate and sodium hydroxide [27,28]. Also, this reaction is mainly influenced by the raw material characteristics [29], activator concentration, and curing process (drying time and temperature) [30]. In this study, fly ash was used as the main raw material, and different percentages of glass powder and/or natural aggregates were introduced in the binder as reinforcement particles. Their chemical composition was analyzed by using X-ray fluorescence (XRF) involving XRF S8 Tiger equipment (Bruker, Karlsruhe, Germany).

2.1. Materials

Geopolymers consist of two main components: the liquid component (activator solution) and the solid component (the material rich in aluminum and silicon oxides and the reinforcing particles).

2.1.1. Indigenous Fly Ash

Fly ashes are, generally, solid torque spheres which result from coal combustion in power plants burning chambers [31]. This micrometric powder ends up being deposited in huge areas near many cities all over the world. Because different dumps present different chemical compositions, the activation solution, as well as the ratio between constituents, must be calculated. The performance of fly ash in geopolymers is strongly influenced by its physical, chemical, and mineralogical properties, and moreover, by its particle dimensions. While the mineralogical and chemical composition (Table 1) depends mainly on the coal composition, the particles can be ground or sifted (Figure 1).

Table 1. Indigenous fly ash oxide chemical composition.

Oxide	SiO_2	Al_2O_3	Fe_xO_y	CaO	K_2O	MgO	TiO_2	Na_2O	P_2O_5	Oth [1]
%, weight	47.80	28.60	10.20	6.40	2.40	2.00	1.30	0.60	0.40	0.30
Stat. error, %	0.32	0.27	0.95	0.77	0.71	1.09	1.81	0.63	0.24	-

[1] Sum of chemical elements lower than 0.1%.

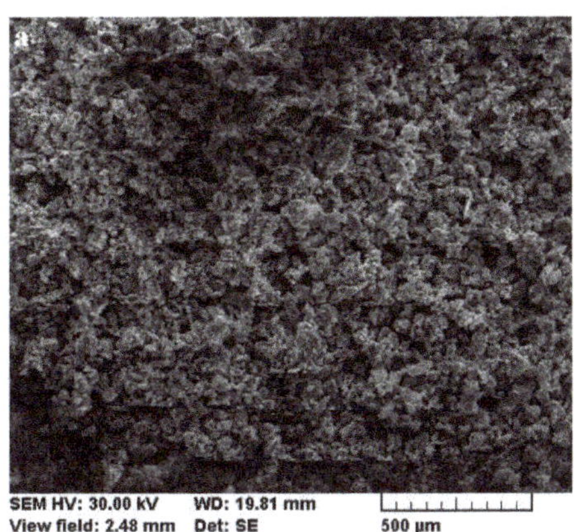

Figure 1. Scanning Electron Microscope (SEM) micrographs of indigenous fly ash after sifting: (**a**) 100X magnification; (**b**) 750X magnification.

In Romania, there are large areas covered by industrial waste from coal burning in the city's power plants. The indigenous thermal power plant ash used for the geopolymer tests comes from CET II—Holboca Iasi Romania ash dumps, which occupied an area of approximately 50 hectares in 2013 [32].

According to the Standard ASTM C618-92a, indigenous fly ash belongs to class F because it has a main oxides (silicon, aluminum and iron) sum that is higher than 70% (Equation (1)):

$$SiO_2 + Al_2O_3 + Fe_2O_3 = 47.8\% + 28.6\% + 10.2\% = 86.6\%. \tag{1}$$

2.1.2. Glass Powder

Another waste that appears in large quantities due to industrialization is glass. This inert material does not decompose naturally, producing negative effects on the environment following storage in landfills. Therefore, the use of glass particles in the manufacturing of environmentally friendly materials has become a worldwide concern. Due to the incorporation ability of geopolymer paste, the introduction of glass powder into the composition of these materials can be done using simple methods.

The glass powder (Figure 2) used as a reinforcing element in geopolymer samples contains only particles smaller than 100 μm (SR EN 933-1/2012) and is obtained by conducting glass waste grinding in a local factory.

By comparing the chemical composition (Table 2) with that of the thermal power plant ash (Table 1), the glass powder contained much higher SiO_2, CaO, and Na_2O, but much lower Al_2O_3. However, according to several studies [33], glass powder reacts in alkaline environments. Therefore, this should increase the geopolymerization rate.

Figure 2. Glass powder.

Table 2. Glass powder oxide chemical composition.

Oxide	SiO_2	Al_2O_3	Fe_xO_y	CaO	MgO	Na_2O	Oth [1]
%, weight	70–71	1.5–2	0.8–1	9–11	2–3	12–14	<0.1

[1] Sum of chemical elements lower than 0.1%.

By using glass particles for geopolymers manufacturing, two main advantages emerge: the first is related to waste recycling and second refers to improving mechanical properties by introducing particles with high mechanical properties.

2.1.3. Natural Aggregates

In order to improve the mechanical characteristics of geopolymers based on local powerplant ash, different quantities or types of aggregates can be added to the composition. Besides the use of waste, another category of reinforcing elements studied [25,34] worldwide is natural aggregates (sand). Depending on the geometric peculiarities of the particles, by introducing them in the geopolymer matrix, compressive strength increases of up to 150% can be obtained. The quantity and type of aggregate used is chosen according to the particle size distribution of the sand, because it may affect the homogeneity of the samples, but also their porosity.

The sand granulometric characteristics analysis conducted by using sifting (SR EN 933-1/2012) was performed after drying the aggregates, in order to reduce the measurement errors due to the fine particles sticking or adhesion to the sieve surface. According to the particle size distribution, close to 30% of particles had a diameter higher than 1.25 mm, and 50% (d50) had a diameter lower or equal to 0.19 mm. Therefore, the type of sand used belongs to the 0/4 aggregate class because all particles pass through the 4 mm mesh sieve (SR ISO 3310-3). The XRF analysis of sand indicated the following composition: 98.8% SiO_2, 0.57% Al_2O_3, 0.33% Fe_xO_y, and the rest being CaO, Na_2O, and other materials as traces.

2.1.4. Sodium Silicate

Sodium silicate is made by a sand (SiO_2) fusion with sodium or potassium carbonate (Na_2CO_3 or K_2CO_3) at temperatures above 1100 °C and dissolving the high-pressure vapor product in a semi-viscous liquid known as silicate. Silicate is rarely used as an independent activator because it does not have a sufficient activation capacity to initiate a geopolymerization reaction.

A commercially purchased high purity Na_2SiO_3 solution (Scharlab S.L., Barcelona, Spain) with a density of 1.37 g/cm^3 and a lower pH than 11.5 was used in this study.

2.1.5. Sodium Hydroxide

The NaOH solution concentration and molarity strongly influence the final properties of the geopolymers. The high concentrations of the NaOH solution result in high resistance to the early reaction stages. NaOH-activated geopolymers possess high crystallinity, having better stability in acidic or sulfate environments [35].

The NaOH solution was prepared at a 10-molar concentration by dissolving the high purity (99%) NaOH flakes in distilled water for 24 h before use (mixing).

2.1.6. Sample Preparation

The samples mixture was prepared according to the BS EN 196-1:1995 by means of a variable speed mixer. In order to increase the homogeneity of the samples, firstly, the solid component was poured into the mixer and stirred in a dry state for 4 min. Secondly, the liquid component was added gradually and mixed for 10 min until a homogeneous binder was obtained. The mix proportion of liquid and solid component of each sample are presented in Table 3, and the process flow diagram is shown in Figure 3.

Table 3. Samples components mix proportion.

Sample	Liquid Component, % Weight		Solid Component, % Weight		
	Na_2SiO_3	NaOH	Fly Ash	Glass Powder	Sand
100FA	60	40	100	0	0
70FA_30PG	60	40	70	30	0
30FA_70S	60	40	30	0	70
15FA_15PG_70S	60	40	15	15	70

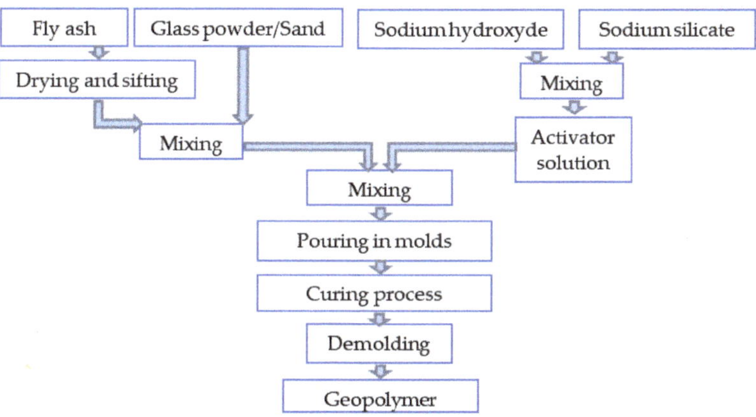

Figure 3. Process flow diagram.

Because the geopolymer characteristics and properties depend on multiple factors, it is essential to set the optimal parameters to be specific to the raw material, the activation solution, and the curing process. Therefore, the following parameters were used in this study:

- a raw materials relative humidity of 0;
- fly ash particles lower than 80 μm;
- glass powder particles lower than 100 μm;
- sand particles lower than 4 mm;
- curing temperature of 70 °C;

- curing time of 8 h.

During the geopolymerization process, minerals rich in aluminum and silicon pass through several phases. In the first phase, these are dissolved by the alkaline solution, forming a gel whose viscosity is given by the ratio between solid and liquid (Equation (2)). In the second phase, the reorganization of the molecules takes place, while water is removed and the material hardening begins.

$$\frac{\text{g of solid (powerplant ash)}}{\text{g of activating solution (sodium silicate + sodium hydroxide)}} = 1.5 \tag{2}$$

In order to evaluate the effect on thermal behavior and compounds mineralogy of replacing different percentages of fly ash with reinforcing elements, four types of geopolymers samples were obtained and studied.

2.2. Methods

Simultaneous thermal analysis consisting of thermogravimetric analysis (TGA) and Differential thermal analysis (DTA) was performed on the obtained samples in order to evaluate their thermal behavior. Because the DTA curve showed multiple peaks in the temperature range where the evaluation was made, X-ray diffraction (XRD) was performed to confirm the phase transition during heating.

2.2.1. Simultaneous Thermal Analysis

The sample's mass evolution by TGA was analyzed simultaneously with the phase transformations analysis using DTA by means of a STA PT-1600 equipment (Linseis, Selb, Germany). The analysis was performed in the 25–1000 °C temperature range, with a heating rate of 10 °C/min on samples and a mass lower than 50 mg, in a static air atmosphere.

Materials analysis conducted using TG-DTA emphasized their thermal stability and the content/type of volatile compounds through two curves simultaneously plotted based on the temperature.

2.2.2. X-ray Diffraction

X-ray diffraction (XRD) is a technique used to identify crystalline phases in different materials and for quantitative analysis of these phases. XRD is used, in particular, due to the superior highlighting of the three-dimensional atomic structure that directly influences the properties and characteristics of the materials. In order to analyze the mineralogical composition of the obtained geopolymers, an X'Pert Pro MPD equipment (Malvern Panalytical Ltd., Eindhoven, The Netherlands) equipped with a copper x-ray tube and a single channel detector was used. The diffractograms between the X-ray intensity on the ordinate and the Bragg angle, θ, on the abscissa were performed on a θ–2θ angle range between 5° and 90° through continuous scanning at a step size of 0.013° at every 60 s, with a scan speed of 0.054 (°/s) at a 45 KV voltage and 40 mA current intensity.

The mineralogical changes produced by the alkaline activator on the fly ash was analyzed on powder obtained by grinding the samples maintained in normal atmosphere conditions (clean air, ≤20 °C) for 90 days, and after being analyzed by TG-DTA.

3. Results and Discussion

3.1. Thermal Behavior Evaluation

The TG-DTA simultaneous thermal analysis was used to evaluate the thermal stability of geopolymers after replacing high percentages of fly ash with two types of particles. By monitoring the mass change during the heating of samples, the fraction of volatile compounds could be determined, so if the DTA curve is plotted at the same time, the mass change at specific temperatures could confirm the quantity of a specific compound.

The DTA curves of samples show multiple peaks at 123–130 °C, 185 °C, 232–240 °C, 312–358 °C, 490–497 °C, and 572–576 °C, respectively. These peaks correspond to the removal of water molecules, which are free or bound are with the structural compounds. In totally inorganic materials, such as geopolymers, water can be found in two main forms:

(i) hygroscopic (free) water, which is removed at rising temperatures up to 120 °C [36]. This water is absorbed into the structure due to the hygroscopicity of geopolymers [37].

(ii) strong physically bonded water which is removed in the 120–300 °C temperature range. This type of water can be divided into three sub-types:

- crystallization water (anionic and cationic or coordinative) which is removed from the structure in the 120–200 °C temperature range. This sub-type of water molecules are bonded in the structure during the formation of crystals from aqueous solution [38].
- water from hydrogels that can be intercrystalline and network types that interact with the crystallization water. This sub-type of water is removed during heating in the 180–300 °C temperature range [39].
- zeolitic water from cavities and channels, which is removed from the structure in the 200–300 °C temperature range [37,40].

When the temperatures exceed 300 °C, the (iii) chemically bound water starts being removed. The peaks on the DTA curve above this temperature corresponded to the decomposition of M (metal) and OH groups compounds [39,41,42]. These compounds exist in the fly-ash based geopolymers structure in different forms, such as:

- Acids: $M\text{-}O^{-}H^{+}$ (Si(IV), Ti(IV), Fe(III))
- Basics: $M^{+}HO^{-}$ hydroxide (Na, Ca (II), K, Mg (II))
- Neutral: M-OH hydroxyl (Al (III), Mn (III)).

The DTA curves (Figure 4a) of the analyzed samples showed an endothermic peak whose minimum was positioned at 123 °C for an 100FA sample, 115 °C for an 70FA_30PG sample, and 130 °C for 30FA_70S and 15FA_15PG_70S, respectively. The peak "A" corresponds to the overlapping of the removing of hygroscopic water evaporation and crystallization water removal [37]. By comparing the peaks broadening, it can be seen that by increasing the percentage of compact particles, the amount of water in these forms is lower. Because the used particles are compact bodies (Figure 5), the porosity of the sample can be related only with the percentage of fly ash. Therefore, high fly ash content ensures a highly porous structure which will increase the amount of absorbed water.

The "B" peaks which are in the temperature range of hydrogel water removal are higher in the case of the 100FA sample. This can be related to the hydrogel-forming capability of fly ash during geopolymerization [43].

Close to 230 °C, another peak, "C", appeared. During this endothermic reaction, the water molecules were removed from the calcium silicate hydrate (C-S-H), C-S-H with Al in its structure (C-A-S-H), and sodium aluminosilicate hydrate (N-A-S-H) channels and pores [44,45].

The "D" peaks corresponded to the iron oxides transition from FeO(OH) amorphous phase (Goethite) into the α-Fe_2O_3 (Hematite) crystalline phase (Equation (3)) [46–49]. The transformation reaction of Fe compounds occurred at around 300 °C but could be moved to higher temperatures due to the presence of silica and aluminum [50].

The "E" peaks represented an endothermic reaction in the 490–497 °C temperature range and corresponded to calcium hydroxide $Ca(OH)_2$ (Portlandite) decomposition following a reaction with carbon from the atmosphere, resulting in $CaCO_3$ and H_2 (Equation (4)) [51–53].

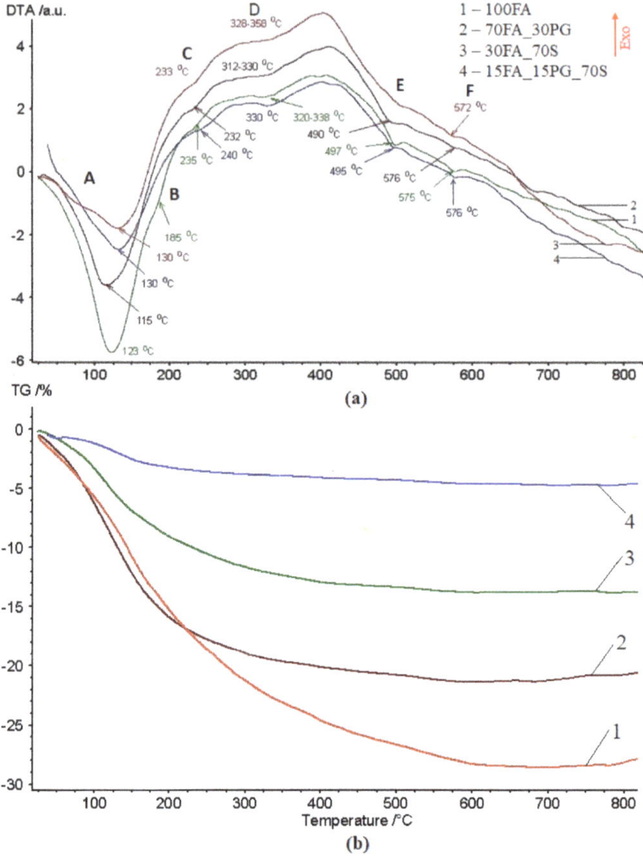

Figure 4. TG-DTA curves in the 22–820 °C temperature range: (**a**) DTA curves; (**b**) TG curves.

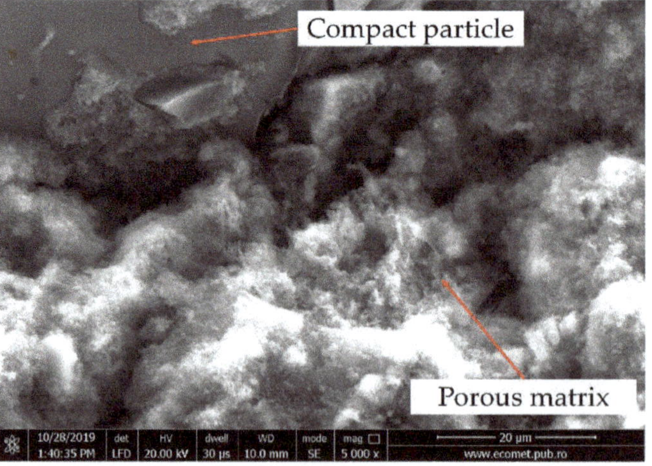

Figure 5. SEM micrographs of geopolymers with particles.

Also, at up to 570 °C, the "F" peaks which appeared on the DTA curve corresponded to the α-quartz to β-quartz conversion and the reaction between the unreacted particles and the activator caught in gel pores [54]. However, in the same temperature range, aluminum hydroxide, ($Al(OH)_3$) decomposition occured (Equation (5)) [55–58].

$$FeO(OH)(s) + 3H^+(aq) = Fe^{2+}(aq) + 2H_2O(aq) \tag{3}$$

$$Ca(OH)_2(s) + CO(g) \rightarrow CaCO_3(s) + H_2(aq) \tag{4}$$

$$2Al(OH)_3(s) \rightarrow Al_2O_3(s) + 3H_2O(aq) \tag{5}$$

In addition, in the same temperature range, the water resulting from the silicon or aluminum hydroxide groups condensation could appear. According to references [5,59], this chemical reaction consists of (Equation (6)):

$$\equiv M\text{-}OH + HO\text{-}M\equiv \rightarrow \equiv M\text{-}O\text{-}M\equiv + H_2O \tag{6}$$

As can be seen in Figure 6a, the obtained geopolymers presented large pores distributed on the entire analyzed surface. After introducing the aggregates, the large pores especially decreased in number (Figure 6b).

Figure 6. Optical micrographs of: (**a**) sample 100FA; (**b**) sample 30FA_70S.

The samples mass loss (Figure 4b) caused by hygroscopic water evaporation was close to 10% for the 100FA sample, 8% for 70FA_30PG, 5% for 30FA_70S, and 2% for 15FA_15PG_70S, respectively. The water absorbance capacity of samples was related to the calcium oxides and silica gel concentration. Up to the temperature when the hydrogels water is removed, the samples mass decreased to close to 16% in the case of 100FA and 70FA_30PG samples, while the samples with sand show lower than 10% mass reduction. However, up to 250 °C, the mass decreases reached close to 18% in the case of 100FA and 70FA_30PG samples, 12% in the case of 30FA_70S, and only 3% in the case of 15FA_15PG_70S.

Even if the percentage of mass loss up to this temperature is relatively high, because these types of water molecules are free or physically bonded, their influence on the mechanical properties is insignificant. However, if these materials are subjected to freeze-thaw cycles, cracks formation may occur due to the water (ice) from the expansion of the pores, which reduces the mechanical resistance of the geopolymers [60].

In the 360–700 °C temperature range, the mass loss is due to the removing of chemical bound water molecules. Therefore, an increase of between 460 °C and 515 °C corresponds to a 1% mass reduction of the 100FA sample and close to a 0.2% mass reduction of 15FA_15PG_70S sample, which is related to the CaOH decomposition. Also, in the $Al(OH)_3$ and FeO(OH) decomposition temperature range, the samples mass loss are lower than 1%.

Furthermore, above this temperature range, the DTA curves still show small peaks. These endothermic or exothermic reactions correspond to the decomposition of $CaCO_3$ at close to 750 °C [61]

(Equation (7)), $Ti(OH)_4$ close to 790 °C [62], or $Mg(OH)_2$ close to 670 °C [63]. Yet, these compounds exist only at the tracks level. Therefore, the effects on sample characteristics are low. At over 700 °C, a mass gain can be observed, which appears to be due to the oxidation of oxygen-poor iron species or pure iron [64].

$$CaCO_3(s) \rightarrow CaO(s) + CO_2(g) \quad (7)$$

The 100FA sample shows four peaks with the largest area. Therefore, the compounds that decompose in the analyzed temperature range come from the power plant fly ash particles.

3.2. Mineralogical Evaluation

The initial phases specific to the raw material and the transition in other phases that is specific to the zeolites are governed by the characteristics of the geopolymerization reaction. This transition is based on the raw material dissolution under alkaline conditions, resulting in reactive precursors of $Si(OH)_4$ and $Al(OH)_4$, and the polymerization and precipitation of the system, resulting in condensation of Si-O-Al molecules in various compounds.

The fly ash diffractogram (Figure 7) shows multiple peaks specific to the main chemical components oxides, such as Q—quartz, C—corundum, M—mullite or H—hematite, but also other more complex crystalline phases including calcium, titanium, or magnesium, such as A—anorthite, G—goethite, Al—albite, Ca—calcite, P—portlandite, Gb—gibbsite, magnesium hydroxide, etc. Both in the case of the raw material diffractogram and in the case of the obtained geopolymers, most of the peaks were positioned between 20° and 45° (2θ). Moreover, the peaks with the highest intensities specific to the quartz and corundum were positioned between 25° and 30° (2θ).

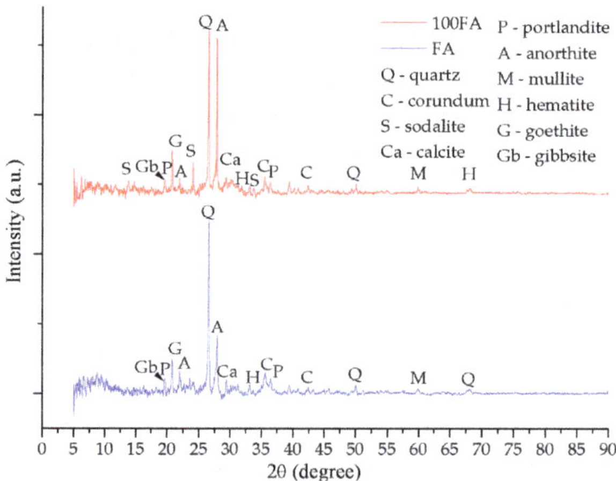

Figure 7. XRD patterns of fly ash powder and sample 100FA. (the Gb and P peaks with the highest intensity are overlapping).

The detected quartz or silicon dioxide is a mineral with a tetrahedral structure formed between silicon atoms and oxygen, crystallizing in the hexagonal system. Its concentration positively influences the mechanical properties of geopolymers, due to the quartz particles capacity of creating barriers for crack propagation.

The corundum detected crystallizes in the rhombohedral system, which is known as one of the main aluminum oxides. This compound is essential for geopolymers due to its hardness being close to that of diamond.

Mullite crystallizes in the orthorhombic system, and is a less commonly encountered compound that forms between aluminum, silicon, and oxygen. Due to its very high melting temperature, 1840 °C, the presence of this mineral produces the increase of the geopolymers refractivity.

Hematite crystallizes in the rhombohedral system due to being a compound of iron with oxygen. It has the same crystallographic structure as that of the corundum, being frequently encountered with it.

Augite crystallizes in the monoclinic system being a complex compound of calcium, magnesium, silica and oxygen. It has a stone-like structure and color, and is rarely encountered with a shiny surface.

The anorthite crystallizes in the triclinic (anorthic) crystalline system, as it has the richest calcium content in the group of plagioclase feldspars. It is found in several colors and consists of calcium, aluminum, silicon, oxygen, but also potassium, sodium, iron, and titanium at trace levels.

Sodalite crystallizes in the cubic system as a mineral complex formed by the reaction between sodium and chlorine with the main elements of the raw material (aluminum, silicon, and oxygen). The natural sodalite consists of an Al-O-Si network that encompasses Cl^+ cations, but the one resulting from geopolymerization shows inter-structural Na^+ cations, similar to zeolites [3].

Following the geopolymerization chemical reaction between the fly ash and the activation solution, the main phase specific to the raw material, the quartz, whose peak is positioned at 26.62°, 2θ, decreases in intensity as a result of the decrease of the glass phase, but there is a significant increase in the anorthite intensity, 28.03°, 2θ, while new peaks specific to the phases created as a result of the reaction between Na and the other compounds also appear.

The diffractogram specific to the sample 100FA (Figure 7) shows the formation of the most important phase specific to the geopolymerization, i.e., sodalite, which shows three peaks between 8° and 35°, with the highest intensity at 24.50°, 2θ. The appearance of such a phase specific to zeolites suggests the formation of a mesoporous material (contains small pores with a diameter between 20 and 50 nm) of semi-crystalline nature [65]. The sodalite content formed is directly proportional to the cation exchange capacity between the raw material and the activation solution. However, prior to and after activation, secondary phases, such as corundum with the highest intensity peak at 35.47°, 2θ, portlandite with the highest intensity peak at 36.48°, 2θ, mullite with the highest intensity peak at 60.76°, 2θ, hematite with the highest intensity peak at 33.69°, 2θ, goethite with the highest intensity peak at 21.09°, 2θ, and calcite were confirmed.

After replacing 30% of the fly ash quantity with glass particles, the sample 70FA_30PG results (Figure 8a) showed a decrease in intensity of the phases characteristic of the chemical reaction between the activation solution and ash.

To confirm the portlandite and goethite decomposition during heating, XRD analysis has been performed after TG-DTA tests. Therefore, the heated samples diffractograms—Figures 8b, 9b and 10b—shows high peaks intensity modifications, especially for calcite and hematite, due to the chemical reactions (Equations (4) and (5)) follow the portlandite and goethite decomposition. Moreover, because the geopolymerization continues during heating, anorthite reacts with the Na^+ cations, creating new phase albite with the highest intensity peak at 27.85°, 2θ.

These type of geopolymers have multiple phases which present similar XRD patterns, and, according to the database, one peak corresponded to multiple phases in this study we have presented, which was the phase with the highest intensity on the diffractograms. Therefore, after STA analysis, some peaks were changed to other phases because the phase intensity is different, e.g., the highest intensity peak at 21.09°, 2θ prior STA corresponds to goethite, yet, after STA it appears at 20.85°, 2θ and corresponds to quartz.

Following the replacement of 70% of the fly ash powder specific to the 100FA sample with sand particles, it was found that the intensity of some phases increased exponentially (Figure 9) while the specific phases of activation decrease significantly as a result of the reduction of the Al content available in the system.

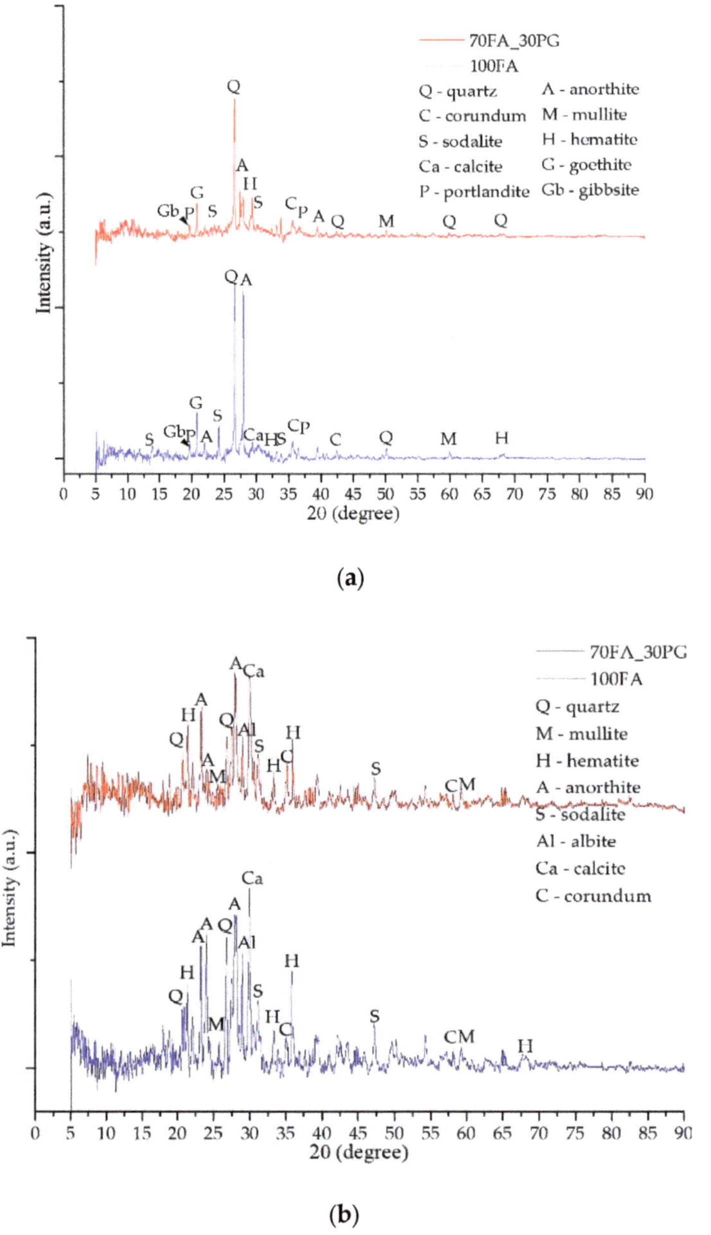

Figure 8. XRD patterns of sample 100FA and sample 70FA_30PG: (**a**) prior TG-DTA analysis; (**b**) after TG-DTA analysis.

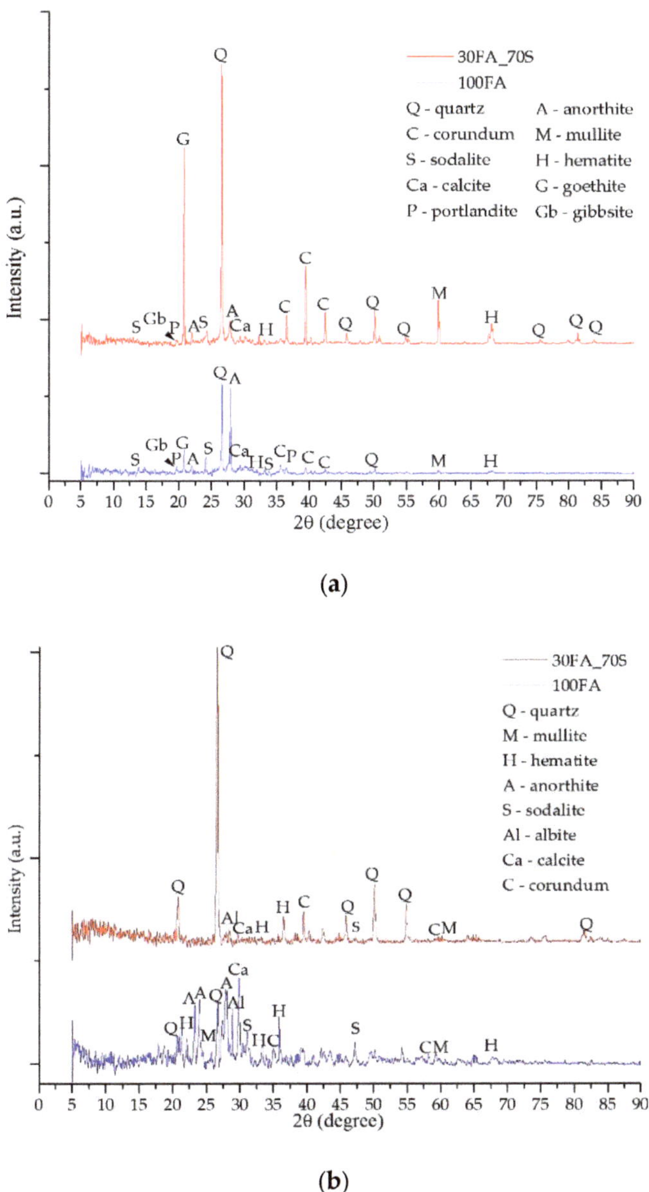

Figure 9. XRD patterns of sample 100FA and sample 30FA_70S: (**a**) prior TG-DTA analysis; (**b**) after TG-DTA analysis.

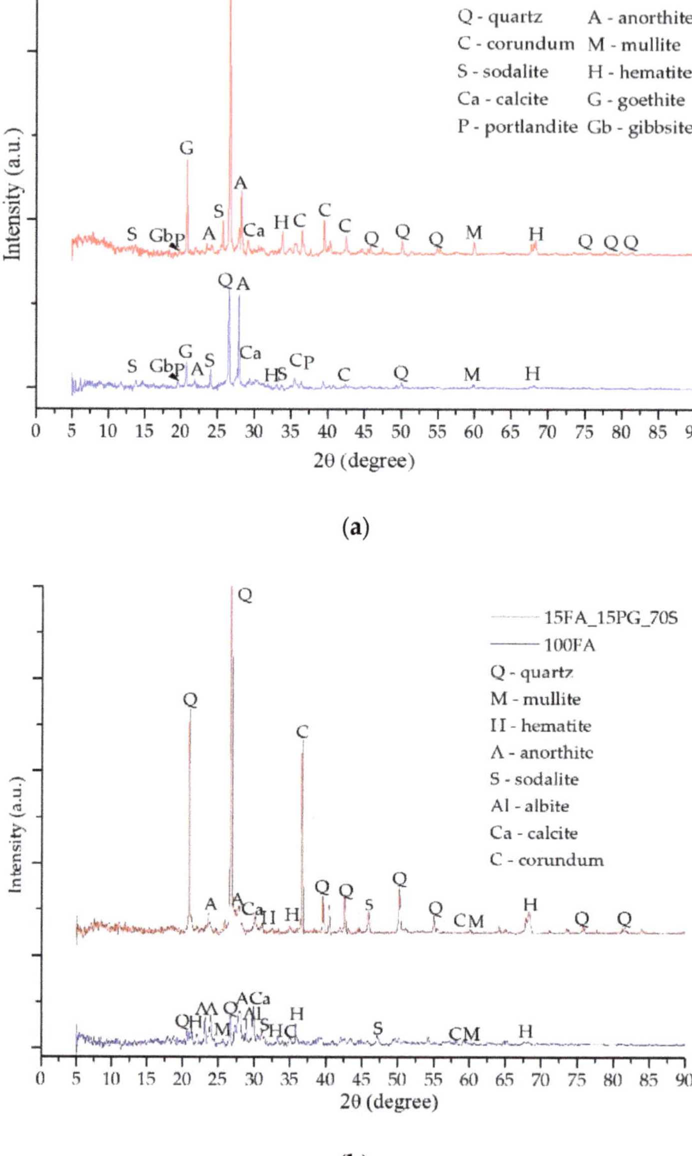

Figure 10. XRD patterns of sample 100FA and sample 15FA_15PG_70S: (**a**) prior TG-DTA analysis; (**b**) after TG-DTA analysis.

The XRD diffractogram of the sample 15FA_15PG_70S (Figure 10b) shows a decrease in peaks intensity specific to the aluminum-containing compounds due to the sample ash content reduction.

Because the differences between the diffractograms specific to the raw material and those of the geopolymer samples are small, we can consider that the geopolymers have a granular structure whose surface is covered by phases resulting from the geopolymerization, which helps to bind them, resulting in a semi-crystalline structure.

The diffractogram of the thermal power plant fly ash shows mainly phases specific to the compounds of Al and Si, and at the trace level Fe, Ca, or Ti compounds can be observed. Following activation, an additional phase specific to zeolites appeared known as sodalite, while the initial phases showed minimal changes.

4. Conclusions

In the analyzed temperature range, the fly ash-based geopolymers exhibited high mass loss due to the removal of free and physically bound water molecules at up to 300 °C. Above this temperature multiple compounds, such as goethite, portlandite, gibbsite, etc. decomposed due to the OH groups (chemically bound water) removed. Yet, it was observed that the mass lose percentage depends on the sample fly ash content. Therefore, the hygroscopicity, as well as the concentration of unstable compounds, are strongly related to the matrix structure.

Taking into account the high number of chemical elements and so many possibilities of compounds formation during heating, such phase transition and structural modifications will result in geopolymers characteristics changing. Therefore, in order to obtain a much thermally stable geopolymer, based on indigenous fly ash, a high percentage of natural aggregates should be introduced in the matrix.

The glass powder introduction in the matrix will result in a denser sample, yet, due to the high calcium content, the thermal stability at high temperatures decreases.

Author Contributions: Writing original draft and investigation, D.D.B.N.; data curation and validation, M.M.A.B.A.; Validation and writing—reviewing and editing, A.V.S.; Validation and writing, project administration and scientific supervision, P.V. All authors have read and agreed to the published version of the manuscript.

Funding: This research received no external funding.

Conflicts of Interest: The authors declare no conflict of interest.

References

1. Nergis, D.D.B.; Abdullah, M.M.A.B.; Vizureanu, P.; Tahir, M.F.M. Geopolymers and Their Uses: Review. *IOP Conf. Ser. Mater. Sci. Eng.* **2018**, *374*, 12019. [CrossRef]
2. Vogt, O.; Ukrainczyk, N.; Ballschmiede, C.; Koenders, E. Reactivity and Microstructure of Metakaolin Based Geopolymers: Effect of Fly Ash and Liquid/Solid Contents. *Materials* **2019**, *12*, 3485. [CrossRef] [PubMed]
3. Papa, E.; Medri, V.; Amari, S.; Manaud, J.; Benito, P.; Vaccari, A.; Landi, E. Zeolite-geopolymer composite materials: Production and characterization. *J. Clean. Prod.* **2018**, *171*, 76–84. [CrossRef]
4. North, M.R.; Swaddle, T.W. Kinetics of Silicate Exchange in Alkaline Aluminosilicate Solutions. *Inorg. Chem.* **2000**, *39*, 2661–2665. [CrossRef] [PubMed]
5. Duxson, P.; Fernández-Jiménez, A.; Provis, J.L.; Lukey, G.C.; Palomo, A.; van Deventer, J.S.J. Geopolymer technology: The current state of the art. *J. Mater. Sci.* **2006**, *42*, 2917–2933. [CrossRef]
6. Farhana, Z.F.; Kamarudin, H.; Rahmat, A.; Al Bakri, A.M.M. A Study on Relationship between Porosity and Compressive Strength for Geopolymer Paste. *Key Eng. Mater.* **2013**, *594–595*, 1112–1116. [CrossRef]
7. Ding, Y.; Dai, J.-G.; Shi, C.-J. Mechanical properties of alkali-activated concrete: A state-of-the-art review. *Constr. Build. Mater.* **2016**, *127*, 68–79. [CrossRef]
8. Rowles, M.; O'Connor, B. Chemical optimisation of the compressive strength of aluminosilicate geopolymers synthesised by sodium silicate activation of metakaolinite. *J. Mater. Chem.* **2003**, *13*, 1161–1165. [CrossRef]
9. Ukrainczyk, N.; Muthu, M.; Vogt, O.; Koenders, E. Geopolymer, Calcium Aluminate, and Portland Cement-Based Mortars: Comparing Degradation Using Acetic Acid. *Materials* **2019**, *12*, 3115. [CrossRef]
10. Rangan, B.V. Design, Properties, and Applications of Low-Calcium Fly Ash-Based Geopolymer Concrete. In *Developments in Porous, Biological and Geopolymer Ceramics*; John Wiley & Sons, Inc.: Hoboken, NJ, USA, 2007; pp. 347–362.
11. Ismail, N.; El-Hassan, H. Development and Characterization of Fly AshSlag Blended Geopolymer Mortar and Lightweight Concrete. *J. Mater. Civ. Eng.* **2018**, *30*, 4018029. [CrossRef]

12. Reig, L.; Soriano, L.; Borrachero, M.V.; Monzó, J.; Payá, J. Influence of calcium aluminate cement (CAC) on alkaline activation of red clay brick waste (RCBW). *Cem. Concr. Compos.* **2016**, *65*, 177–185. [CrossRef]
13. Scrivener, K.L.; Capmas, A. Calcium Aluminate Cements. In *Leas Chemistry of Cement and Concrete*; Elsevier: Amsterdam, The Netherlands, 1998; pp. 713–782.
14. Vafaei, M.; Allahverdi, A. Influence of calcium aluminate cement on geopolymerization of natural pozzolan. *Constr. Build. Mater.* **2016**, *114*, 290–296. [CrossRef]
15. Nath, P.; Sarker, P.K. Use of OPC to improve setting and early strength properties of low calcium fly ash geopolymer concrete cured at room temperature. *Cem. Concr. Compos.* **2015**, *55*, 205–214. [CrossRef]
16. Azmi, A.A.; Abdullah, M.M.A.B.; Ghazali, C.M.R.; Sandu, A.V.; Hussin, K.; Sumarto, D.A. A Review on Fly Ash Based Geopolymer Rubberized Concrete. *Key Eng. Mater.* **2016**, *700*, 183–196. [CrossRef]
17. Burduhos Nergis, D.D.; Vizureanu, P.; Corbu, O. Synthesis and characteristics of local fly ash based geopolymers mixed with natural aggregates. *Rev. Chim.* **2019**, *70*, 1262–1267.
18. Ma, H.; Qian, S.; Li, V.C. Influence of Fly Ash Type on Mechanical Properties and Self-Healing Behavior of Engineered Cementitious Composite (ECC). In Proceedings of the 9th International Conference on Fracture Mechanics of Concrete and Concrete Structures, Berkeley, CA, USA, 28 May–1 June 2016.
19. Ibrahim, W.M.W.; Hussin, K.; Abdullah, M.M.A.B.; Kadir, A.A.; Binhussain, M. A Review of Fly Ash-Based Geopolymer Lightweight Bricks. *Appl. Mech. Mater.* **2015**, *754–755*, 452–456. [CrossRef]
20. Davidovits, J. Geopolymers Based on Natural and Synthetic Metakaolin a Critical Review. In Proceedings of the 41st International Conference on Advanced Ceramics and Composites, Daytona Beach, FL, USA, 22–27 January 2017; pp. 201–214.
21. Available online: http://www.anpm.ro/documents/839616/33839940/Raport_Amplasament_CET2+Holboca+Rev.+Sep2017.pdf/7931f8ce-1e21-4af7-9abf-a16c602acc5b (accessed on 12 December 2019).
22. Zhang, S.; Keulen, A.; Arbi, K.; Ye, G. Waste glass as partial mineral precursor in alkali-activated slag/fly ash system. *Cement Concrete Res.* **2017**, *102*, 29–40. [CrossRef]
23. Shaikh, F.U.A. Mechanical and durability properties of fly ash geopolymer concrete containing recycled coarse aggregates. *Int. J. Sustain. Built Environ.* **2016**, *5*, 277–287. [CrossRef]
24. Embong, R.; Kusbiantoro, A.; Shafiq, N.; Nuruddin, M.F. Strength and microstructural properties of fly ash based geopolymer concrete containing high-calcium and water-absorptive aggregate. *J. Clean. Prod.* **2016**, *112*, 816–822. [CrossRef]
25. Joseph, B.; Mathew, G. Influence of aggregate content on the behavior of fly ash based geopolymer concrete. *Sci. Iran.* **2012**, *19*, 1188–1194. [CrossRef]
26. Dave, S.; Bhogayata, A.; Arora, D.N. Impact Resistance of Geopolymer Concrete Containing Recycled Plastic Aggregates. *EasyChair* **2017**, *1*, 137–143.
27. Provis, J.L. Alkali-activated Binders and Concretes: The Path to Standardization. In *Geopolymer Binder Systems*; ASTM International: West Conshohocken, PA, USA, 2013; pp. 185–195.
28. Nordin, N.; Abdullah, M.M.A.B.; Fakri, W.M.N.R.W.; Tahir, M.F.M.; Sandu, A.V.; Hussin, K.; Zailani, W.W.A. Exploration on fly ash waste as global construction materials for dynamics marketability. In Proceedings of the Applied Physics of Condensed Matter (APCOM 2019), Pleso, Slovak, 19–21 June 2019; AIP Publishing: Melville, NY, USA, 2019.
29. Xu, H.; van Deventer, J.S.J. Effect of source materials on geopolymerization. *Ind. Eng. Chem. Res.* **2003**, *42*, 1698–1706. [CrossRef]
30. Mo, B.H.; Zhu, H.; Cui, X.M.; He, Y.; Gong, S.Y. Effect of curing temperature on geopolymerization of metakaolin-based geopolymers. *Appl. Clay Sci.* **2014**, *99*, 144–148. [CrossRef]
31. Abdullah, M.M.A.B.; Yahya, Z.; Tahir, M.F.M.; Hussin, K.; Binhussain, M.; Sandhu, A.V. Fly Ash Based Lightweight Geopolymer Concrete Using Foaming Agent Technology. *Appl. Mech. Mater.* **2014**, *679*, 20–24. [CrossRef]
32. Harja, M.; Barbuta, M.; Rusu, L. Obtaining and Characterization of the Polymer Concrete with Fly Ash. *J. Appl. Sci.* **2009**, *9*, 88–96. [CrossRef]
33. Corbu, O.; Ioani, A.M.; Al Bakri Abdullah, M.M.; Meita, V.; Szilagyi, H.; Sandu, A.V. The pozzoolanic activity level of powder waste glass in comparisons with other powders. *Key Eng. Mater.* **2015**, *660*, 237–243. [CrossRef]
34. Lee, W.K.W.; van Deventer, J.S.J. Chemical interactions between siliceous aggregates and low-Ca alkali-activated cements. *Cem. Concr. Res.* **2007**, *37*, 844–855. [CrossRef]

35. Ismail, I.; Bernal, S.A.; Provis, J.L.; Hamdan, S.; van Deventer, J.S.J. Microstructural changes in alkali activated fly ash/slag geopolymers with sulfate exposure. *Mater. Struct.* **2012**, *46*, 361–373. [CrossRef]
36. Wuddivira, M.N.; Robinson, D.A.; Lebron, I.; Bréchet, L.; Atwell, M.; de Caires, S.; Oatham, M.; Jones, S.B.; Abdu, H.; Verma, A.K.; et al. Estimation of soil clay content from hygroscopic water content measurements. *Soil Sci. Soc. Am. J.* **2012**, *76*, 1529–1535. [CrossRef]
37. Rico, P.; Adriano, A.; Soriano, G.; Duque, J.V. Characterization of Water Absorption and Desorption properties of Natural Zeolites in Ecuador. In *International Symposium on Energy*; Puerto Rico Energy Center-Laccei: Puerto Rico, 2013.
38. Longhi, M.A.; Zhang, Z.; Rodríguez, E.D.; Kirchheim, A.P.; Wang, H. Efflorescence of alkali-activated cements (geopolymers) and the impacts on material structures: A critical analysis. *Front. Mater.* **2019**, *6*. [CrossRef]
39. Smart Nanoconcretes and Cement-Based Materials: Properties, Modelling and Applications. Available online: https://books.google.ro/books?id=oaq-DwAAQBAJ&pg=PA173&lpg=PA173&dq=strong+physically+bound+water&source=bl&ots=F2VLsZWYHy&sig=ACfU3U11W_kRxqwL2yooFg6auXynm-80ww&hl=ro&sa=X&ved=2ahUKEwjjptPehrLmAhV_wAIHHVmYCG8Q6AEwDnoECAoQAQ#v=onepage&q=strong%20physically%20bound%20water&f=false (accessed on 13 December 2019).
40. Van Reeuwijk, L.P. *The Thermal Dehydration of Natural Zeolites (with a Summary in Dutch)*; H. Veenman & Zonen: Wageningen, The Netherlands, 1974.
41. Valdiviés-Cruz, K.; Lam, A.; Zicovich-Wilson, C.M. Chemical interaction of water molecules with framework Al in acid zeolites: A periodic ab initio study on H-clinoptilolite. *Phys. Chem. Chem. Phys.* **2015**, *17*, 23657–23666. [CrossRef] [PubMed]
42. Calero, S.; Gómez-Álvarez, P. Hydrogen bonding of water confined in zeolites and their zeolitic imidazolate framework counterparts. *RSC Adv.* **2014**, *4*, 29571–29580. [CrossRef]
43. Glad, B.E.; Kriven, W.M. Geopolymer with Hydrogel Characteristics via Silane Coupling Agent Additives. *J. Am. Ceram. Soc.* **2014**, *97*, 295–302. [CrossRef]
44. Palomo, A.; Krivenko, P.; Garcia-Lodeiro, I.; Kavalerova, E.; Maltseva, O.; Fernández-Jiménez, A. A review on alkaline activation: New analytical perspectives. *Mater. Constr.* **2014**, *64*, 22. [CrossRef]
45. Criado, M.; Aperador, W.; Sobrados, I. Microstructural and mechanical properties of alkali activated Colombian raw materials. *Materials* **2016**, *9*, 158. [CrossRef] [PubMed]
46. Musicá, S.M.; Krehula, S.; Popovićb, S.; Popovićb, P. Thermal decomposition of h-FeOOH. *Mater. Lett.* **2004**, *58*, 444–448.
47. Derie, R.; Ghodsi, M.; Calvo-Roche, C. DTA study of the dehydration of synthetic goethite αFeOOH. *J. Therm. Anal.* **1976**, *9*, 435–440. [CrossRef]
48. Rendon, J.L.; Cornejo, J.; de Arambarri, P.; Serna, C.J. Pore structure of thermally treated goethite (α-FeOOH). *J. Colloid Interface Sci.* **1983**, *92*, 508–516. [CrossRef]
49. Giovanoli, R.; Brütsch, R. Kinetics and mechanism of the dehydration of γ-FeOOH. *Thermochim. Acta* **1975**, *13*, 15–36. [CrossRef]
50. Walter, D.; Buxbaum, G.; Laqua, W. The mechanism of the thermal transformation from goethite to hematite*. *J. Therm. Anal. Calorim.* **2001**, *63*, 733–748. [CrossRef]
51. Cheng-Yong, H.; Yun-Ming, L.; Abdullah, M.M.A.B.; Hussin, K. Thermal Resistance Variations of Fly Ash Geopolymers: Foaming Responses. *Sci. Rep.* **2017**, *7*, 45355. [CrossRef] [PubMed]
52. Cornejo, M.H.; Togra, B.; Baykara, H.; Soriano, G.; Paredes, C.; Elsen, J. Effect of Calcium Hydroxide and Water to Solid Ratio on Compressive Strength of Mordenite-Based Geopolymer and the Evaluation of Its Thermal Transmission Property. In Proceedings of the ASME International Mechanical Engineering Congress and Exposition (IMECE), Pittsburgh, PA, USA, 9–15 November 2018; American Society of Mechanical Engineers (ASME): New York, NY, USA, 2018; Volume 12.
53. Abdullah, M.M.A.B.; Ming, L.Y.; Yong, H.C.; Tahir, M.F.M. Clay-Based Materials in Geopolymer Technology. In *Cement Based Materials*; InTechOpen: London, UK, 2018. [CrossRef]
54. Bajare, D.; Vitola, L.; Dembovska, L.; Bumanis, G. Waste stream porous alkali activated materials for high temperature application. *Front. Mater.* **2019**, *6*. [CrossRef]
55. Dehydration Reactions and Kinetic Parameters of Gibbsite—Science Direct. Available online: https://www.sciencedirect.com/science/article/pii/S0272884210002592 (accessed on 13 December 2019).
56. MacKenzie, K.J.D.; Temuujin, J.; Okada, K. Thermal decomposition of mechanically activated gibbsite. *Thermochim. Acta* **1999**, *327*, 103–108. [CrossRef]

57. Zhu, B.; Fang, B.; Li, X. Dehydration reactions and kinetic parameters of gibbsite. *Ceram. Int.* **2010**, *36*, 2493–2498. [CrossRef]
58. Redaoui, D.; Sahnoune, F.; Heraiz, M.; Raghdi, A. Mechanism and Kinetic Parameters of the Thermal Decomposition of Gibbsite $Al(OH)_3$ by Thermogravimetric Analysis. *Acta Phys. Pol. Ser. A* **2017**, *131*, 562–565. [CrossRef]
59. Hao, H.; Lin, K.-L.; Wang, D.; Chao, S.-J.; Shiu, H.-S.; Cheng, T.-W.; Hwang, C.-L. Elucidating Characteristics of Geopolymer with Solar Panel Waste Glass. *Environ. Eng. Manag. J.* **2015**, *14*, 79–87.
60. Grawe, S.; Augustin-Bauditz, S.; Clemen, H.C.; Ebert, M.; Eriksen Hammer, S.; Lubitz, J.; Reicher, N.; Rudich, Y.; Schneider, J.; Staacke, R.; et al. Coal fly ash: Linking immersion freezing behavior and physicochemical particle properties. *Atmos. Chem. Phys.* **2018**, *18*, 13903–13923. [CrossRef]
61. Duan, P.; Yan, C.; Zhou, W. Compressive strength and microstructure of fly ash based geopolymer blended with silica fume under thermal cycle. *Cem. Concr. Compos.* **2017**, *78*, 108–119. [CrossRef]
62. Paunović, P.; Petrovski, A.; Načevski, G.; Grozdanov, A.; Marinkovski, M.; Andonović, B.; Makreski, P.; Popovski, O.; Dimitrov, A. Pathways for the production of non-stoichiometric titanium oxides. In *Nanoscience Advances in CBRN Agents Detection, Information and Energy Security*; Springer: Dordrecht, The Netherlands, 2015; pp. 239–253. ISBN 9789401796972.
63. Anderson, P.J.; Horlock, R.F. Thermal decomposition of magnesium hydroxide. *Trans. Faraday Soc.* **1962**, *58*, 1993–2004. [CrossRef]
64. Zulkifly, K.; Yong, H.C.; Abdullah, M.M.A.B.; Ming, L.Y.; Panias, D.; Sakkas, K. Review of Geopolymer Behaviour in Thermal Environment. In Proceedings of the IOP Conference Series: Materials Science and Engineering, Bali, Indonesia, 26–27 July 2016; Institute of Physics Publishing: Bristol, UK, 2017; Volume 209.
65. Alvarez-Ayuso, E.; Querol, X.; Plana, F.; Alastuey, A.; Moreno, N.; Izquierdo, M.; Font, O.; Moreno, T.; Diez, S.; Vázquez, E.; et al. Environmental, physical and structural characterisation of geopolymer matrixes synthesised from coal (co-)combustion fly ashes. *J. Hazard. Mater.* **2008**, *154*, 175–183. [CrossRef]

 *materials*

Article

Strength Development and Elemental Distribution of Dolomite/Fly Ash Geopolymer Composite under Elevated Temperature

Emy Aizat Azimi [1,*], Mohd Mustafa Al Bakri Abdullah [1,*], Petrica Vizureanu [2,*], Mohd Arif Anuar Mohd Salleh [1], Andrei Victor Sandu [1,2,3,4], Jitrin Chaiprapa [5], Sorachon Yoriya [6], Kamarudin Hussin [1] and Ikmal Hakem Aziz [1]

1 Center of Excellence Geopolymer and Green Technology, School of Materials Engineering, Universiti Malaysia Perlis (UniMAP), P.O. Box 77, D/A Pejabat Pos Besar, 01000 Kangar, Perlis, Malaysia; arifanuar@unimap.edu.my (M.A.A.M.S.); sav@tuiasi.ro (A.V.S.); kamarudin@unimap.edu.my (K.H.); ikmalhakem@gmail.com (I.H.A.)

2 Faculty of Materials Science and Engineering, "Gheorghe Asachi" Technical University, Blvd. D. Mangeron 71, 700050 Iasi, Romania

3 Romanian Inventors Forum, Str. Sf. P. Movila 3, Iasi 700089, Romania

4 National Institute for Research and Development in Environmental Protection, 294 Splaiul Independenței Blv, 060031 Bucharest, Romania

5 Synchrotron Light Research Institute (SLRI), 111 University Avenue, Muang District, Nakhon Ratchasima 30000, Thailand; jitrin@slri.or.th

6 National Metal and Materials Technology Center (MTEC), 114 Thailand Science Park, Phaholyothin Road, Klong 1, Klongluang, Pathumthani 12120, Thailand; sorachy@mtec.or.th

* Correspondence: emyaizat@gmail.com (E.A.A.); mustafa_albakri@unimap.edu.my (M.M.A.B.A.); peviz2002@yahoo.com (P.V.)

Received: 5 February 2020; Accepted: 21 February 2020; Published: 24 February 2020

Abstract: A geopolymer has been reckoned as a rising technology with huge potential for application across the globe. Dolomite refers to a material that can be used raw in producing geopolymers. Nevertheless, dolomite has slow strength development due to its low reactivity as a geopolymer. In this study, dolomite/fly ash (DFA) geopolymer composites were produced with dolomite, fly ash, sodium hydroxide, and liquid sodium silicate. A compression test was carried out on DFA geopolymers to determine the strength of the composite, while a synchrotron Micro-Xray Fluorescence (Micro-XRF) test was performed to assess the elemental distribution in the geopolymer composite. The temperature applied in this study generated promising properties of DFA geopolymers, especially in strength, which displayed increments up to 74.48 MPa as the optimum value. Heat seemed to enhance the strength development of DFA geopolymer composites. The elemental distribution analysis revealed exceptional outcomes for the composites, particularly exposure up to 400 °C, which signified the homogeneity of the DFA composites. Temperatures exceeding 400 °C accelerated the strength development, thus increasing the strength of the DFA composites. This appears to be unique because the strength of ordinary Portland Cement (OPC) and other geopolymers composed of other raw materials is typically either maintained or decreases due to increased heat.

Keywords: dolomite/fly ash; geopolymer; strength development; temperature exposure; Micro-XRF

1. Introduction

Geopolymers refer to binder materials that are generated through the activation of aluminosilicate materials with alkali or alkali–silicate solutions [1,2]. Geopolymers that are otherwise inorganic polymers or alkali-activated binders have garnered interest at the global level [3]. Generally, geopolymers are amorphous to semi-crystalline and three-dimensional silica alumina-based materials. Geopolymers can be produced by mixing aluminosilicate precursor materials, such as fly ash, kaolin or metakaolin, metal slag, and dolomite, with strong alkali solutions. The common alkali solutions used to produce geopolymers are sodium hydroxide (NaOH), potassium hydroxide (KOH), sodium silicate, and potassium silicate [4]. The geopolymer paste can be cured either at room temperature or an elevated temperature. The aluminosilicate source dissolves and forms free Si^{4+}, Al^{3+} tetrahedral units and Ca^{2+}, under strong alkali solution. The progress of the reaction continues with the slow removal of water. Next, SiO_4, AlO_4 tetrahedral, and CaO clusters are integrated to produce polymeric precursors through sharing all oxygen atoms. This will eventually form amorphous or semi-amorphous geopolymers. The primary block of the geopolymer chain is Si-O-Al [5]. Basically, geopolymers are a man-made material that offer several advantages, including good mechanical strength (similar or more than ordinary Portland Cement (OPC)) and the capacity to encapsulate hazardous waste, as well as being water- and fire-resistant [6]. The strength of geopolymers is a critical factor within the building and construction domain. The development of strength in geopolymers strongly depends on the raw materials and the alkali activator solutions [7–16]. Even though geopolymers with slow strength development generate materials with lower strength, this drawback can be addressed by increasing heat and aging time [17].

The compressive strength of materials after being exposed to elevated temperatures appears to be one of the aspects of concern for application in the building and construction domain. A prior study assessed the strength of geopolymers after they were exposed to elevated temperatures using Class F fly ash [18]. The strength of geopolymer composites reduced after the temperature was increased from 100 to 800 °C. The strength went far lower than the strength of the geopolymer that was not exposed to an elevated temperature. Another study on exposure to elevated temperature was performed by comparing the strength of geopolymer (fly ash) with OPC composites [19]. As a result, the compressive strength of both materials deteriorated gradually. It is noteworthy to highlight that the OPC composites displayed the worst decline in strength, when compared to that of the geopolymer composite. Sodium hydroxide (NaOH) concentration (part of the alkali solution) greatly affected the dissolution process of silica and alumina from fly ash, wherein increased molarity led to an increased dissociation of the active species of raw material, apart from yielding the formation of more geopolymer gel network [20]. Another crucial factor in generating geopolymers refers to the ratio of NaOH to sodium silicate (Na_2SiO_3). Morsy et al., reported that increments in compressive strength were affected by the NaOH/Na_2SiO_3 ratio, which escalated from 1.0 to 2.5 [21].

In a previous study, dolomite was only used as a filler and replacement for other raw materials in alkali-activated composites and geopolymer composites. However, no study has applied dolomite as the major or main raw material in geopolymer composites. Zarina et al., [22] used dolomite only as an addition to boiler ash, which turned into geopolymer paste, while Yip et al., [23] applied dolomite merely as a carbonate mineral addition to a metakaolin-based geopolymer. Dolomite has the potential to be used as geopolymer raw material due to its Al, Si, and Ca contents. Nonetheless, the incorporation of dolomite in geopolymer is still new and in its infancy within the research domain. Information concerning dolomite geopolymer is in scarce, especially in terms of strength development under exposure to elevated temperature. Therefore, this study investigated the effect of elevated temperature exposure on geopolymer composites based on dolomite/fly ash (DFA). The study outcomes enhance comprehension regarding the mechanical properties for the future improvement and application of dolomite geopolymers, especially those exposed to elevated temperatures.

2. Experimental

2.1. Materials

Dolomite was applied as the geopolymer raw material in this study. The dolomite was supplied by Perlis Dolomite Industries Sdn. Bhd., Perlis, Malaysia. The chemical composition of dolomite was tabulated in Table 1. The size of the dolomite was set to below 63 μm. The hardness of the dolomite ranged from 3.5 to 4.0 (Mohs hardness) with a specific gravity between 2.8 and 2.9. The solid dolomite was ground to obtain its powder form with irregular particle shapes. The resultant dolomite was used as a raw material in the investigation carried out in this study.

Table 1. Chemical composition of dolomite.

Chemical Compound	Composition (wt %)
CaO	80.21
Al_2O_3	1.52
SiO_2	2.50
Fe_2O_3	0.15
MgO	15.50
CuO	0.07
MnO	0.02

F class fly ash was gathered from a coal combustion plant located in Manjung, Perak, Malaysia. Fly ash refers to waste generated from YTL Corporation Berhad. The size of fly ash was fixed at below 63 μm. The collected fly ash was in a fine powder form with a generally spherical shape. The fly ash was applied as a raw material for this study purpose. The chemical composition of fly ash was tabulated in Table 2.

Table 2. Chemical composition of fly ash.

Chemical Compound	Composition (wt %)
SiO_2	55.3
Al_2O_3	25.8
Fe_2O_3	5.5
CaO	2.9
MgO	0.8
SO_3	0.3

The NaOH used in this study refers to caustic soda flakes called Formosoda-P supplied by Formosa Plastic Corporation, Taiwan. The NaOH in flake form was diluted in water to form alkaline solution. The molecular weight of the NaOH was 40 g/mol with 99.0% purity. The use of flake-type sodium hydroxide gave a solution with high purity (can be controlled during diluting process), when compared to that in liquid form.

The technical grade of liquid Na_2SiO_3 was supplied by South Pacific Chemical Industries Sdn. Bhd. (SPCI), Malaysia. The liquid Na_2SiO_3 is colorless and dissolves readily in 60.5% water. Sodium silicate, which is readily soluble in water, appears to be the most silicon-rich when compared to its powder form. The Na_2SiO_3 is composed of 30.1% of silica and 9.4% of sodium oxide. The molecular weight of Na_2SiO_3 is 122.06 g/mol. Both the silica content and the viscosity of this sodium silicate are suitable for application in geopolymer. Sodium silicate was applied as part of the alkali activator solution in this study.

2.2. Sample Preparation

NaOH solution with 22M concentration was prepared in a volumetric flask, which was put in the circulate water bath to ensure the cooling down of the solution. The NaOH solution was mixed with Na_2SiO_3 solution with a Na_2SiO_3/NaOH ratio of 2.5 to formulate the alkali activator solution 24 hours prior to analysis. Next, DFA with a 60/40 ratio and an alkali activator solution was mixed with a solid to liquid ratio of 2.5 and stirred well by using a mechanical mixer. The fresh paste was rapidly poured into steel mold and was compressed into each cube compartment at each layer by adhering to ASTM C109. The samples were oven-dried for 24 hours at 80 °C for curing purposes. During the curing process, the samples were sealed with thin plastic at the exposed part of the mold. The samples were exposed to elevated temperatures from 200 to 1000 °C after 28 days of curing, in order to achieve the study objective.

2.3. Testing and Chracterization Method

Instron machine series 5569 Mechanical Tester was employed to assess the compressive strength of all specimens. The specimens referred to geopolymer composites that were taken out from the oven after 24 hours of curing and were placed at room temperature until the day of testing for control sample, wherein the samples were already exposed to elevated temperature (200–1000 °C). The compressive test was performed to examine the strength development of the specimens. The samples were tested after seven days of curing, in which three specimens were tested for each parameter.

Microstructural characterization of DFA geopolymer composites was performed using the JSM-6460LA model Scanning Electron Microscope (JEOL) with secondary electron detectors. In microstructural analysis, both dolomite and fly ash powder was sprinkled onto double-sided carbon taped prior to analysis, whereby the blower was used to discard loosely held powder. In microstructural analysis, the samples were taken from the surface of the internal structure of the geopolymer breakage prior to compressive test. The samples was were prepared with a size of up to 50 mm × 50 mm × 10 mm and coated with palladium. The coated samples were placed in SEM chamber for characterization process.

A Micro X-Ray Fluorescence (XRF) machine using a source from synchrotron radiation at beam line 6b (BL6b) of the Synchrotron Light Research Institute (SLRI) was applied to determine the chemical composition and the elemental distribution of the reaction products. The BL6b exploited the continuous synchrotron radiation that emitted from the bending magnet. The specimen was positioned on a motorized stage with three-degree freedom. This Micro-XRF employed an 8 keV radiation from synchrotron, along with three gate valves. The Micro-XRF station used a capillary half-lens for X-ray focusing. This optic focused on X-ray beam from 5 × 2 mm (H × V) at the entrance of the lens down to a diameter of 50 µm measured at a sample positioned 22 mm (lens focal point) downstream of the lens exit. AMPTEK single-element Si (PIN) solid-state was installed to detect the fluorescence element.

Perkin Elmer FTIR Spectrum RX1 Spectrometer was used to determine the functional groups of a DFA geopolymer composite. The samples were analyzed using the attenuated total reflectance (ATR) technique. The samples were crushed into powder form using a mortar. Initially, the specimens were placed on the sample slot (ATR crystal) located at the sample platform of the machine. The pressure tower that contained the compression tip was moved to the top of the specimens and closed tightly. The specimens were scanned between 500 and 4000 cm^{-1} with a resolution of 4 cm^{-1}. The sample spectrum was collected after gathering the background spectrum.

An XRD-6000 Shimadzu X-Ray diffractometer was applied to characterize the DFA geopolymer composite. The x-ray diffractometer (XRD) was equipped with auto-search/match software as standard to identify the crystalline phases. The specimens were in powder form. The samples were prepared in a size that ranged from 1 to 63 µm. Extremely fine grain specimens were required for the analysis of powders via XRD to attain good signal to noise ratio, avoid fluctuations in intensity and spottiness, and to reduce the preferred orientation. The XRD analysis was performed using Cu-Kα radiation with X-ray tube that operated at 40 kV and 35 mA. The XRD data were collected at 2θ values in the range of 10° to 90° at a scan rate of 2° per minute and scan steps of 0.02°(2θ). Good phase identification may be made if at least three of the main diffraction peaks of the unknown phase match the standard diffraction pattern of a known crystal phase retrieved from the literature of powder diffraction database.

3. Results and Discussions

3.1. Microstructure Analysis

Figure 1 illustrates the microstructure of DFA geopolymer composites for (a) before temperature exposure, as well as after temperature exposure, at (b) 200, (c) 400, and (d) 1000 °C. In Figure 1a, the microstructure signifies that the geopolymer matrix was not fully developed in the system, mainly because some raw materials retained their original shape (e.g., spherical fly ash). The microstructure in Figure 1b displays no occurrence of crack on the surface of the geopolymer composite. The growth of geopolymer matrices began to continue when heat was supplied from the exposure. Based on Figure 1c, the microstructure exhibits the occurrence of small voids. The geopolymer that served as a binder in the DFA composite was completely cured. In Figure 1d, cracks are noted on the structure of the DFA composite, which also signifies porosity. The microstructure in Figure 1 exemplifies the development of the geopolymer matrix and the increased density of the geopolymer structure from (a) to (c), whereas porosity and density decreased from (c) to (d).

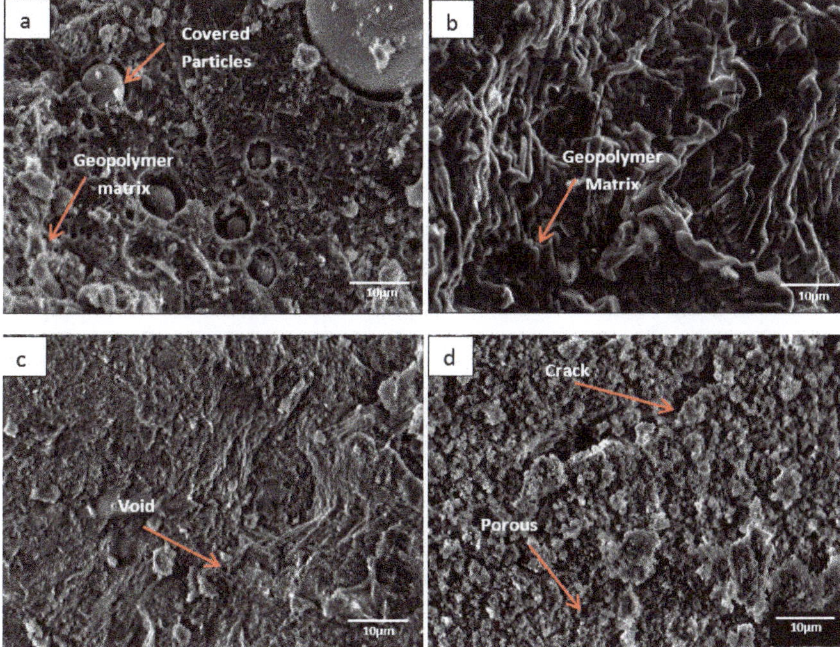

Figure 1. Microstructure of dolomite/fly ash geopolymer composite (**a**) before temperature exposure, after fire exposure at (**b**) 200, (**c**) 400, and (**d**) 1000 °C.

In Figure 1b, the growth of the geopolymer matrix began to take place as heat was continuously supplied. Dolomite has a high amount of calcium (Ca) content. During the dissolution process of geopolymer, the CaO content in dolomite was attacked by the active species from alkali solution and turned into active species of Ca^{2+} and O^{2-}. The Ca^{2+} ions, along with Si^{4+} and Al^{3+} that reacted to OH^- ions from alkali solution, formed a geopolymer composite. At this stage, the Ca content in dolomite began promoting the optimum reaction within the geopolymer system. Increments in temperature accelerated the continuous development of geopolymer matrices, thus increasing the strength of DFA composites.

As the heat increased due to the increasing temperature, the geopolymer matrix in Figure 1c accelerated in its growth. The structure turned out to be dry and smooth, along with some void due to water evaporation. As the polycondensation reaction continued, a gel network underwent the rearrangement process in geopolymerisation. This promoted the geopolymer matrices to continue in its growth and to fill the space within the geopolymer composite structure (space-filling gel) [24].

In Figure 1d, a crack is noted and the structures of the geopolymer composite turned porous. This was due to the increase in heat supply to the sample. At this point, shrinkage started to take place due to dehydration and dehydroxylation. The dehydroxylation process included the heating process, through which the hydroxyl group (OH) was released by forming a water molecule. Distortion and buckling of the polymeric structure of aluminosilicates appeared to be the consequence of the release of structural water via dihydroxylation, which resulted in a disordered structure [25]. Equation (1) shows the general mechanism of dehydroxylation in geopolymers.

$$\begin{gathered} OH^- \leftrightarrow H^+ + O^{2-} \\ H^+ + OH^- \leftrightarrow H_2O \end{gathered} \tag{1}$$

The high stresses in the void wall developed the movement of non-combined water, thus the greater degradation occurred with rising temperature due to fine pores and shrinkage [26]. This was also due to breakage of the bonding and phase changing from an amorphous to a crystalline phase.

3.2. Elemental Distribution Analysis

In order to gain elemental maps, Micro-XRF that used a synchrotron source was applied. In this investigation, the Ca, Si, Al, Mg, and Fe maps were determined. Micro-XRF maps of a ~961 × 50 μm overview of DFA geopolymer composites before (control) and after exposure to 400 °C are presented in Figures 2 and 3, respectively. The sample exposed to 400 °C was selected as it recorded the highest strength amongst the rest. Generally, all the elements in DFA geopolymer composites are well distributed. Some variances were noted in the maps of elemental distribution between composites that were not exposed and those exposed to 400 °C. Based on Figure 3, the elemental distribution of composite that was exposed to 400 °C displayed an increment in the amount of distribution region, when compared to the composite unexposed to elevated temperature. The elements of Ca, Si, and Al in the maps portrayed some increment in the red region (high concentration) after being exposed to 400 °C. The combination of Si and Al maps showcased the formation of Si-O-Al, which emerged as one of the most important bonds that determined the strength of the geopolymer. The green and yellow regions (medium concentration), along with a high-concentration region for Si and Al elements, had more spread in the map of the composite that was exposed to 400 °C. The magnesium (Mg) element also increased after exposure to 400 °C, while the Fe element displayed slight changes in the concentration region. The blue region (low concentration) indicates no element or void in the composite. The green and yellow regions (medium concentration) signify a reaction between elements within the composite. The high-concentration region or the red region reflects a reaction between the investigated elements that leads to the formation of a phase.

A high amount of Ca in geopolymer usually produces a good setting time. Despite material with high Ca content obtaining a better setting time compared to other geopolymers with low Ca content, the strength development was found to be slow. Strength is gained with an increase in curing time and curing temperature. The occurrence of Si and Al elements in geopolymer composites affects the strength development. Increasing Si and Al elements increases the strength development of the geopolymer composites as more geopolymer chains are created. The geopolymer's main structure consists of Si-O-Al, which clearly shows the important of Si and Al elements in producing good strength development. However, the occurrence of Mg in the geopolymer retarded the strength development of the geopolymer. The occurrence of this has disturbed the backbone structure of Ca-Si-O-Al, thus disturbing the strength development of the geopolymer. With enough Ca, the increased bulk of Mg promotes the formation of low Al C-(A)-S-H due to the formation of hydrotalcite group phases and a reduction in the available Al element. Hydrotalcite group phase formation is linked to the increase in C-(N)-A-S-H gel polymerization, decreased gel Al uptake and increased formation of the third aluminate hydrate. MgO or Mg(OH) content affects the properties of geopolymer [27,28].

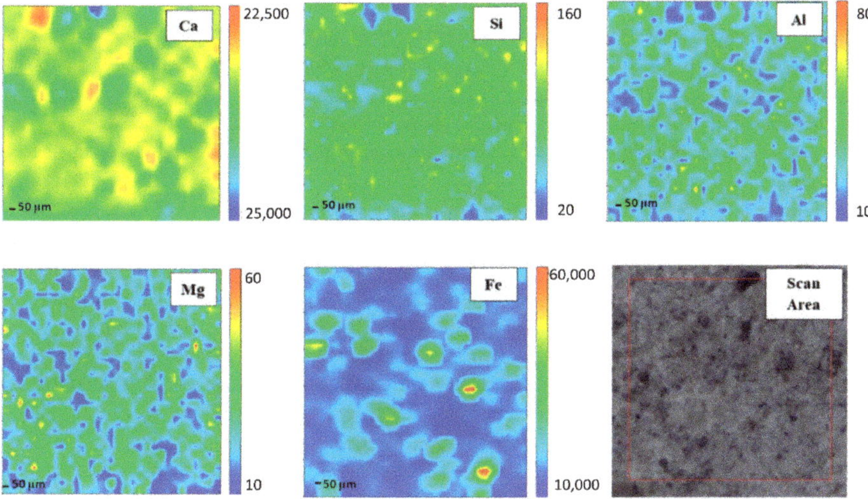

Figure 2. Micro-XRF elemental mapping of Ca, Si, Al, Mg and Fe in a ~961 × 50 μm overview of dolomite/fly ash geopolymer composites unexposed at elevated temperature.

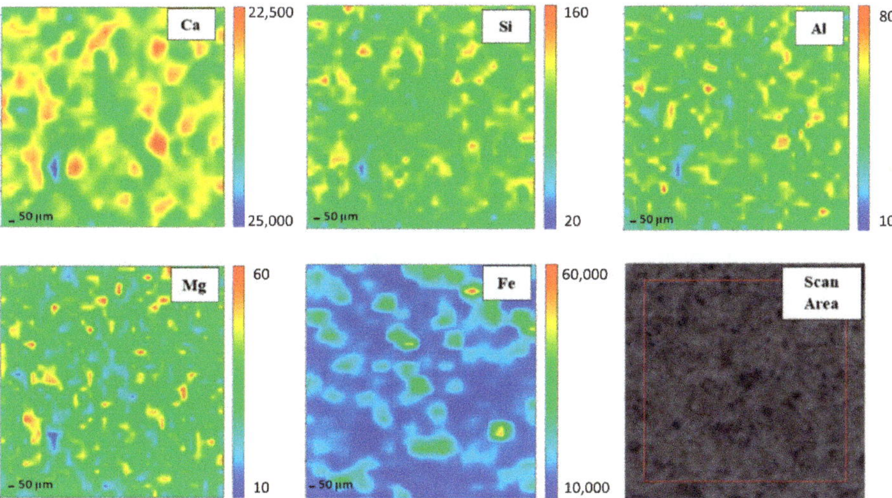

Figure 3. Micro-XRF elemental mapping of Ca, Si, Al, Mg and Fe in a ~961 × 50 μm overview of dolomite/fly ash geopolymer composites after exposure at 400 °C.

Well-distributed elements in the geopolymer composite samples indicated the production of a homogenous geopolymer composite. A homogenous mixture of geopolymer composites enhances the strength of the material. The combination of Ca, Si, and Al maps led to the formation of a calcium aluminate hydrate (C-A-S-H) phase [29]. An increased high concentration region for Ca, Si, and Al in the map of the composite after exposure to 400 °C reflects an increment in the C-A-S-H phase in the composites. Geopolymer composites that contained Ca element generated the C-A-S-H phase after they went through the geopolymerisation process. Heat supply enhanced the strength development of geopolymer composites. With increased strength development, the formation of the C-A-S-H phase increased. This is proven by the formation of more geopolymer matrices in the microstructure of samples exposed to 400 °C.

The increasing green and yellow regions (medium concentration), along with the high-concentration region for Si and Al elements of the composite exposed to 400 °C, signified the increase in the Si-O-Al bond in geopolymer. Upon increments in the exposed temperature, more geopolymer chains were formed, offering more strength to the composites. The geopolymer consists of repeating units of Si-O-Al. The formation of more Si-O-Al bonds confers more strength to the geopolymer.

The element of Mg increased after the increment in exposed temperature. The occurrence of Mg led to the formation of a hydrotalcite group that may weaken geopolymer strength. Nevertheless, the strength of the composite continued to increase due to the increment in Si and Al elements. At this stage, the Mg element began to grow within the composites. This Mg reacted with other element(s) that occurred from the breakage of the geopolymer structure due to an increase in temperature from 400 to 1000 °C. Due to the increasing pressure in the geopolymer sample as the exposed temperature was increased, both dehydration and dehydroxylation processes took place. Dehydroxylation in clay-occurred at temperatures that ranged between 500 and 700 °C [30]. This led to the breakage of several geopolymer bonds, hence the occurrence of more Si and Al elements [31]. These free elements reacted with Mg and formed new crystalline phases that comprised Mg, Si, and Al, which is further discussed in later phase analysis. The new crystalline phases, such as Akermanite, Juanite, Clinohumite and Epsomite, which consisted of Mg, occurred after the dehydroxylation process in the geopolymer composite. Some crystalline phases weakened the DFA geopolymer composite's strength. The Fe element exhibited slight changes in the concentration region, mainly because only a small amount of Fe was produced from the raw material (fly ash) [32].

3.3. Phase Analysis

Figure 4 illustrates the XRD pattern of the DFA geopolymer composite before (Control) and after being exposed to 400 and 1000 °C. By referring to Figure 4, y axis was indicated to the various intensity obtained while x-axis represented to diffraction angle of mineral phase appearance. The original sample of the XRD patterns portrayed the semi-crystalline phase of the DFA geopolymer composite. The crystalline C-A-S-H peak (ICDD 01-081-1448) was detected in the original geopolymer composite, while the binders of semi-crystalline, Na-geopolymer, Ca-geopolymer, and C-A-S-H gel were discovered in the DFA geopolymer composite system. The control composites were composed of calcite (ICDD 00-003-0670), quartz (ICDD 01-079-1906), mullite (ICDD 01-079-1906), and magnetite (ICDD 01-089-0950).

As the DFA composite was heated to 400 °C, the intensity of the C-A-S-H peak ($2\theta = 32°$) was increased. This was due to the growth of the C-A-S-H gel, along with the removal of free water. Upon increments in exposure temperature, the C-A-S-H gel continued to grow and develop. The increased exposure temperature accelerated the strength development of the composites. This is attributable to the microstructure of composites exposed to 400 °C, which displayed a greater development of geopolymer matrices that was fully cured. At this point, the phase of the geopolymer remains the same, except for the intensity of the C-A-S-H peak. The formation of a new crystalline reaction phase that still did not occur signified the structural evolution of the C-A-S-H gel, hence the increased composite strength that resulted later. Calcite was formed due to the excessive amount of CaO that reacted with CO_2, while the magnetite phase took place due to the small amount of Fe contained in the chemical composition of fly ash [33]. The presence of the hydrosodalite phase indicated some geopolymerisation, but with low strength [34].

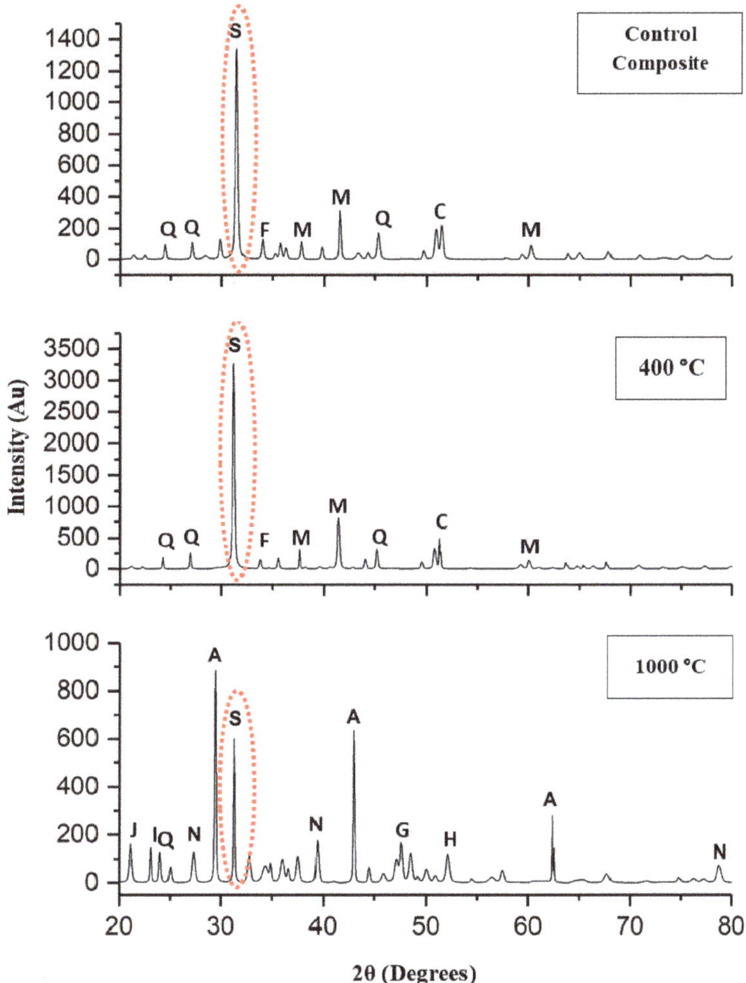

C-A-S-H, N = Nepheline ($NaAlSiO_4$), A = Akermanite ($Ca_2MgSi_2O_7$), C = Calcite, F = Magnetite (Fe_3O_4), M = Muliite ($Al_6Si_2O_{13}$), G = Juanite ($Ca_{10}Mg_4Al_2Si_{11}O_{39}4H_2O$), H = Clinohumite ($(MgFe)_9(SiO_4)_4(FOH)_2$), I = Dickite ($Al_2Si_2O_5(OH)_4$), J = Epsomite ($MgSO_47\ (H_2O)$).

Figure 4. X-Ray Diffraction (XRD) pattern of dolomite/fly ash geopolymer before (Control) and after exposure at 400 and 1000 °C.

After the DFA, the geopolymer composite was exposed up to 1000 °C, and the formation of a new crystalline product or peak was clearly noted. Both akermanite ($Ca_2MgSi_2O_7$) (ICDD 01-074-0990) and nepheline ($NaAlSiO_4$) (ICDD 01-076-1858) phases were predominantly formed. Akermanite ($Ca_2MgSi_2O_7$) refers to a melilite mineral of the sorosilicate group that consists of calcium, magnesium, silicon, and oxygen. Akermanite is a product of contact metamorphism (a change in mineral in existing rocks that occurs preliminarily due to heat and pressure) of siliceous limestones and dolostones (dolomite). The crystalline phases of clinohumite ($(MgFe)_9(SiO_4)_4(FOH)_2$) and epsomite ($MgSO_4(H_2O)$) also occurred. Clinohumite is an uncommon member of the humite group that contains magnesium silicate. Its empirical formula reflects four olivines (Mg_2SiO_4) and a brucite ($Mg(OH)_2$). Clinohumite has brittle properties that normally lead to weak strength if occuring in the composite. Epsomite was originally a hydrous magnesium sulphate that is classified in the orthorhombic system of crystallization and normally forms as efflorescence. Epsomite absorbs water from the air and converts it to hexahydrate with the loss of a water molecule and a switch to a monoclinic structure (unequal length system). Both of these crystalline phases that occurred after the exposure of the composite to 1000 °C contained Mg element.

The occurrence of Mg, along with the high composition of Ca, disrupted the geopolymer chain structure and decreased its strength [35,36]. Figure 5 illustrates the interference of Mg content in the geopolymer system. Crystalline phases of nepheline are commonly detected from XRD diffractograms of geopolymer samples after being exposed to elevated temperature. Nepheline is known as a thermally stable material. The presence of crystalline phases that are thermally stable is critical for the thermal stability of the geopolymer structure. Nevertheless, the presence of nepheline in the DFA geopolymer composite was minimal and had low intensity, when compared to the akermanite phase.

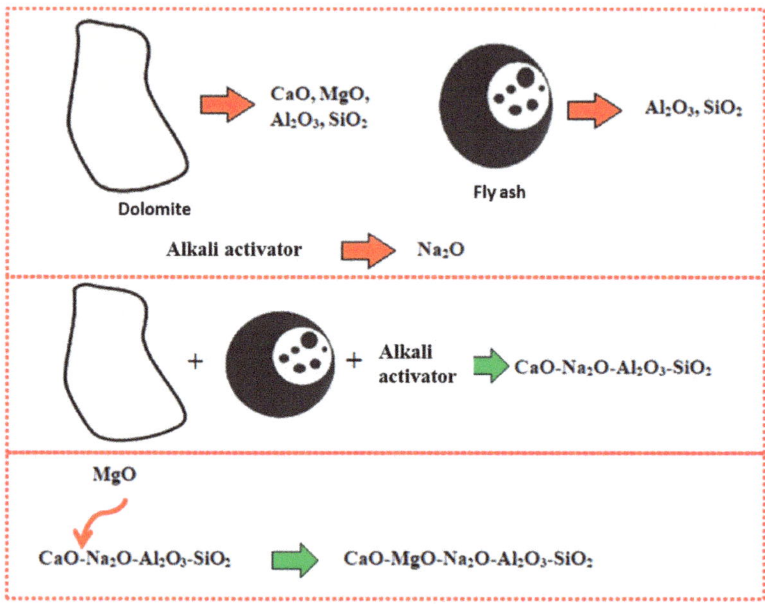

Figure 5. Schematic of Magnesium Oxide (MgO) interference in geopolymer system.

The intensity of the C-A-S-H peak decreased due to the decomposition of some geopolymer phases via dehydroxylation process. The dehydration and dehydroxylation caused shrinkage to DFA geopolymer composites. This is closely linked with the occurrence of the crack noted in the composite exposed to 1000 °C in the microstructure analysis result presented earlier. The peak of nepheline ($NaAlSiO_4$) was observed as the geopolymer composite was exposed to from 900 to 1000 °C [37]. The peak of dickite ($Al_2Si_2O_5(OH)_4$) was also formed from the breakage structure of the DFA geopolymer composite. As the temperature was elevated up to 1000 °C, some new crystalline phases were simultaneously produced, such as akermanite in geopolymer-based metakaolin/GGBS [38].

3.4. Functional Group Analysis

Figure 6 portrays the Fourier Transform Infrared (FTIR) spectra of DFA geopolymer composites before (control) and after exposure to 200, 400, and 1000 °C. Based on the spectrum displayed in Figure 6, peak changes occurred at bands 3600 and 1690 cm^{-1} for all samples exposed to a high temperature. The peak at 3600 cm^{-1} became broader due to high transmittance and, at 1690 cm^{-1}, the peak turned almost flat. The peak at 1400 cm^{-1} increased and shifted to 1450 cm^{-1} as the temperature increased to 1000 °C. Peaks at 970 and 790 cm^{-1} also experienced changes in the control DFA composite as the temperature increased. A change was noted at band 450 cm^{-1} after the composite was exposed to an elevated temperature.

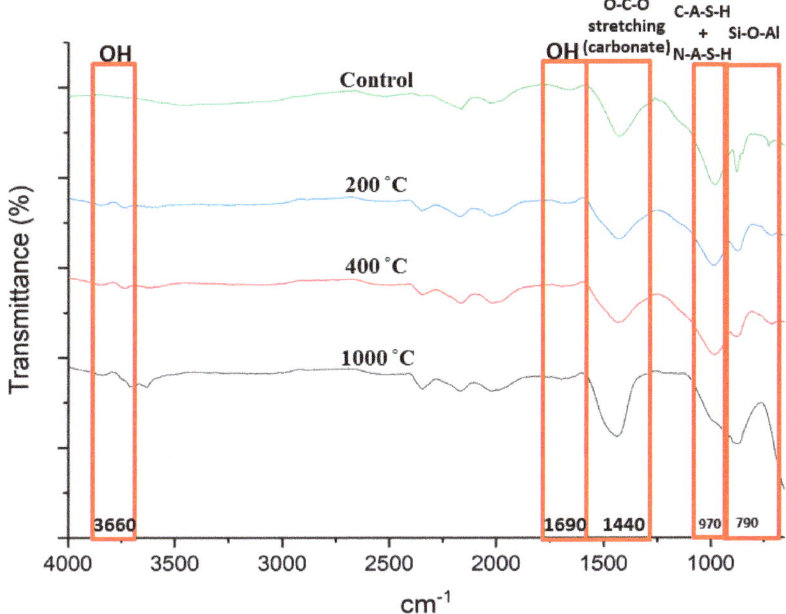

Figure 6. FTIR spectra of dolomite/fly ash geopolymer before (control) and after exposure to 200, 400, 1000 °C.

Changes in peaks occurred at bands 3600 and 1690 cm^{-1} for all samples exposed to a high temperature, mainly due to the removal of free water in the DFA. Bands ranging from 3550 to 3700 and 1690 cm^{-1} corresponded to OH^- stretching vibrations, which represented the water molecules present in the material [39].

A shift in the peak from 1400 to 1450 cm^{-1} was observed as the temperature was increased to 1000 °C. As the peak shifted to 1450 cm^{-1}, the strength of the composite decreased. The band at 1450 cm^{-1} reflects the characteristic of the asymmetric O-C-O stretching mode, which suggests the presence of sodium carbonate as a result of the reaction between excessive sodium and atmospheric carbon dioxide. This formation was due to the atmospheric carbonation on the surface of the matrix, as it reacted with CO_2 due to the exposure to heat. Zaharaki et al., [40] and Assaedi et al., [41] also reported the occurrence of a peak at 1440 cm^{-1} in the spectra of the samples subjected to heat, which is attributed to the presence of atmospheric carbonation.

The change in peaks at 970 and 790 cm^{-1} from the original DFA composite due to an increase in temperature was clearly observed. As the temperature was elevated, some geopolymer bonds were broken and resulted in the reduction in composite strength. This explains the low compressive strength of the geopolymer composite under 1000 °C heat, which is discussed later. This is also associated with the decreased intensity of the C-A-S-H peak due to the decomposition of some geopolymer phases in the phase analysis result. A band at 970 cm^{-1} was attributed to the formation of C-A-S-H with N-A-S-H gel, while the 790 cm^{-1} bands were assigned to Si–O–Si symmetrical stretching. The magnitude of these bands is attributable to the amorphous nature of the material. The Si-O-Al stretching vibration band is also located at 970 cm^{-1}. The Si-O-Al was determined by the peaks found between 700 and 1000 cm^{-1} [42].

The band at 450 cm^{-1} experienced a change after the composite was exposed to an elevated temperature. This band is related to Al-O or Si-O in bending and plane mode. The clear change was noted at composites exposed to 1000 °C, which led to the breakage of the Si-O-Al bond that later turned into either Si-O or Al-O.

3.5. Compressive Strength

The result of the compressive strength noted in the composite before (control) and after the temperature exposure at 200, 400, 600, 800, and 1000 °C is shown in Figure 7. The graph displays that the compressive strength of DFA composites increased as the exposed temperature increased to 400 °C. At 400 °C, the compressive strength of the DFA geopolymer composites was 23% higher than that of the original (without temperature exposure) DFA composites. However, the strength of the composites began to decrease when the exposure temperature was elevated from 600 to 1000 °C. The DFA composite at 1000 °C exhibited lower compressive strength than those recorded at other temperatures. The highest strength was 74.48 MPa, after exposure to 400 °C, while the lowest strength was 32.54 MPa, upon exposure to 1000 °C. The difference between the highest and the lowest strength was 51.31%.

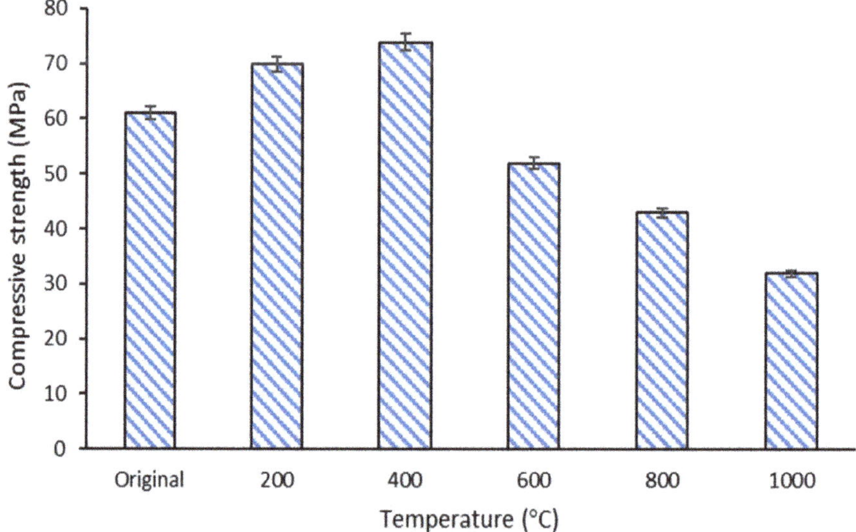

Figure 7. Compressive strength of dolomite/fly ash geopolymer composites before and after fire exposure at 200, 400, 600, 800, 1000 °C.

The strength of the DFA geopolymer increased as the temperature increased, attaining a peak strength of 74.48 MPa at 400 °C. At this stage, dolomite, which contains a high Ca content, promoted the optimum reaction to the geopolymerisation process. This is proven by the microstructure of composites, whereby all the raw materials had already reacted in the microstructure analysis earlier. The geopolymer matrix that had already undergone an optimum reaction was indicated by all the materials that lost their original shape. The strength development of the geopolymer using material that contained a high Ca composition was slow. The strength development in geopolymer could be accelerated by increasing the aging time, or when subjected to temperature exposure, or both. Thus, further introduction to heat (by increasing the temperature) up to a certain temperature accelerated the strength development of the geopolymer that contained high Ca materials. This further explains the increasing compressive strength after exposure to 400 °C.

The increase of 23% strength at 400 °C is attributed to the promotion of polycondensation between chain-like geopolymer gels. This can be attributed to the stiffening of gel and an increment in surface forces between the gel particles due to the release of adsorbed moisture or due to the promotion of polycondensation between chain-like geopolymer gels [43]. This is closely linked with the dehydration of free or weakly bound water content entrapped in the geopolymer during mixing and curing processes. Dehydration and oxidation often act, starting from 500 °C for OPC and other geopolymer materials. In some cases, the dehydration of the geopolymer matrix can lead to a significant reduction in strength, and some phase transformations occur after being exposed to temperatures ranging between 450 and 850 °C [6]. Regarding the geopolymer that contains dolomite with slow reactivity, the action occurred as early as 400 °C. At this stage, the DFA geopolymer was dry and did not contain entrapped water content, thus enhancing the strength of the composites. Heating the geopolymer composite to this temperature generated more high-density C-A-S-H with sodium aluminate silicate (N-A-S-H) phase. As the formation of C-A-S-H with N-A-S-H gel increased, the strength of the composites increased as well. More geopolymer phases were produced without the entrapped water content, resulting in the strength improvement of the geopolymer sample [44].

As the temperature increased from 600 to 1000 °C, the shrinkage and deterioration of the geopolymer sample started to occur for this composite. With increasing temperature, the pressure also increased in the geopolymer sample due to the expanding water vapour, which is also known as the dehydration and dehydroxylation of the binders. Differential thermal analysis of minerals revealed that dehydroxylation in the clay-like based occurred at varied temperatures, starting from ~500 or ~700 °C, or between 500 and 700 °C [30,45]. This pressure leads to shrinkage and cracking, thus decreasing the strength of the sample [37]. This is also proven by the change in peaks at 970 and 790 cm^{-1} in functional group analysis of the composite under 1000 °C from the control DFA composite. The strength deterioration from 600 to 1000 °C is attributable to the $Ca(OH)_2$ decomposition that occurs, starting at 600 °C. Basically, $Ca(OH)_2$ was generated from the decomposition of the C-A-S-H phase that contained Ca element due to exposure to elevated temperature [46]. The decomposition of C-A-S-H gels that took place in the DFA composite was due to the introduction of CaO. The CaO, which was derived from dolomite, may also cause severe strength reduction at high temperatures. Although the strength of DFA composites decreased after exposure to 1000 °C, the strength was still higher and better when compared to OPC-based composites.

4. Conclusions

The dolomite has the potential to be used as a raw material for geopolymer. However, it produces a low reactivity in the geopolymerisation process, which leads to slow strength development. Applying heat or temperature exposure to the DFA geopolymer composites appears to be successful in increasing the reactivity of the geopolymerisation process, thus helping to improve strength development. Surprisingly, the composite strength increased after being exposed to 400 °C. This outcome is considered unique, mainly because OPC, or a geopolymer using other materials, would start to decrease or retain their original strength. Based on the elemental distribution analysis, the elements in the geopolymer composites were distributed well, signifying a homogenous composite sample.

Author Contributions: Conceptualization and writing, E.A.A.; supervision and resources, M.M.A.B.A.; writing—review and editing, P.V.; validation, M.A.A.M.S.; methodology, A.V.S.; formal analysis, J.C.; investigation, S.Y.; resources, K.H.; visualization, reviewing and editing, I.H.A. All authors have read and agree to the published version of the manuscript.

Funding: The authors gratefully acknowledge the Centre of Excellent Geopolymer and Green Technology (CeGeoGTech), UniMAP for the financial support. The authors would like to extend their gratitude to the European Union for sponsoring the "Partnership for Research in Geopolymer Concrete" (PRI-GeoC-689857) grant.

Acknowledgments: A special thanks is dedicated to Synchrotron Light Research Institute (SLRI), Thailand for essential testing accommodation.

Conflicts of Interest: The authors declare no conflict of interest.

References

1. Sarker, P.K.; Kelly, S.; Yao, Z. Effect of fire exposure on cracking, spalling and residual strength of fly ash geopolymer concrete. *Mater. Des.* **2014**, *63*, 584–592. [CrossRef]
2. Zeng, L.; Dan-yang, C.; Xu, Y.; Chun-wei, F.; Xiao-qin, P. Novel Method for Preparation of Calcined Kaolin Intercalation Compound Based Geopolymer. *Appl. Clay Sci.* **2014**, *101*, 637–642. [CrossRef]
3. Duxson, P.; Fernández-Jiménez, A.; Provis, J.L.; Lukey, G.C.; Palomo, A.; van Deventer, J.S.J. Geopolymer Technology: The Current State of The Art. *J. Mater. Sci.* **2007**, *42*, 2917–2933. [CrossRef]
4. Scrivener, K.L.; John, V.M.; Gartner, E.M. Eco-efficient cements: Potential economically viable solutions for a low-CO2 cement-based materials industry. *Cem. Concr. Res.* **2018**, *114*, 2–26. [CrossRef]
5. Davidovits, J. Geopolymer Chemistry and Application. *Geopolym. Inst.* **2008**, *2*, 585.
6. Temuujin, J.; Rickard, W.; Lee, M.; van Riessen, A. Preparation and Thermal Properties of Fire Resistant Metakaolin Based Geopolymer Type Coatings. *J. Non-Cryst. Solids* **2011**, *357*, 1399–1404. [CrossRef]
7. Coppola, L.; Coffetti, D.; Crotti, E. Pre-packed alkali activated cement-free mortars for repair of existinf masonry buildings and concrete structures. *Constr. Build. Mater.* **2018**, *173*, 111–117. [CrossRef]
8. Burduhos Nergis, D.D.; Abdullah, M.M.A.B.; Sandu, A.; Vizureanu, P. XRD and TG-DTA Study of New Alkali Activated Materials Based on Fly Ash with Sand and Glass Powder. *Materials* **2020**, *13*, 343. [CrossRef]
9. Ming, L.Y.; Sandu, A.V.; Yong, H.C.; Tajunnisa, Y.; Azzahran, S.F.; Bayuji, R.; Abdullah, M.M.A.B.; Vizureanu, P.; Hussin, K.; Jin, T.S.; et al. Compressive Strength and Thermal Conductivity of Fly Ash Geopolymer Concrete Incorporated with Lightweight Aggregate, Expanded Clay Aggregate and Foaming Agent. *Rev. Chim.* **2019**, *70*, 4021–4028. [CrossRef]
10. Burduhos Nergis, D.D.; Vizureanu, P.; Corbu, O. Synthesis and Characteristics of Local Fly Ash Based Geopolymers Mixed with Natural Aggregates. *Rev. Chim.* **2019**, *70*, 1262–1267. [CrossRef]
11. Shahedan, N.F.; Abdullah, M.M.A.B.; Mahmed, N.; Kusbiantoro, A.; Hussin, K.; Sandu, A.V.; Naveed, A. Thermal Insulation Properties of Insulated Concrete. *Rev. Chim.* **2019**, *70*, 3027–3031. [CrossRef]
12. Burduhos Nergis, D.D.; Abdullah, M.M.A.B.; Vizureanu, P. The Effect of Fly Ash/Alkaline Activator Ratio in Class F Fly Ash Based Geopolymers. *Eur. J. Mater. Sci. Eng.* **2017**, *2*, 111–118.
13. Bouaissi, A.; LI, L.Y.; Moga, L.M.; Sandu, I.G.; Abdullah, M.M.A.B.; Sandu, A.V. A Review on Fly Ash as a Raw Cementitious Material for Geopolymer Concrete. *Rev. Chim.* **2018**, *69*, 1661–1667. [CrossRef]
14. Burduhos Nergis, D.D.; Abdullah, M.M.A.B.; Vizureanu, P.; Faheem, M.T.M. Geopolymers and Their Uses: Review. *IOP Conf. Ser. Mater. Sci. Eng.* **2018**, *374*, 012019. [CrossRef]
15. Burduhos Nergis, D.D.; Vizureanu, P.; Andrusca, L.; Achitei, D. Performance of local fly ash geopolymers under different types of acids. *IOP Conf. Ser. Mater. Sci. Eng.* **2019**, *572*, 012026. [CrossRef]
16. Jun, N.H.; Minciuna, M.G.; Abdullah, M.M.A.; Jin, T.S.; Sandu, A.V.; Ming, L.Y. Mechanism of Cement Paste with Different Particle Sizes of Bottom Ash as Partial replacement in Portland Cement. *Rev. Chim.* **2017**, *68*, 2367–2372. [CrossRef]
17. Al-Majidi, M.H.; Lampropoulos, A.; Cundy, A.; Meikle, S. Development of geopolymer mortar under ambient temperature for in situ applications. *Constr. Build. Mater.* **2016**, *120*, 198–211. [CrossRef]
18. Talha Junaid, M.; Kayali, O.; Khennane, A. Response of alkali activated low calcium fly-ash based geopolymer concrete under compressive load at elevated temperatures. *Mater. Struct.* **2016**, *50*, 50. [CrossRef]
19. Sitarz, M.; Hager, I.; Kochanek, J. Effect of High Temperature on Mechanical Properties of Geopolymer Mortar. *MATEC Web Conf.* **2018**, *163*, 06004. [CrossRef]
20. Hamidi, R.M.; Man, Z.; Azizli, K.A. Concentration of NaOH and the Effect on the Properties of Fly Ash Based Geopolymer. *Procedia Eng.* **2016**, *148*, 189–193. [CrossRef]
21. Morsy, M.; Alsayed, S.; Al-Salloum, Y.; Almusallam, T. Effect of Sodium Silicate to Sodium Hydroxide Ratios on Strength and Microstructure of Fly Ash Geopolymer Binder. *Arab. J. Sci. Eng.* **2014**, *39*, 4333–4339. [CrossRef]
22. Zarina, Y.; Kamarudin, H.; Al Bakri, A.M.M.; Khairul Nizar, I.; Rafiza, A.R. Influence of Dolomite on The Mechanical Properties of Boiler Ash Geopolymer Paste. *Key Eng. Mater.* **2014**, *594*, 8–12. [CrossRef]

23. Yip, C.K.; Provis, J.L.; Lukey, G.C.; van Deventer, J.S.J. Carbonate Mineral Addition to Metakaolin-Based Geopolymers. *Cem. Concr. Compos.* **2008**, *30*, 979–985. [CrossRef]
24. Valencia Saavedra, W.G.; Mejía de Gutiérrez, R. Performance of Geopolymer Concrete Composed of Fly Ash After Exposure to Elevated Temperatures. *Constr. Build. Mater.* **2017**, *154*, 229–235. [CrossRef]
25. Ahmari, S.; Zhang, L. The properties and durability of mine tailings-based geopolymeric masonry blocks. In *Eco-Efficient Masonry Bricks and Blocks*; Pacheco-Torgal, F., Lourenço, P.B., Labrincha, J.A., Kumar, S., Chindaprasirt, P., Eds.; Woodhead Publishing: Oxford, UK, 2015.
26. Zulkifly, K.; Yong, H.; Abdullah, M.; Ming, L.; Panias, D.; Sakkas, K. Review of Geopolymer Behaviour in Thermal Environment. *IOP Conf. Ser. Mater. Sci. Eng.* **2017**, *209*, 012085. [CrossRef]
27. Walkley, B.; San Nicolas, R.; Sani, M.-A.; Bernal, S.A.; van Deventer, J.S.J.; Provis, J.L. Structural Evolution of Synthetic Alkali Activated CaO-MgO-Na2O-Al2O3-SiO2 Materials is Influenced by Mg Content. *Cem. Concr. Res.* **2017**, *99*, 155–171. [CrossRef]
28. Bernal, S.A.; San Nicolas, R.; Myers, R.J.; Mejía de Gutiérrez, R.; Puertas, F.; van Deventer, J.S.J.; Provis, J.L. MgO Content of Slag Controls Phase Evolution and Structural Changes Induced by Accelerated Carbonation in Alkali Activated Binders. *Cem. Concr. Res.* **2014**, *57*, 33–43. [CrossRef]
29. Sudbrink, B.; Khanzadeh Moradllo, M.; Hu, Q.; Ley, M.T.; Davis, J.M.; Materer, N.; Apblett, A. Imaging the presence of silane coatings in concrete with micro X-ray fluorescence. *Cem. Concr. Res.* **2017**, *92*, 121–127. [CrossRef]
30. Heller-Kallai, L. Chapter 10.2—Thermally Modified Clay Minerals. In *Developments in Clay Science*; Bergaya, F., LAGALY, G., Eds.; Elsevier: Jerusalem, Israel, 2013.
31. Clegg, F.; Breen, C.; Carter, M.A.; Ince, C.; Savage, S.D.; Wilson, M.A. Dehydroxylation and Rehydroxylation Mechanisms in Fired Clay Ceramic: A TG-MS and DRIFTS Investigation. *J. Am. Ceram. Soc.* **2012**, *95*, 416–422. [CrossRef]
32. Salleh, M.M.; McDonald, S.; Terada, Y.; Yasuda, H.; Nogita, K. Development of a microwave sintered TiO2 reinforced Sn–0.7 wt% Cu–0.05 wt% Ni alloy. *Mater. Des.* **2015**, *82*, 136–147. [CrossRef]
33. Giergiczny, Z. Fly ash and slag. *Cem. Concr. Res.* **2019**, *124*, 105826. [CrossRef]
34. Phoo-ngernkham, T.; Maegawa, A.; Mishima, N.; Hatanaka, S.; Chindaprasirt, P. Effects of Sodium Hydroxide and Sodium Silicate Solutions on Compressive and Shear Bond Strengths of FA–GBFS Geopolymer. *Constr. Build. Mater.* **2015**, *91*, 1–8. [CrossRef]
35. Machner, A.; Zajac, M.; Ben Haha, M.; Kjellsen, K.O.; Geiker, M.R.; De Weerdt, K. Chloride-binding capacity of hydrotalcite in cement pastes containing dolomite and metakaolin. *Cem. Concr. Res.* **2018**, *107*, 163–181. [CrossRef]
36. Ke, X.; Bernal, S.A.; Provis, J.L. Controlling the reaction kinetics of sodium carbonate-activated slag cements using calcined layered double hydroxides. *Cem. Concr. Res.* **2016**, *81*, 24–37. [CrossRef]
37. Kuenzel, C.; Grover, L.M.; Vandeperre, L.; Boccaccini, A.R.; Cheeseman, C.R. Production of Nepheline/Quartz Ceramics from Geopolymer Mortars. *J. Eur. Ceram. Soc.* **2014**, *33*, 251–258. [CrossRef]
38. Zhang, Y.J.; Li, S.; Wang, Y.C.; Xu, D.L. Microstructural and Strength Evolutions of Geopolymer Composite Reinforced by Resin Exposed to Elevated Temperature. *J. Non-Cryst. Solids* **2012**, *358*, 620–624. [CrossRef]
39. Abo Sawan, S.E.; Zawrah, M.F.; Khattab, R.M.; Abdel-Shafi, A.A. Fabrication, sinterabilty and characterization of non-colored and colored geopolymers with improved properties. *Mater. Res. Express* **2019**, *6*, 075205. [CrossRef]
40. Zaharaki, D.; Komnitsas, K.; Perdikatsis, V. Use of Analytical Techniques for Identification of Inorganic Polymer Gel Composition. *J. Mater. Sci.* **2010**, *45*, 2715–2724. [CrossRef]
41. Assaedi, H.; Shaikh, F.U.A.; Low, I.M. Effect of Nano-clay on Mechanical and Thermal Properties of Geopolymer. *J. Asian Ceram. Soc.* **2016**, *4*, 19–28. [CrossRef]
42. Peyne, J.; Gautron, J.; Doudeau, J.; Rossignol, S. Development of Low Temperature Lightweight Geopolymer Aggregate, From Industrial Waste, in Comparison With High Temperature Processed Aggregates. *J. Clean. Prod.* **2018**, *189*, 47–58. [CrossRef]
43. Part, W.K.; Ramli, M.; Cheah, C.B. An overview on the influence of various factors on the properties of geopolymer concrete derived from industrial by-products. *Constr. Build. Mater.* **2015**, *77*, 370–395. [CrossRef]
44. Kumar, R.; Singh, S.; Singh, L.P. Studies on Enhanced Thermally Stable High Strength Concrete Incorporating Silica Nanoparticles. *Constr. Build. Mater.* **2017**, *153*, 506–513. [CrossRef]

45. Yuan, P. Chapter 7—Thermal-Treatment-Induced Deformations and Modifications of Halloysite. In *Developments in Clay Science*; Yuan, P., Thill, A., Bergaya, F., Eds.; Elsevier: Guangzhou, China, 2016.
46. Kong, D.L.Y.; Sanjayan, J.G. Effect of Elevated Temperatures on Geopolymer Paste, Mortar and Concrete. *Cem. Concr. Res.* **2010**, *40*, 334–339. [CrossRef]

MDPI

Article

Influence of TiO_2 Nanoparticles on the Resistance of Cementitious Composite Materials to the Action of Bacteria

Andreea Hegyi [1], Adrian-Victor Lăzărescu [1,*], Henriette Szilagyi [1], Elvira Grebenişan [1], Jana Goia [2] and Andreea Mircea [3,*]

[1] NIRD URBAN-INCERC Cluj-Napoca Branch, 117 Calea Florești, 400524 Cluj-Napoca, Romania; andreea.hegyi@incerc-cluj.ro (A.H.); henriette.szilagyi@incerc-cluj.ro (H.S.); elvira.grebenisan@incerc-cluj.ro (E.G.)

[2] Municipal Hospital, 14–16 1 Mai Street, 405200 Dej, Romania; janagoia@yahoo.com

[3] Facultatea de Construcţii, Technical University of Cluj-Napoca, 28 Memorandumului, 400114 Cluj-Napoca, Romania

* Correspondence: adrian.lazarescu@incerc-cluj.ro (A.-V.L.); andreea.mircea@ccm.utcluj.ro (A.M.)

Abstract: The formation of biofilms on cementitious building surfaces can cause visible discoloration and premature deterioration, and it can also represent a potential health threat to building occupants. The use of embedded biofilm-resistant photoactivated TiO_2 nanoparticles at low concentrations in the cementitious composite matrix is an effective method to increase material durability and reduce maintenance costs. Zone of inhibition studies of TiO_2-infused cementitious samples showed efficacy toward both Gram-negative and Gram-positive bacteria.

Keywords: cementitious composites; TiO_2 nanoparticles; photocatalysis; bactericidal effect

Citation: Hegyi, A.; Lăzărescu, A.-V.; Szilagyi, H.; Grebenişan, E.; Goia, J.; Mircea, A. Influence of TiO_2 Nanoparticles on the Resistance of Cementitious Composite Materials to the Action of Bacteria. *Materials* **2021**, *14*, 1074. https://doi.org/10.3390/ma14051074

Academic Editor: Stefano Lettieri

Received: 20 January 2021
Accepted: 23 February 2021
Published: 25 February 2021

1. Introduction

Worldwide, a shift in human lifestyles over the last half century has seen a growing number of daily activities move from the outdoors to enclosed inner spaces [1]. At the present time, it is known that the growth of micro-organisms (fungi, bacteria, viruses, algae, lichen, dust mites) on building surfaces (floors, walls, ceilings) has a detrimental effect on population health, particularly with contaminated interior surfaces. Maintaining good air quality and appearance of homes and workplaces has resulted in increased maintenance and repair costs. The existence of a so-called "Sick Building Syndrome (SBS)" manifests itself in the population that operates, partially or totally, inside buildings affected by mold or colonies (biofilms) of bacteria [1–3]. Pathogenic bacteria readily develop and survive on surfaces, especially in conditions of high humidity (min. 97%) and at temperatures ranging from −5 to +60 °C. The development of simple, inexpensive, and preventative antibacterial treatment methods of these surfaces are critical to maintain good public health [4,5]. In particular, areas of intensive medical use require the regular and thorough disinfection of surfaces in order to reduce the number of bacteria and prevent bacterial transmission to patients [4–7]. These untreated surfaces can act as reservoirs of microorganisms, which in turn could lead to the spread of infections [8]. Applications of the photocatalytic process of nano-TiO_2 provides a conceptually simple and promising preventative technology for inhibiting contamination from bacteria, as well as an alternative to the constant use of chemical disinfectants [5,9–11].

The specific properties of TiO_2 were discovered in the 1950s, but their exploitation in practical applications began only in 1972, when Fujishima and Honda used it for water splitting. Another very important property of TiO_2 is its photoinduced superhydrophilicity, which was unexpectedly discovered in 1995 in SiO_2/TiO_2 composites illuminated by ultraviolet (UV) light.

An important feature of TiO_2-SiO_2 compounds is that, unlike TiO_2, whose photocatalytic activity ceases without UV illumination, their photocatalytic effect continues for hours or even days after the removal of the UV source. In 1997, the first publication by Luigi Cassar et al. focused on the production of cementitious materials with self-cleaning properties [12]. A significant amount of research has shown the benefits of the introduction of TiO_2 nanoparticles on the performance of cement composites: reduction of intake time, increase of the degree of cement hydration, increase of mechanical resistances (bending, compression, abrasion), adhesion to the reinforcement and resistance to its corrosion, and improvement of durability and freeze–thaw resistance [13–16].

In 1985, Matsunaga et al. first demonstrated the photocatalytic cytotoxic mechanisms in microbial cells from *Saccharomyces cerevisiae* (yeast), *Lactobacillus acidophilus* and *Escherichia coli* (bacteria), and *Chlorella vulgaris* (green algae) in water [17–21]. It is currently known that when the supplied photon energy is greater than the energy difference between the valence and conduction band edges of TiO_2 (which occurs for UV radiation), photogenerated electrons (e−) and holes (h+) react with O_2 and H_2O, forming anionic radicals (O^{2-}) and (OH). These oxidative species (h^+, O^{2-}, and OH) are highly reactive, contributing to the destruction of the cells of microorganisms [1,22–26]. The mechanism of destruction can be synthesized in the following sequence of reactions:

$$TiO_2 + h\nu \rightarrow TiO_2\,(e^-_{cb} + h^+_{vb}) \tag{1}$$

$$O_2 + e^-_{cb} \rightarrow O^{2-} \tag{2}$$

$$H_2O + h^+_{vb} \rightarrow \bullet OH + H^+ \tag{3}$$

$$\bullet OH + \bullet OH \rightarrow H_2O_2 \tag{4}$$

$$O^{2-} + H_2O_2 \rightarrow \bullet OH + OH^- + O_2 \tag{5}$$

$$\bullet OH + \text{Organic} + O_2 \rightarrow CO_2 + H_2O. \tag{6}$$

A number of studies have been carried out in this direction. Sunada et al. [27] have shown that cell membranes are photocatalytically destroyed in the case of *Escherichia coli* bacteria. Saito et al. [28] proposed a mechanism for the destruction of bacteria by inhibiting their respiration function once they come into contact with TiO_2. Oguma et al. [29] discussed a mechanism for the destruction of bacteria through the destruction of the cell wall and the induction of disorder at the cellular level following the contact of the microorganism with TiO_2. Evidences highlight that wavelengths in the range of 320–400 nm are the most efficient to activate the photo-cytotoxic activity of TiO_2. Gogniat et al. [30] showed that the adsorption capacity of TiO_2 is positively correlated with its biocidal effect. Adsorption has been consistently associated with a reduction in the integrity of the bacterial membrane, as indicated by flow cytometry. The authors suggested that the adsorption of cells on photoactivated TiO_2 is followed by a loss of membrane integrity, which was key to the biocidal effect. Mazurkova et al. [31] analyzed the effect of nano-TiO_2 on the influenza virus, indicating the destruction of the virus in the presence of nanoparticles. After 15 min of incubation, the nanoparticles adhered to the outer surface of the virus, the surface spinules of the virus were glued together, and its outer membrane, of a lipoprotein nature, was torn. After 30 min, the degree of destruction increased, and after 1–5 h of incubation, the virus that came into contact with the nanoparticles was completely destroyed. It is considered that this effect depends on the duration of exposure/incubation, the viral concentration, and the concentration of nano-TiO_2. The tests were carried out in three lighting conditions: dark, UV radiation, and natural light. Adams et al. [32] showed that the concentration of *Bacillus subtilis* and *Escherichia coli* were reduced upon contact with a suspension of nano-TiO_2 under natural light illumination. A similar effect on *Bacillus subtilis* has been reported by Armelao et al. [33]. Research conducted by Dedkova et al. [34] on samples of kaolin composites containing nano-TiO_2 indicated their biocidal effect in the presence of *Escherichia coli*, *Enterococcus faecalis*, and *Pseudomonas Aeruginosa*, after 2 days of exposure to artificial light. Results were also consistent with those of Gurr [35], who assessed that the antibacterial effect of TiO_2 composites is manifested in the presence of natural light, without necessarily requiring UV photoactivation. The study carried out by Hamdani [36] has shown that based on the behavior of the two cementitious mixtures with 3% and 5% nano-TiO_2-based type-Aeroxide P25, when in contact with *E. coli*, they were able to reduce the viability of bacteria after 24 h of exposure by 60–70%. The main observation, based on results in the literature [37–39], is that cementitious composite surfaces containing nano-TiO_2 have the ability to inhibit biofilm growth, destroying the *E-coli bacteria* with which they come in contact. This is also supported by results obtained by Daly et al. [40], Carre et al. [41], and Kubacka et al. [42], which confirm that by the formation of free radicals and anions strongly oxidized by the photoactivation of nano-TiO_2 (OH• and O_2^-), at the cellular level, plasma components such as DNA, RNA, lipids, and proteins are destroyed, and cell membranes are broken.

The easiest method of studying the resistance capacity of various building materials to the attack of microorganisms is the adapted antibiogram method, which is already used in medicine. This method is known as the halo inhibition method [43] or the Kirby–Bauer method, and it is currently standardized according to AATCC TM147 and AATCC TM30 (American Association of Textile Chemists and Colorists).

Although a large number of authors confirm the anti-bactericidal capacity of cement compounds with nano-TiO_2, there is still controversy about maintaining the bactericidal capacity in the absence of light (during the night), as well as the influence of the type of bacteria that contaminate the surface. Results show that their destruction begins after only 20 min of UV radiation exposure and 60–120 min is sufficient to destroy all the bacteria [44], since hydroxyl radicals are the main factors responsible for the bactericidal capacity of semiconductor's photocatalysts. They also possess a destruction capacity of *Escherichia coli* bacteria which is 10^3–10^4 times more effective than chemical disinfection products [45].

The aim of this work is to analyze the potential of inhibiting the growth of bacterial films on the surface of cementitious composites by adding different amounts of nano-TiO_2, using the zone of inhibition to get a quick test of the antibacterial efficiency, and to identify the nanoparticle-functionalized regions in relation to the amount of cement, which ensure a successful effect of resistance from a biological point of view. For this purpose, four types of bacteria were used, which were chosen because of the frequency with which they are encountered in the building environment: *Escherichia coli, Pseudomonas Aeruginosa, Staphylococcus Aureus,* and *Streptococcus Pyogenes.*

2. Materials and Methods

In order to study the self-cleaning and anti-bacterial properties of cementitious composites, the literature shows three types of concrete modifications: concrete covered with a thin layer of TiO_2, concrete covered by a thick layer of photoactive concrete on the top, and finally, different weight percentages of TiO_2 in the concrete mass (when TiO_2 substitutes cement or is present as additive) [46]. Each of these approaches has advantages and disadvantages in relation to the compatibility and adhesion to the substrate (in the case of thin films on the surface), durability, and impact on physical and mechanical performances (in the case of the introduction of TiO_2 nanoparticles into the cementitious mass) and even economic impact in terms of costs [47]. The methodology used was based on the studies presented by Meija-De Gutierrez [43].

The preparation of the cementitious composites was performed by using white Portland cement as binder and by adding TiO_2 nanoparticles as addition in different mass percentages, relative to cement quantity. Preliminary results obtained on the same mix-design ratios, when subjected to the method of staining with Rhodamine B and methylene blue, extended even for the situation of staining with exhaustion gas particles on the surface of the samples [48,49], have demonstrated the photocatalytic activity of nano-TiO_2 addition in the cementitious matrix and thus the self-cleaning capacity of the samples. Furthermore, the influence of added TiO_2 nanoparticles on the hydrophilicity of the cementitious composites surfaces was studied on the same mixtures [50].

Commercial Aeroxide P25 TiO_2 nanoparticles, 99.5% purity (Evonik Industries AG, Hanau, Germany), containing more than 70% anatase, with a minor amount of rutile and a small amount of amorphous phase anatase and rutile crystallites, with a reported ratio of 70:30 or 80:20, were used in the production of the samples. Both phases play an important role in industrial applications and contribute to the photoactivation mechanism [3]. According to the manufacturer's data sheet, the mean size of the TiO_2 particles is 21 nm, with a specific surface of 35–65 m^2/g. The test specimens were prepared and conditioned as shown in Table 1 in order to conduct the specific tests.

Table 1. Mix ratio and conditioning of cementitious composites.

Mixture Number	P1	P2	P3	P4	P5	P6	P7	P8	P9	P10
Amount of nanoparticles relative to the amount of cement (%)	0	1	2	3	3.6	4	5	6	10	12
CEM I 52,5R white cement, HOLCIM (g)	500	500	500	500	500	500	500	500	500	500
Amount of water relative to the amount of total dry mixture (water/(cement + nano-TiO_2)) (g)	0.5									
Conditioning	- 24 h in molds, 90% RH, 20 °C, without light; - demolding; - 27 days complete water immersion, 20 °C, without light.									

Then, the cementitious composite paste was cast into rectangular molds from which small, 17.4 mm circular samples were cut. Then, the samples were subjected to photoactivation, undergoing a 24 h UV ray treatment by using a light source with 400–315 nm spectrum emission (corresponding to the UVA band), which was located at a distance of 10 cm above the surface of the specimens, which determined a luminous flux intensity of 860 lux.

For testing resistance to the action of bacteria of the nano-TiO_2 cementitious composite samples, *Escherichia coli* (ATCC 25922), *Pseudomonas Aeruginosa* (ATCC 27853), *Staphylococcus Aureus* (ATCC 25923), and *Streptococcus Pyogenes* (ATCC 19615) solutions have been used. These solutions were prepared by harvesting them from reference bacteria cultures and introducing 2 colonies (2 loops of 1 μL each) of biological material into 1 mL of physiological serum. The biological load of the prepared solutions was semi-quantified by the contact plate method, using the specifications of MicroKount® microbilogic load test (Figure 1), determining a concentration of 10^7 CFUs for each type of bacteria used.

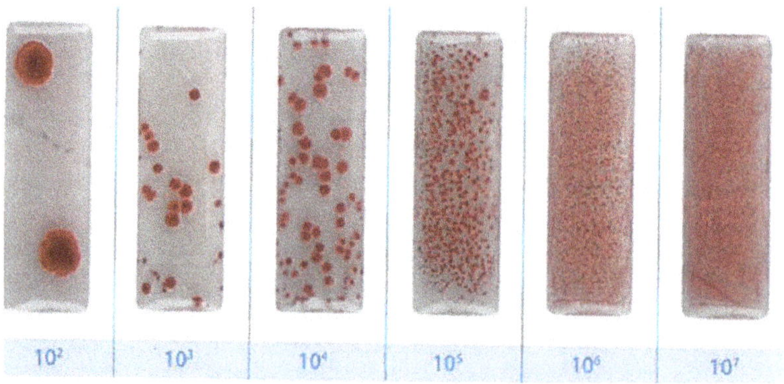

Figure 1. Quantification of biological load according to MicroKount® specifications.

At the same time, φ = 90 mm Petri dishes were prepared, in which suitable nutrient substrates were placed in order to develop the bacteria cultures, such as agar for bacteria cultures *Escherichia coli*, *Pseudomonas Aeruginosa*, and *Staphylococcus Aureus*, respectively, and blood agar for bacteria culture *Streptococcus Pyogenes*. All of them were sterilized under UV rays.

In each sterilized Petri dish with a suitable nutrient substrate, depending on the type of bacteria, 1 mL of bacterium suspension was applied and distributed so that the entire surface of the nutrient substrate was covered. Consequently, the photoactivated cementitious composite samples were centrally placed in the Petri dishes without any cross-contamination of the system.

Subsequently, 0.5 mL of biological material–bacterium suspension was applied on the samples, and the Petri dish lid was sealed by isolating the whole system on the edge to prevent cross contamination. One sample for each type of bacteria used was made without using any cementitious composite sample (P0). This was considered the primary control sample and was used to demonstrate the viability of the used bacteria in the suspensions.

Romanian STAS 12718 offers the possibility of semi-quantitative quantification of the microbiological load of the system, providing a quantification grid as follows: 0 (−) no growth (sterile); 1 (+) 1–10 colonies of microorganisms; 2 (++) over 10 colonies of microorganisms; 3 (+++) areas with confluent colonies; 4 (++++) growth throughout the surface.

Then, the farming systems were placed in the laboratory, at a temperature of 30 °C, under natural light conditions. As a result of exposure to natural light, the photoactivation process initially induced by UV exposure was continuously refreshed throughout 21 days, similarly to the alternation of day/night periods. Moreover, the initial UV exposure, for 24 h, ensured the sterilization of the cementitious samples so that there was no pre-existing contamination. At regular intervals of time, i.e., 2, 3, 4, 6, 7, 14, and 21 days, the development of the systems was examined visually and microscopically for signs of growth/development of the material (the colonies). The presence/development of the halo of inhibition was also studied. The quantification of the behavior of the bacteria contaminated samples was carried out qualitatively, according to STAS 12718 (Table 2) and using the Kirby–Bauer technique (AATCC TMII47), which is commonly used in medicine and has been known since 1966 as the antibiogram method. This method was adapted to the present requirements and conditions, by which the diameter of the inhibition halo, φ (mm) was measured.

Table 2. Quantification of the biological load according to STAS 12718.

0 (-)	no growth (sterile)
1 (+)	1–10 colonies of microorganisms
2 (++)	over 10 colonies of microorganisms
3 (+++)	areas with confluent colonies
4 (++++)	growth throughout the surface

The effectiveness of the antibacterial effect induced by nano-TiO_2 photoactivation was assessed quantitatively by introducing a quantifiable parameter, EEA, which represents the percentage change in the diameter of the inhibition halo of the analyzed sample (with nano-TiO_2 content in the matrix >0, relative to the control sample (with 0% nano-TiO_2 content in the matrix), according to Equation (7).

$$EEA = (\varphi_{\%TiO2} - \varphi_{control})/\varphi_{control} \times 100\ (\%) \quad (7)$$

where $\varphi_{\%TiO2}$—inhibition halo diameter measured for the sample with $TiO_2\% > 0$ (mm); $\varphi_{control}$—inhibition halo diameter measured for TiO_2 0% control sample (mm).

3. Results and Discussion

Experimental results on the behavior of cementitious composites with the addition of TiO_2 in an environment contaminated with *Escherichia coli* are presented in Table 3 and in Figures 2 and 3. For the environment contaminated with *Pseudomonas Aeruginosa*, results are presented in Table 4 and in Figures 3 and 4. For the environment contaminated with *Staphylococcus Aureus*, results are shown in Table 5 and in Figures 3 and 5 and for the samples contaminated with *Streptococcus Pyogenes*, results are shown in Table 6 and in Figures 3 and 6.

Table 3. Quantification of the microbiological load of the system according to STAS 12718 for samples exposed to *Escherichia coli*.

Exposure Period (Days)	P0 (without Composite Sample)	P1 (0% TiO_2)	P2 (1% TiO_2)	P3 (2% TiO_2)	P4 (3% TiO_2)	P5 (3.6% TiO_2)	P6 (4% TiO_2)	P7 (5% TiO_2)	P8 (6% TiO_2)	P10 (12% TiO_2)
2	1 (+)	1 (+)	1 (+)	1 (+)	1 (+)	1 (+)	0 (-)	0 (-)	0 (-)	0 (-)
3	1 (+)	1 (+)	1 (+)	1 (+)	1 (+)	1 (+)	0 (-)	0 (-)	0 (-)	0 (-)
4	3 (+++)	1 (+)	1 (+)	1 (+)	1 (+)	1 (+)	0 (-)	0 (-)	0 (-)	0 (-)
6	3 (+++)	1 (+)	1 (+)	1 (+)	1 (+)	1 (+)	0 (-)	0 (-)	1 (+)	1 (+)
7	3 (+++)	1 (+)	1 (+)	1 (+)	1 (+)	1 (+)	0 (-)	1 (+)	1 (+)	1 (+)
14	3 (+++)	1 (+)	1 (+)	1 (+)	1 (+)	1 (+)	1 (+)	1 (+)	1 (+)	1 (+)
21	3 (+++)	1 (+)	3 (+++)	1 (+)	3 (+++)	3 (+++)	2 (++)	1 (+)	1 (+)	1 (+)

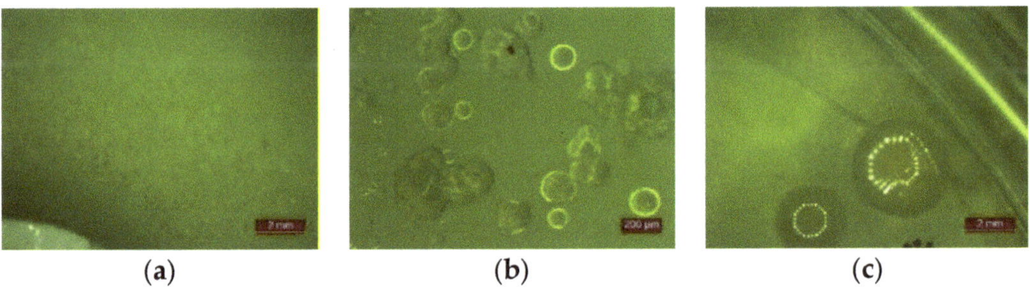

Figure 2. Microscopic examination of the behavior of samples exposed to *Escherichia coli*: (**a**) Identification of the inhibition halo area (1:2 mm); (**b**) Growth colonies (1:200 µm); (**c**) Detail of growth colonies (1:2 mm).

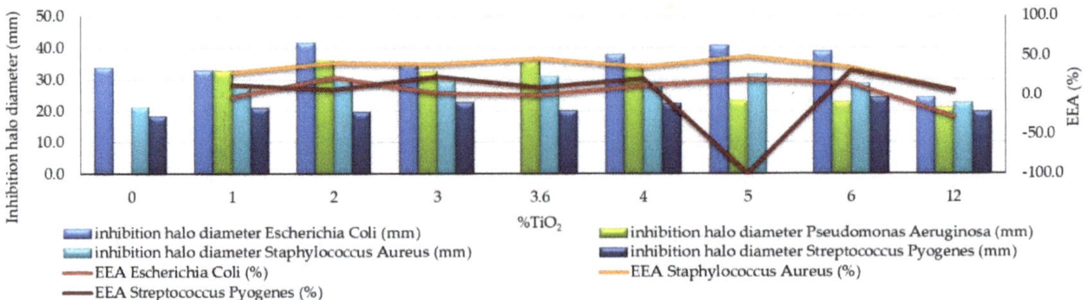

Figure 3. Inhibition halo size development after 2 days of exposure to bacteria *Escherichia coli*, *Pseudomonas Aeruginosa*, *Staphylococcus Aureus*, and *Streptococcus Pyogenes*. (In the case of *Pseudomonas Aeruginosa* contamination, the EEA could not be quantified, because the control sample (0% TiO_2) did not show an inhibition halo). EEA: the percentage change in the diameter of the inhibition halo of the analyzed sample, indicating the effectiveness of the antibacterial effect.

Table 4. Quantification of the microbiological load of the system according to STAS 12718 for samples exposed to *Pseudomonas Aeruginosa*.

Exposure Period (Days)	P0 (Without Composite Sample)	P1 (0% TiO_2)	P2 (1% TiO_2)	P3 (2% TiO_2)	P4 (3% TiO_2)	P5 (3.6% TiO_2)	P6 (4% TiO_2)	P7 (5% TiO_2)	P8 (6% TiO_2)	P10 (12% TiO_2)
2	3 (+++)	3 (+++)	2 (++)	2 (++)	2 (++)	2 (++)	2 (++)	2 (++)	2 (++)	2 (++)
3	3 (+++)	3 (+++)	2 (++)	2 (++)	2 (++)	2 (++)	2 (++)	2 (++)	2 (++)	2 (++)
4	3 (+++)	3 (+++)	2 (++)	2 (++)	2 (++)	2 (++)	2 (++)	2 (++)	2 (++)	2 (++)
6	3 (+++)	3 (+++)	2 (++)	2 (++)	2 (++)	2 (++)	2 (++)	2 (++)	2 (++)	2 (++)
7	3 (+++)	3 (+++)	3 (+++)	2 (++)	2 (++)	2 (++)	2 (++)	2 (++)	2 (++)	2 (++)
14	3 (+++)	3 (+++)	3 (+++)	3 (+++)	2 (++)	2 (++)	2 (++)	2 (++)	2 (++)	2 (++)
21	3 (+++)	3 (+++)	3 (+++)	3 (+++)	3 (+++)	2 (++)	2 (++)	2 (++)	2 (++)	2 (++)

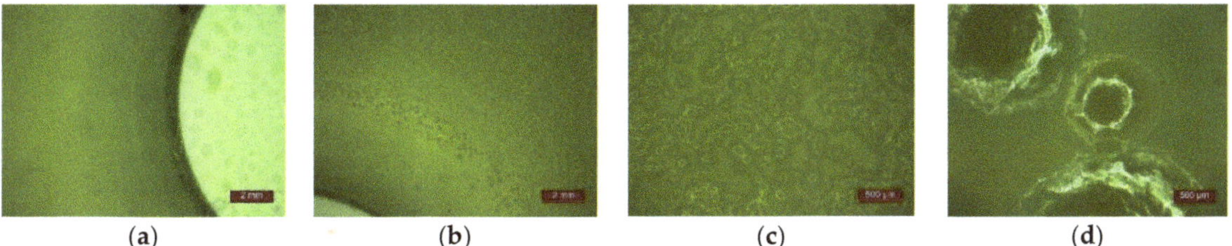

Figure 4. Microscopic examination of the behavior of samples exposed to *Pseudomonas Aeruginosa*: (**a**) The presence of biological material on the surface of the composite specimen at the time of sowing (1:2 mm); (**b**) Identification of the inhibition halo zone (1:2 mm); (**c**) Growth of colonies (1:500 μm); (**d**) Detail growth of colonies (1:500 μm).

Table 5. Quantification of the microbiological load of the system according to STAS 12718 for samples exposed to *Staphylococcus Aureus*.

Exposure Period (Days)	P0 (without Composite Sample)	P1 (0% TiO_2)	P2 (1% TiO_2)	P3 (2% TiO_2)	P4 (3% TiO_2)	P5 (3.6% TiO_2)	P6 (4% TiO_2)	P7 (5% TiO_2)	P8 (6% TiO_2)	P10 (12% TiO_2)
2	0 (-)	0 (-)	0 (-)	0 (-)	0 (-)	0 (-)	0 (-)	0 (-)	0 (-)	0 (-)
3	1 (+)	1 (+)	0 (-)	0 (-)	0 (-)	0 (-)	0 (-)	1 (+)	0 (-)	0 (-)
4	1 (+)	1 (+)	0 (-)	0 (-)	1 (+)	0 (-)	0 (-)	1 (+)	0 (-)	1 (+)
6	1 (+)	1 (+)	1 (+)	0 (-)	1 (+)	0 (-)	1 (+)	1 (+)	0 (-)	1 (+)
7	1 (+)	1 (+)	1 (+)	0 (-)	1 (+)	1 (+)	1 (+)	3 (+++)	0 (-)	1 (+)
14	1 (+)	1 (+)	1 (+)	0 (-)	1 (+)	1 (+)	1 (+)	3 (+++)	1 (+)	1 (+)
21	1 (+)	1 (+)	1 (+)	1 (+)	1 (+)	1 (+)	1 (+)	3 (+++)	1 (+)	1 (+)

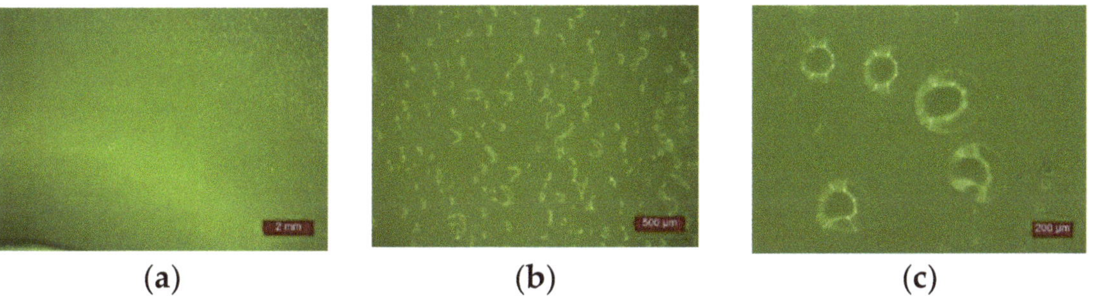

Figure 5. Microscopic examination of the behavior of samples exposed to *Staphylococcus Aureus*: (**a**) Identification of the inhibition halo area (1:2 mm); (**b**) Growth of colonies (1:500 μm); (**c**) Detail growth of colonies (1:200 μm).

Table 6. Quantification of the microbiological load of the system according to STAS 12718 for samples exposed to *Streptococcus Pyogenes*.

Exposure Period (Days)	P0 (without Composite Sample)	P1 (0% TiO_2)	P2 (1% TiO_2)	P3 (2% TiO_2)	P4 (3% TiO_2)	P5 (3.6% TiO_2)	P6 (4% TiO_2)	P7 (5% TiO_2)	P8 (6% TiO_2)	P10 (12% TiO_2)
2	2 (++)	2 (++)	1 (+)	1 (+)	1 (+)	1 (+)	1 (+)	1 (+)	1 (+)	2 (++)
3	3 (+++)	2 (++)	2 (++)	1 (+)	1 (+)	1 (+)	1 (+)	1 (+)	1 (+)	2 (++)
4	3 (+++)	2 (++)	2 (++)	2 (++)	2 (++)	2 (++)	2 (++)	2 (++)	2 (++)	2 (++)
6	3 (+++)	3 (+++)	3 (+++)	2 (++)	2 (++)	2 (++)	2 (++)	2 (++)	3 (+++)	3 (+++)
7	3 (+++)	3 (+++)	3 (+++)	2 (++)	2 (++)	2 (++)	2 (++)	3 (+++)	3 (+++)	3 (+++)
14	4 (++++)	4 (++++)	3 (+++)	3 (+++)	3 (+++)	3 (+++)	3 (+++)	3 (+++)	4 (++++)	4 (++++)
21	4 (++++)	4 (++++)	4 (++++)	3 (+++)	3 (+++)	3 (+++)	3 (+++)	3 (+++)	4 (++++)	4 (++++)

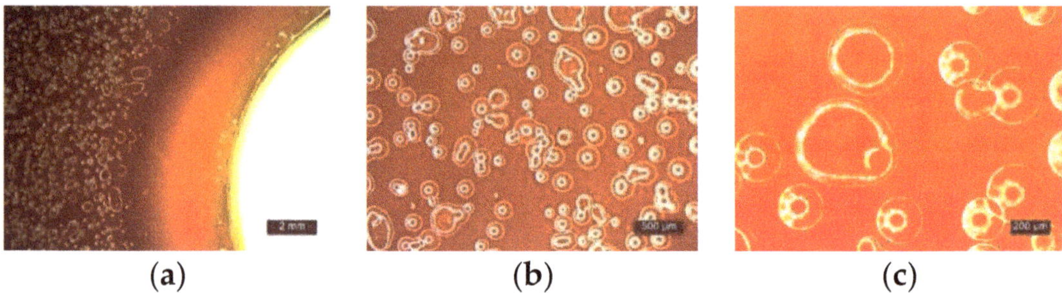

Figure 6. Microscopic examination of the behavior of samples exposed to *Streptococcus Pyogenes*: (**a**) Identification of the inhibition halo area (1:2 mm); (**b**) Growth of colonies (1:500 μm); (**c**) Detail growth of colonies (1:200 μm).

By analyzing the obtained results, several general features have been observed on all systems, regardless of the type of contaminant:

- No traces of contamination/development of bacterial colonies were observed on the surface of any of the tested cementitious material samples during the entire test period (21 days).
- In the first 48 h after exposure in the contaminated environment, the formation of inhibition haloes was observed, which remained constant in size and shape throughout the test. The systems presented a concentric shape: the cementitious composite sample being surrounded by a circular area with microbiological load, evaluated according to STAS 12718/1989, in Class 0 (-). This was followed by a zone of growth and the development of biological material. This increase was more intense as the distance from the edge of the cementitious composite increased (Figures 2–7). The only exception was observed for samples tested with *Pseudomonas Aeruginosa*, for which no inhibition halo was identified in the cementitious composite control system.
- Sample P10 (12% TiO_2) had in general a smaller halo diameter than samples with lower nanoparticle content. This behavior can be attributed to the inhomogeneity and the improper dispersion of nanoparticles in the cementitious matrix, which may tend to agglomerate.
- The P0 system, without the cementitious composite sample, had the most intense and rapid development of colonies of bacteria.
- Microscopic analysis revealed the presence and development of colonies of bacteria in the areas outside the inhibition halo, which again indicate the viability of the suspension used for seeding, the right choice of nutrient substrate, and exposure conditions.
- The formation of the inhibition halo for the control composite system (0% TiO_2) also indicated resistance to the development of bacteria. This mainly happened because of the chemical composition of white Portland cement, which usually contains a certain quantity of TiO_2.

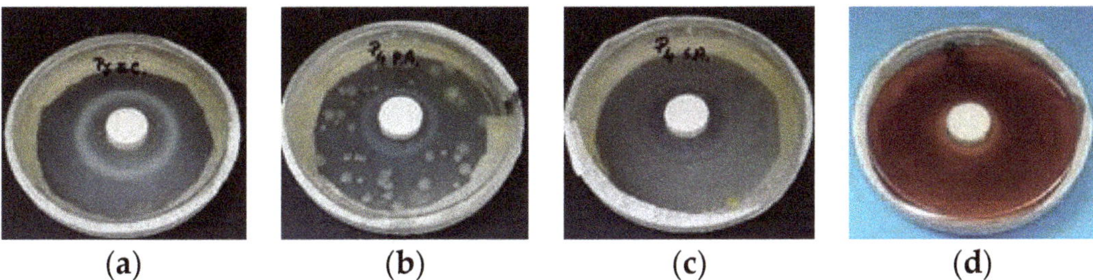

Figure 7. Nano-TiO_2 systems evaluation: (**a**) 5% TiO_2, (14 days exposure to *Escherichia coli*); (**b**) 3.6% TiO_2 (6 days exposure to *Pseudomonas Aeruginosa)*; (**c**) 3% TiO_2 (14 days exposure to *Staphylococcus Aureus*); (**d**) 6% TiO_2 (21 days exposure to *Streptococcus Pyogenes*).

In case of contamination with *Escherichia coli*, the following aspects could be identified:

- Cementitious composites with nano-TiO_2 content in the range of 2–6% had the most effective behavior, with an efficiency of the antibacterial effect (EEA) of more than 10% (Figure 3). The highest value of this parameter (23%) was reported for the samples with 2% nano-TiO_2 addition.
- When evaluating the entire system by quantifying the microbiological load of the system, according to STAS 12718 (Table 3), Classes 0 (-) or 1 (+) were observed/maintained for a longer period. It was also noticed that the samples with 4–12% nano-TiO_2—(Class 0 (-))—maintain this sterile behavior longer (even after 7 days of exposure in contaminated environment—sample P6 (4% TiO_2)). In addition, the P3 sample (2% TiO_2) had a distinguished behavior, by keeping Class 1 (+) constant until the end of the test period.
- In the case of the P0 system, the formation of zones with confluent colonies (Class 3 (+++)) was observed earlier, after only 4 days at exposure in the contaminated environment. This confirmed the viability of the inoculated bacterial material.

In the case of contamination with *Pseudomonas Aeruginosa*, the following aspects can be identified:

- Due to the lack of visible and measurable inhibition halo in the control sample, the effectiveness of the antibacterial effect (EEA) could not be calculated, thus indicating a resistance effect to these bacteria of the cementitious composite matrix (Figure 3). However, for samples with 3.6% and 4% nano-TiO_2, large inhibition halos have been observed.
- When evaluating the entire system by quantifying the microbiological load of the system, according to STAS 12718 (Table 4), Class 2 (++) was observed and maintained for a longer period. This happened due to the higher content of nanoparticles in the cementitious composite mass. For composite samples with 3.6–12% nano-TiO_2, the framing Class 2 (++) was maintained throughout the 21 days of testing. This also happened for sample P4 (3% TiO_2), whose framing class changed from 2 (++) to 3 (+++) only at the last stage of testing (during 14–21 days of exposure in the contaminated environment).
- In the case of the P0 system and the P1 control sample system (0% TiO_2), the formation of areas with Class 3 (+++) confluent colonies was rapidly observed after 2 days of exposure in the contaminated environment, which on the one hand indicates the pre-viability of the bacterial inoculum material and on the other hand indicates the lack of antibacterial activity in the case of the P1 control composite matrix (0% TiO_2) (Table 4).

In the case of contamination with *Staphylococcus Aureus*, the following aspects could be identified:

- Samples containing nano-TiO_2 in the range of 1% to 5% had the most satisfactory behavior, i.e., an efficiency of the antibacterial effect (EEA) of more than 25% (Figure 3). The maximum effectiveness of the antibacterial effect (EEA) was achieved by the samples with 5% nano-TiO_2, for which this indicator was 49%.
- The development of *Staphylococcus Aureus* colonies occurred less readily compared to the other types of bacteria analyzed in the study. The identified colonies were visible to the naked eye after only 2–3 days of exposure in the contaminated environment.
- In the case of the P0 system, colony formation Class 1 (+) was observed after 3 days of exposure, which confirms the viability of the inoculated bacterial material (Table 5).

In the case of contamination with *Streptococcus Pyogenes*, the following aspects could be identified:

- Samples with nano-TiO_2 content in the range of 3–6% showed better behavior in terms of the ability to inhibit colony growth.
- When evaluating the entire system by quantifying the microbiological load of the system, according to STAS 12718 (Table 6), Classes 1 (+) or 2 (++) were observed and maintained during the first 2–3 days after exposure to the contaminated environment. For samples with 2–6% nano-TiO_2, this classification was kept constant for up to 3 days. In these cases, the EEA quantifiable parameter reached the maximum value, i.e., 31%, for the 6% nano-TiO_2 composition (Figure 3);
- In the case of the P0 system, the formation of more than 10 colonies, Class 2 (++) was observed after 2 days, Class 3 (+++) confluent colonies were observed after 3 days, and also Class 4 (++++) growth was observed throughout the surface (Table 6). Almost the same behavior, differentiated only by the delay in their development, was observed for P1 (0% TiO_2), P2 (1% TiO_2), and even the system with the maximum nanoparticle content, P10 (12% TiO_2).

4. Conclusions

From the research performed, we can draw the following conclusions:

- The viability of the contaminants, selection of nutrients, and temperature conditions were proven. Therefore, the identification, quantification, and comparison between their action and the results regarding the growth of the biological material, when subjected to the cementitious composites, was demonstrated, based on the retention time of the samples in the contaminated environment and the content of nano-TiO_2 in the samples.
- The effect of the development of the inhibition halo, when subjected to *Escherichia coli*, *Pseudomonas Aeruginosa*, and *Staphylococus Aureus* bacteria, namely, *Streptococcus Pyogenes* has also been confirmed for samples containing nano-TiO_2 in the range of 2% to 5%. However, the introduction of large quantities of nanoparticles in the matrix of the composite may be on one hand beneficial in terms of antibacterial effects but, on the other hand, it is harmful as a result of the tendency of agglomeration of the nanoparticles in the matrix of the composite. Therefore, the effect of the antibacterial agent is considerably reduced.
- It was considered that for a good inhibiting activity against the development of contaminants of type *Escherichia coli*, *Pseudomonas Aeruginosa*, *Staphylococcus Aureus*, and *Streptococcus Pyogenes*, the content of TiO_2 nanoparticles in the cementitious composite matrix should be at least 2% and not more than 5% relative to the amount of cement. The possibility remains open that composite samples with more than 5% nano-TiO_2 are antibacterial effective if adequate nanoparticle dispersion is ensured. This range of identified nano-TiO_2 amount is consistent with reports in the literature [36–42,51–56].

The results and observations presented in this paper can be a starting point for further research, consisting of a relatively fast and inexpensive semi-quantitative method for evaluating the antibacterial performance of cementitious composites with nano-TiO_2. Through the work process adopted, it was also possible to highlight the antibacterial efficiency of cementitious composites with different nano-TiO_2 percentages and to open new perspectives in the development of self-cleaning construction materials.

Author Contributions: Conceptualization, A.H., A.M. and H.S.; methodology, E.G. and J.G.; validation, A.H., H.S., A.-V.L. and A.M.; investigation, A.H., E.G., J.G. and A.-V.L.; resources, H.S., J.G. and A.M.; data curation, A.H., E.G., H.S. and A.-V.L.; writing—original draft preparation, A.H.; writing—review and editing, A.H., E.G., H.S. and A.-V.L.; visualization, A.-V.L.; supervision, A.H., H.S., A.M. All authors have read and agreed to the published version of the manuscript.

Funding: This research received no external funding.

Institutional Review Board Statement: Not applicable.

Informed Consent Statement: Not applicable.

Data Availability Statement: The data presented in this study are available on request from the corresponding authors.

Acknowledgments: The authors would like to thank Michael Grantham of Sandberg LLP (former Visiting Professor at The University of Leeds and Queen's University Belfast and a Past President of the Institute of Concrete Technology, UK) for his support and contribution in technical editing, language editing and proofreading.

Conflicts of Interest: The authors declare no conflict of interest.

References

1. Haleem Khan, A.A.; Mohan Karuppayil, S. Fungal pollution of indoor environments and its management. *Saudi J. Biol. Sci.* **2012**, *19*, 405–426. [CrossRef]
2. Ebbehoj, N.E.; Hansen, M.O.; Sigsgaard, T.; Larsen, L. Building-related symptoms and molds: A two-step intervention study. *Indoor Air* **2002**, *12*, 273–277. [CrossRef]
3. Zeliger, H.I. Toxic effects of chemical mixtures. *Arch. Environ. Health* **2003**, *58*, 23–29. [CrossRef]
4. Yadav, H.M.; Kim, J.S.; Pawar, S.H. Developments in photocatalytic antibacterial activity of nano TiO_2: A review. *Korean J. Chem. Eng.* **2016**, *33*, 1989–1998. [CrossRef]
5. Wang, L.; Hu, C.; Shao, L. The antimicrobial activity of nanoparticles: Present situation and prospects for the future. *Int. J. Nanomed.* **2017**, *12*, 1227–1249. [CrossRef]
6. Kühn, K.P.; Chaberny, I.F.; Massholder, K.; Stickler, M.; Benz, V.W.; Sonntag, H.G.; Erdinger, L. Disinfection of surfaces by photocatalytic oxidation with titanium dioxide and UVA light. *Chemosphere* **2003**, *53*, 71–77. [CrossRef]
7. Drugă, B.; Ukrainczyk, N.; Weise, K.; Koenders, E.; Lackner, S. Interaction between wastewater microorganisms and geopolymer or cementitious materials: Biofilm characterization and deterioration characteristics of mortars. *Int. Biodeterior. Biodegrad.* **2018**, *134*, 58–67. [CrossRef]
8. Machida, M.; Norimoto, K.; Kimura, T. Antibacterial Activity of Photocatalytic Titanium Dioxide Thin Films with Photodeposited Silver on the Surface of Sanitary Ware. *J. Am. Ceram. Soc.* **2005**, *88*, 95–100. [CrossRef]
9. Page, K.; Wilson, M.; Parkin, I.P. Antimicrobial surfaces and their potential in reducing the role of the inanimate environment in the incidence of hospital-acquired infections. *J. Mater. Chem.* **2009**, *19*, 3819. [CrossRef]
10. Watts, R.J.; Kong, S.; Orr, M.P.; Miller, G.C.; Henry, B.E. Photocatalytic inactivation of coliform bacteria and viruses in secondary wastewater effluent. *Water Res.* **1995**, *29*, 95–100. [CrossRef]
11. Vohra, A.; Goswami, D.Y.; Deshpande, D.A.; Block, S.S. Enhanced photocatalytic inactivation of bacterial spores on surfaces in air. *J. Ind. Microbiol. Biotechnol.* **2005**, *32*, 364–370. [CrossRef]
12. Cassar, L. Nanotechnology and photocatalysis in cementitious materials. In *NICOM 2: 2nd International Symposium on Nanotechnology in Construction*; de Miguel, Y., Porro, A., Bartos, P.J.M., Eds.; RILEM Publications SARL: Paris, France, 2006; pp. 277–284.
13. Wang, L.; Zhang, H.; Gao, Y. Effect of TiO_2 nanoparticles on physical and mechanical properties of cement at low temperatures. *Adv. Mater. Sci. Eng.* **2018**, *2018*, 8934689. [CrossRef]
14. Sadawy, M.M.; Elsharkawy, E.R. Effect of Nano-TiO_2 on mechanical properties of concrete and corrosion behavior of reinforcement bars. *Int. J. Eng. Res. Appl.* **2016**, *6*, 61–65.
15. Salemi, N.; Behfarnia, K.; Zaree, S.A. Effect of nanoparticles on frost durability of concrete. *Asian J. Civ. Eng.* **2014**, *15*, 411–420.
16. Shen, W.; Zhang, C.; Li, Q.; Zhang, W.; Cao, L.; Ye, J. Preparation of titanium dioxide nano particle modified photocatalytic self-cleaning concrete. *J. Clean. Prod.* **2015**, *87*, 762–765. [CrossRef]
17. Matsunaga, T.; Tomoda, R.; Nakajima, T.; Nakamura, N.; Komine, T. Continuous-sterilization system that uses photosemiconductor powders. *Appl. Environ. Microbiol.* **1988**, *54*, 1330–1333. [CrossRef]
18. Abdel-Gawwad, H.A.; Mohamed, S.A.; Mohammed, M.S. Recycling of slag and lead-bearing sludge in the cleaner production of alkali activated cement with high performance and microbial resistivity. *J. Clean. Prod.* **2019**, *220*, 568–580. [CrossRef]

19. Guzmán-Aponte, L.; Mejía de Gutiérrez, R.; Maury-Ramírez, A. Metakaolin-Based Geopolymer with Added TiO_2 particles: Physicomechanical Characteristics. *Coatings* **2017**, *7*, 233. [CrossRef]
20. Mohd Adnan, M.A.; Muhd Julkapli, N.; Amir, M.N.I.; Maamor, A. Effect on different TiO_2 photocatalyst supports on photodecolorization of synthetic dyes: A review. *Int. J. Environ. Sci. Technol.* **2019**, *16*, 547–566. [CrossRef]
21. Haider, A.J.; Anbari, R.; Kadhim, G.R.; Salame, C.T. Exploring potential environmental applications of TiO_2 nanoparticles. *Energy Procedia* **2017**, *119*, 332–345. [CrossRef]
22. Hoffmann, M.R.; Martin, S.T.; Choi, W.; Bahnemannt, D.W. Environmental Applications of Semiconductor Photocatalysis. *Chem. Rev.* **1995**, *95*, 69–96. [CrossRef]
23. Mejía, J.M.; Mendoza, J.D.; Yucuma, J.; Mejía de Gutiérrez, R.; Mejía, D.E.; Astudillo, M. Mechanical, in-vitro biological and antimicrobial characterization as an evaluation protocol of a ceramic material based on alkaline activated metakaolin. *Appl. Clay Sci.* **2019**, *178*, 105141. [CrossRef]
24. Syamsidar, D.; Nurfadilla, S. The Properties of Nano TiO_2 -Geopolymer Composite as a Material for Functional Surface Application. *MATEC Web Conf.* **2017**, *97*, 01013. [CrossRef]
25. Sikora, P.; Augustyniak, A.; Cendrowski, K.; Horszczaruk, E.; Rucinska, T.; Nawrotek, P.; Mijowska, E. Characterization of mechanical and bactericidal properties of cement mortars containing waste glass aggregate and nanomaterials. *Materials* **2016**, *9*, 701. [CrossRef] [PubMed]
26. Hamidi, F.; Aslani, F. TiO_2-based Photocatalytic Cementitious Composites: Materials, Properties, Influential Parameters, and Assessment Techniques. *Nanomaterials* **2019**, *9*, 1444. [CrossRef] [PubMed]
27. Sunada, K.; Watanabe, T.; Hashimoto, K. Bactericidal Activity of Copper-Deposited TiO_2 Thin Film under Weak UV Light Illumination. *Environ. Sci. Technol.* **2003**, *37*, 4785–4789. [CrossRef]
28. Saito, T.; Iwase, T.; Horie, J.; Morioka, T. Mode of photocatalytic bactericidal action of powdered semiconductor TiO2 on mutans streptococci. *J. Photochem. Photobiol. B* **1992**, *14*, 369–379. [CrossRef]
29. Oguma, K.; Katayama, H.; Ohgaki, S. Photoreactivation of Escherichia coli after Low- or Medium-Pressure UV Disinfection Determined by an Endonuclease Sensitive Site Assay. *Appl. Environ. Microbiol.* **2002**, *68*, 6029–6035. [CrossRef]
30. Gogniat, G.; Thyssen, M.; Denis, M.; Pulgarin, C.; Dukan, S. The bactericidal effect of TiO_2 photocatalysis involves adsorption onto catalyst and the loss of membrane integrity. *FEMS Microbiol. Lett.* **2006**, *258*, 18–24. [CrossRef]
31. Mazurkova, N.A.; Spitsyna, Y.E.; Shikina, N.V.; Ismagilov, Z.R.; Zagrebel'Nyi, S.N.; Ryabchikova, E.I. Interaction of titanium dioxide nanoparticles with influenza virus. *Nanotechnol. Russia* **2010**, *5*, 417–420. [CrossRef]
32. Adams, L.K.; Lyon, D.Y.; McIntosh, A.; Alvarez, P.J.J. Comparative toxicity of nano-scale TiO_2, SiO_2 and ZnO water suspensions. *Water Sci. Technol.* **2006**, *54*, 327–334. [CrossRef]
33. Armelao, L.; Barreca, D.; Bottaro, G.; Gasparotto, A.; Maccato, C.; Maragno, C.; Tondello, E.; Stangar, U.L.; Bergant, M.; Mahne, D. Photocatalytic and antibacterial activity of TiO_2 and Au/TiO_2 nanosystems. *Nanotechnology* **2007**, *18*, 375709. [CrossRef]
34. Dědková, K.; Matějová, K.; Lang, J.; Peikertová, P.; Kutláková, K.M.; Neuwirthová, L.; Frydrýšek, K.; Kukutschová, J. Antibacterial activity of kaolinite/nanoTiO_2 composites in relation to irradiation time. *J. Photochem. Photobiol. B* **2014**, *135*, 17–22.
35. Gurr, J.R.; Wang, A.S.S.; Chen, C.H. Ultrafine titanium dioxide particles in the absence of photoactivation can induce oxidative damage to human bronchial epithelial cells. *Toxicology* **2005**, *213*, 66–73. [CrossRef]
36. Hamdany, A.H. Photocatalytic Cementitious Material for Self-Cleaning and Anti-Microbial Application. Ph.D. Thesis, Nanyang Technological University, Singapore, 2019.
37. Davidson, H.; Poon, M.; Saunders, R.; Shapiro, I.M.; Hickok, N.J.; Adams, C.S. Tetracycline tethered to titanium inhibits colonization by Gram-negative bacteria. *J. Biomed. Mater. Res. B Appl. Biomater.* **2015**, *103*, 1381–1389. [CrossRef]
38. Lorenzetti, M.; Dogša, I.; Stošicki, T.; Stopar, D.; Kalin, M.; Kobe, S.; Novak, S. The Influence of Surface Modification on Bacterial Adhesion to Titanium-Based Substrates. *ACS Appl. Mater. Interfaces* **2015**, *7*, 1644–1651. [CrossRef] [PubMed]
39. Peng, Z.; Ni, J.; Zheng, K.; Shen, Y.; Wang, X.; He, G.; Jin, S.; Tang, T. Dual effects and mechanism of TiO_2 nanotube arrays in reducing bacterial colonization and enhancing C3H10T1/2 cell adhesion. *Int. J. Nanomed.* **2013**, *8*, 3093–3105.
40. Daly, M.J.; Gaidamakova, E.K.; Matrosova, V.Y.; Vasilenko, A.; Zhai, M.; Leapman, R.D.; Lai, B.; Ravel, B.; Li, S.-M.W.; Kemner, K.M.; et al. Protein Oxidation Implicated as the Primary Determinant of Bacterial Radioresistance. *PLoS Biol.* **2007**, *5*, 92. [CrossRef]
41. Carre, G.; Estner, M.; Gies, J.-P.; Andre, P.; Hamon, E.; Ennahar, S.; Keller, V.; Keller, N.; Lett, M.-C.; Horvatovich, P. TiO_2 Photocatalysis Damages Lipids and Proteins in Escherichia coli. *Appl. Environ. Microbiol.* **2014**, *80*, 2573–2581. [CrossRef] [PubMed]
42. Kubacka, A.; Diez, M.S.; Rojo, D.; Bargiela, R.; Ciordia, S.; Zapico, I.; Albar, J.P.; Barbas, C.; Martins dos Santos, V.A.P.; Fernández-garcía, M.; et al. Understanding the antimicrobial mechanism of TiO2-based nanocomposite films in a pathogenic bacterium. *Sci. Rep.* **2014**, *4*, 4134. [CrossRef] [PubMed]
43. Mejía-de Gutiérrez, R.; Villaquirán-Caicedo, M.; Ramírez-Benavides, S.; Astudillo, M.; Mejía, D. Evaluation of the Antibacterial Activity of a Geopolymer Mortar based on Supplemented with TiO_2 and CuO Particles Using Glass Waste as Fine Aggregate. *Coattings* **2020**, *10*, 157. [CrossRef]
44. Pacheco-Torgal, F.; Jalali, S. Nanotechnology: Advantages and drawbacks in the field of construction and building materials. *Contr. Build. Mater.* **2011**, *25*, 582–590. [CrossRef]

45. Cho, M.; Chung, H.; Choi, W.; Yoon, J. Linear correlation between inactivation of *E. coli* and OH radical concentration in TiO_2 photocatalytic disinfection. *Water Res.* **2004**, *38*, 1069–1077. [CrossRef]
46. Janus, M.; Zajac, K. Concretes with Photocatalytic Activity. In *High Performance Concrete Technology and Applications*; Yilmax, S., Baytan, H., Eds.; IntechOpen: London, UK, 2016; Chapter 7, pp. 141–161.
47. Carmona-Quiroga, P.M.; Martinez-Ramirez, S.; Viles, H.A. Efficiency and durability of a self-cleaning coating on concrete and stones under both natural and artificial ageing trials. *App. Surf. Sci.* **2018**, *433*, 312–320. [CrossRef]
48. Grebenişan, E.; Hegyi, A.; Szilagyi, H.; Lăzărescu, A.V.; Ionescu, B.A. Influence of the Addition of TiO_2 Nanoparticles on the Self-Cleaning Performance of Cementitious Composite Surfaces. *Proceedings* **2020**, *1*, 42. [CrossRef]
49. Grebenişan, E.; Hegyi, A.; Lăzărescu, A.V. Research Regarding the Influence of TiO_2 Nanoparticles on the Performance of Cementitious Materials. In *IOP Conference Series: Materials Science and Engineering, Proceedings of the International Conference on Innovative Research—ICIR EUROINVENT 2020, Iasi, Romania, 21–23 May 2020*; IOP Publishing: Bristol, UK, 2020; Volume 877, p. 012004.
50. Hegyi, A.; Szilagyi, H.; Grebenișan, E.; Sandu, A.V.; Lăzărescu, A.V.; Romila, C. Influence of TiO_2 Nanoparticles Addition on the Hydrophilicity of Cementitious Composites Surfaces. *Appl. Sci.* **2020**, *10*, 4501. [CrossRef]
51. Gopalan, A.-I.; Lee, J.-C.; Saianand, G.; Lee, K.-P.; Sonar, P.; Dharmarajan, R.; Hou, Y.; Ann, K.-Y.; Kannan, V.; Kim, W.-J. Recent progress in the abatement of hazardous pollutants using photocatalytic TiO_2-based building materials. *Nanomaterials* **2020**, *10*, 1854. [CrossRef]
52. Shaaban, I.G.; El-Sayad, H.; El-Ghaly, A.E.; Moussa, S. Effect of micro TiO_2 on cement mortar. *EJME* **2020**, *5*, 58–68. [CrossRef]
53. Daniyal, M.; Azam, A.; Akhtar, S. Application of Nanomaterials in Civil Engineering. In *Nanomaterials and Their Applications. Advanced Structured Materials*; Khan, Z., Ed.; Springer: Singapore, 2018; Volume 84, pp. 169–189.
54. Wang, D.; Geng, Z.; Hou, P.; Yang, P.; Cheng, X.; Huang, S. Rhodamine B Removal of TiO2@ SiO2 Core-Shell Nanocomposites Coated to Buildings. *Crystals* **2020**, *10*, 80. [CrossRef]
55. Guo, M.Z.; Maury-Ramirez, A.; Poon, C.S. Self-cleaning ability of titanium dioxide clear paint coated architectural mortar and its potential in field application. *J. Clean. Prod.* **2016**, *112*, 3583–3588. [CrossRef]
56. Sikora, P.; Cendrowski, K.; Markowska-Szczupak, A.; Horszczaruk, E.; Mijowska, E. The effects of silica/titania nanocomposite on the mechanical and bactericidal properties of cement mortars. *Constr. Build. Mater.* **2017**, *150*, 738–746. [CrossRef]

Article

Innovative Use of Sheep Wool for Obtaining Materials with Improved Sound-Absorbing Properties

Simona Ioana Borlea (Mureşan) [1], Ancuţa-Elena Tiuc [1,*], Ovidiu Nemeş [1,2,*], Horaţiu Vermeşan [1] and Ovidiu Vasile [3]

1 Faculty of Materials and Environmental Engineering, Technical University of Cluj-Napoca, Cluj-Napoca, 28 Memorandumului Street, 400114 Cluj-Napoca, Romania; Ioana.Muresan@staff.utcluj.ro (S.I.B.M.); horatiu.vermesan@imadd.utcluj.ro (H.V.)

2 National Institute for Research and Development in Environmental Protection, 294 Blvd Splaiul Independenței, Sector 6, 060031 Bucharest, Romania

3 Department of Mechanics, Politehnica University of Bucharest, 313 Splaiul Independentei, 060042 Bucharest, Romania; ovidiu.vasile@upb.ro

* Correspondence: ancuta.tiuc@imadd.utcluj.ro (A.-E.T.); ovidiu.nemes@sim.utcluj.ro (O.N.); Tel.: +40-756-102-923 (A.-E.T.); +40-724-072-598 (O.N.)

Received: 28 November 2019; Accepted: 1 February 2020; Published: 4 February 2020

Abstract: In recent years, natural materials are becoming a valid alternative to traditional sound absorbers due to reduced production costs and environmental protection. This study explores alternative usage of sheep wool as a construction material with improved sound absorbing properties beyond its traditional application as a sound absorber in textile industry or using of waste wool in the textile industry as a raw material. The aim of this study was to obtain materials with improved sound-absorbing properties using sheep wool as a raw material. Seven materials were obtained by hot pressing (60 ÷ 80 °C and 0.05 ÷ 6 MPa) of wool fibers and one by cold pressing. Results showed that by simply hot pressing the wool, a different product was obtained, which could be processed and easily manipulated. The obtained materials had very good sound absorption properties, with acoustic absorption coefficient values of over 0.7 for the frequency range of 800 ÷ 3150 Hz. The results prove that sheep wool has a comparable sound absorption performance to mineral wool or recycled polyurethane foam.

Keywords: sheep wool recovery; acoustic materials; sound absorption coefficient

1. Introduction

From a sustainable development perspective, an important goal is to choose raw materials that are easily recyclable and renewable as well as locally available and environmentally friendly. This includes timber, clay, stone, straw, bio-based fibers, and sheep wool, provided that any further processing is carried out with low energy consumption.

The origin of these materials can be vegetable or animal so that their manufacturing has a low environmental impact due to the energy saved in the production process [1]. The use of natural fibers as raw material for acoustic applications have been intensively studied [2–14], especially in recent years. Many industries are moving toward natural material-based, environmentally friendly products [15,16]. This may be due to the fact that the energy required to process these types of materials is lower compared to that required for synthetic materials. For instance, processing 1 m^3 of sheep wool insulation produces almost 5.4 kg of CO_2, whereas the quantity of CO_2 produced is 135 kg in the case of mineral wool [17]. Thus, the environmental impact when using these types of materials is low, and there is no negative effect on the environment [18].

Many natural materials, such as bamboo, kenaf, sisal, flax, hemp, sheep wool, cork, or coconut fibers, show good sound-absorbing performance and can therefore be used as sound absorbers in acoustic rooms and noise barriers [19,20].

Sheep wool is an easily recyclable, easily renewable, and environmentally friendly source of raw material, which consists of 60% animal protein fibers, 10% fat, 15% moisture, 10% sheep sweat, and 5% impurities on average. Zach et al. evaluated the thermal, hygrothermal, and acoustic performance of samples of sheep wool materials. A mixture of sheep wool was mechanically fastened to a reinforcing cloth with varying thickness and density. The results showed that sheep wool was characterized by high hygroscopicity that reached up to 35% and that sheep wool could therefore be an excellent acoustic insulating material [21].

Del Rey et al. studied sheep wool as a sustainable material for acoustic applications. The materials were made from sheep wool by thermofusion with polyester fibers obtained from recycled polyethylene terephthalate (PET) flakes that acted as a binder (PET fibers melt at 140–150 °C). The final material had 80% sheep wool fibers (first quality, second quality, or blend), and the remaining 20% was PET fibers. From the measurement results, it was demonstrated that sheep wool with PET fiber was a good sound-absorbing material at medium and high frequencies, with acoustic absorption coefficient values of over 0.5 for the frequency range of 600 ÷ 3150 Hz for the best material obtained [1].

Until now, sheep wool has traditionally been used in the textile industry for the manufacturing of conventional woolen products, such as carpets, garments, curtains, covers, and bedding. More recently, they have also been used in the building industry due to their thermal properties. For the fabrication of wool-based building materials, coarse fibers or those fibers that cannot be used in the textile industry are generally used [22]. Wool has good thermal characteristics, with the thermal conductivity of wool panels varying between 0.040 and 0.041 W/mK for densities of 25 ÷ 92.5 kg/m^3 [23].

Sheep wool fibers have a similar size as mineral fibers. A 33 ÷ 36 μm sheep wool fiber would roughly be the same size as PET polyester fibers (33 μm) [24] or Kenaf fibers (36 μm) [25]. Unlike synthetic fibers, sheep fibers do not have a fixed thickness. Their thickness range has a standard deviation of 2 μm, according to scientific literature [26]. The fiber diameter also depends on the breed of the sheep.

The surface of wool fibers has many scales, and the fibers can only move on one direction. Under mechanical agitation, friction, and pressure in the presence of moisture and heat, the scale edge of one fiber locks into the interscale gap of another fiber like a "ratchet" mechanism. The fibers interlock and cannot return to their original positions, resulting in irreversible felting shrinkage [27].

Pressed felt is produced from wool or animal hair by mechanical agitation and compression of the fibers in warm, moist conditions [28].

The aim of this study was to analyze the sound absorption coefficient of some materials or structures based on sheep wool as an alternative to the classical (wool felts, mineral wool, or foams) or the new series of improved sound absorbers. The sound absorption capability of sheep wool was measured in an impedance tube. Experimental results indicated the material's excellent performance in the development of building elements for sound absorption with or without the addition of other elements (polyurethane foam, epoxy, or polyester resin).

Compared with the classical acoustic materials existing in the market or in the literature, the ones obtained in this research have the advantage of good properties. They are also environmentally friendly due to the fact that no binders are used, and the working parameters (pressure and temperature) require low energy consumption. These materials with very good acoustic absorption properties can be obtained by hot pressing without the presence of humidity compared to the standard mode of felting.

2. Materials and Methods

2.1. Materials

In order to obtain the desired sound-absorbing materials, black merino sheep wool (different shades of black, including dark brown) was used. Figure 1 shows the raw sheep wool (60 ÷ 80 mm length, 18 ÷ 20 μm fineness, ripple of 100 mm, and density of 3.4578 g/cm^3). Prior to experiments, the raw wool was washed to remove impurities, sand, and dust, and it was then dried and carded.

Figure 1. The raw material used: black merino sheep wool.

2.2. Manufacturing Process

The samples used in this research were obtained by hot and cold pressing. The mold used to obtain the material samples had a cylindrical shape with two aluminum hot plates (top and bottom). The samples were heated from both sides to obtain a uniform temperature in the mold. The mold was equipped with four thermocouples disposed on the outside, which were connected to a temperature regulator and measured the working temperature. The mold was also fitted with a thermostat to maintain a constant temperature. Figure 2 shows the mold that was used to obtain the material samples.

Because the ability of a material to reduce the acoustic energy depends on its thickness, different wool quantities were considered in order to prepare different sample thicknesses [29].

Figure 2. The mold used to obtain material samples by hot pressing [30].

The mold for the cold-pressed samples was made of steel in rectangular shape with a lid. To obtain the hot-pressed samples, hot pressing was done in a mold by applying a pressure of 0.05 ÷ 6 MPa on the material, which was heated to 60 ÷ 80 °C. When wool fibers are heated, they easily fill the new shape, and the pressure in the mold forces the wool fibers to compress. The obtained samples were kept in the mold under pressure until complete cooling. After this, the mold was opened, and the sample was extracted. The main parameters that were followed for this process were pressure, pressing time, and temperature. The required heat was transferred through the mold walls. The required pressure force was obtained from a manually operated hydraulic press (Unicraft WPP 50 E, Stürmer Machinen Gmbh, Hallstadt, Germany).

Eight sound-absorbing materials were obtained following the procedure described above. The obtained materials were divided into three groups (Figure 3) depending on the embodiment: hot-pressed wool moistened with water (WHW, labelled as A), hot-pressed wool (WH, labelled as B), and cold-pressed wool (WC, labelled as C). Table 1 presents the technical parameters of the obtained material samples.

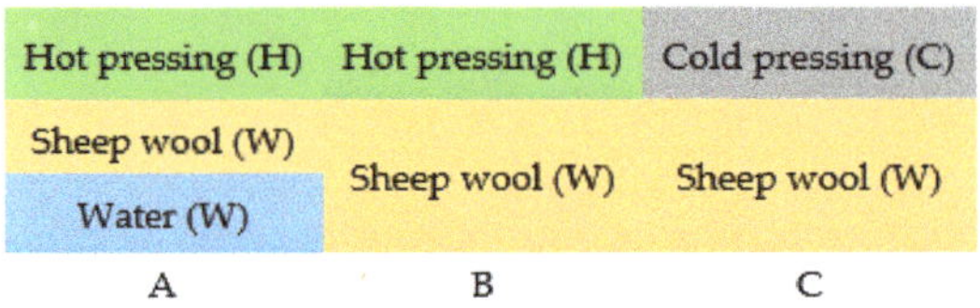

Hot pressing (H)	Hot pressing (H)	Cold pressing (C)
Sheep wool (W) Water (W)	Sheep wool (W)	Sheep wool (W)
A	B	C

Figure 3. Groups of obtained materials.

Table 1. Technical parameters of the materials obtained.

Group	Code	Initial Height (mm)	Final Height (mm)	Temperature (°C)	Pressure (MPa)	Water (ml)
A	WHW40_3_25	40	1	60	3	25
	WHW80_6_50	80	2.5	70	6	50
	WHW80_6_75	80	3	80	6	75
B	WH120_4	120	15	80	4	-
	WH240_4	240	25	80	4	-
	WH120_0.05	124	35	80	0.05	-
	WH240_0.05	240	50	80	0.05	-
C	WC40	40	25	25	0.003	-

The materials in group A were made of wool by hot pressing (pression 3 and 6 MPa and temperature 60, 70, and 80 °C). In order to analyze the influence of humidity on the new materials, three tests were carried out by varying the amount of water used for moistening, i.e., 25, 50, and 75 mL. Results showed that the wool became plastic by wetting, especially at temperatures around 80 °C. It is upon this thermoplastic property that the pressing and elimination of the wrinkles in the wool fiber is based [31]. Thus, sample WHW80_6_75 heated at 80 °C had the consistency of a plywood with portions of glossy faces.

In the case of materials from group B, the wool was not wetted, and the initial wool layer (120 and 240 mm) was hot pressed at 4 and 0.05 MPa at a temperature of 80 °C. Four samples were obtained, the parameters and codes of which are presented in Table 1.

The material constituting group C was obtained by cold pressing a wool layer with an initial height of 40 mm at a pressure of 0.003 MPa.

2.3. Methods

The determination of apparent density was performed according to ISO 845:2006 [32]. The tested samples were cylindrical with a diameter of 63.5 mm and a specific height for each material.

The dimensions of the specimens were measured with a 0.1 mm accuracy. The weight was determined using a laboratory balance with 0.01 g accuracy. The apparent density of specimens was calculated using the following formula:

$$\rho = \frac{m}{V}\left[\mathrm{g/cm^3}\right] \quad (1)$$

where m is the mass of the sample, and V is the volume of the sample.

The sound absorption coefficient at normal incidence (α) is the quotient between the acoustic energy absorbed by the surface of the test sample and the incident acoustic energy for a plane acoustic wave at normal incidence. ISO 10534-2 standard [33] establishes a test procedure to determine the sound absorption coefficient for normal incidence of acoustic absorbers by means of an impedance tube, two microphone positions, and a digital analysis system signal.

The method of measuring the acoustic absorption coefficient by means of the impedance tube is based on the fact that the reflection coefficient at normal incidence (r) can be calculated from the measured transfer function (H12) between two positions of the microphone at different distances from the sample. The transfer function of the incident (H_I) and reflecting waves (H_R) between the microphone positions are defined as follows [34]:

$$H_I = \frac{p_{2I}}{p_{1I}} = e^{-jk\cdot(x_1-x_2)} = e^{-jk\cdot s} \quad (2)$$

$$H_R = \frac{p_{2R}}{p_{1R}} = e^{-jk\cdot(x_1-x_2)} = e^{-jk\cdot s} \quad (3)$$

where s is the distance between the two microphone positions; x_1 and x_2 are the distances from the reference point to microphone position 1 and 2, respectively; p_I and p_R are the sound pressure propagating in the incident and reflected direction, respectively; and jk is a complex-valued wavenumber.

The transfer function (H_{12}) for the total sound field can be calculated with the following formula [34]:

$$H_{12} = \frac{p_2}{p_1} = e^{-jk\cdot(x_1-x_2)} = e^{-jk\cdot s} \quad (4)$$

The reflection coefficient (r) at the sample surface (x = 0) is as follows [33]:

$$r = \frac{H_{12} - H_I}{H_R - H_{12}} e^{2jk\cdot x_1} \quad (5)$$

The sound absorption coefficient at normal incidence is calculated with the following formula [34]:

$$r = 1 - |r|^2 \quad (6)$$

The α values were ascertained by producing standing waves in a tube with 63.5 mm diameter, so the tests were performed on circular samples (Figure 4) with a diameter of 63.5 mm. The circular samples were placed at the end of the Kundt's tube Brüel&Kjaer Type 4206 during each test. Measurements were recorded at the third-octave frequency band within the intervals of 100 ÷ 3200 Hz and conducted at an air temperature of 26 °C, relative humidity of 55%, and pressure of 100.5 kPa.

Figure 4. Samples prepared for sound absorption coefficient measurement.

3. Results and Discussion

3.1. Results for Apparent Density Tests

The material density is an important factor for the acoustic absorption of a material. As the density of the material increases, the sound absorption at medium and high frequencies also increases. Increased number of fibers per unit area increases the apparent density. Energy loss increases with the increase in friction surface, thus increasing the sound absorption coefficient [35].

The apparent density determined for the obtained materials is shown in Figure 5. It can be seen that the hot-pressed materials had a much higher density than the cold-pressed materials. The density of the materials made from sheep wool increased with the increase in pressure.

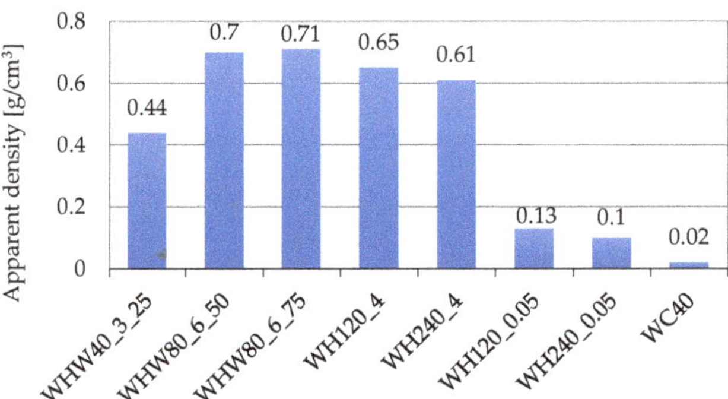

Figure 5. Apparent density values.

3.2. Results of Acoustic Tests

The acoustic characterization of the materials was based on the sound absorption coefficient α [36]. This parameter is the ratio of absorbed sound intensity to incident sound intensity on a surface [37]. The potential for materials to absorb sound energy depends on the following factors: density, thickness, porosity, fiber diameter, airflow resistivity, tortuosity, surface impedance, compression, air gap, and multilayers [19,38].

3.2.1. The Effect of Material Thickness on the Sound Absorption Coefficient

This section highlights and discusses the variation in acoustic absorption coefficient with the thickness of materials obtained from hot-pressed sheep wool at 80 °C and a pressure between 4 and 0.05 MPa. The influence of material thickness obtained from compressed sheep wool at a pressure of 4 MPa on the coefficient of acoustic absorption is presented in Figure 6.

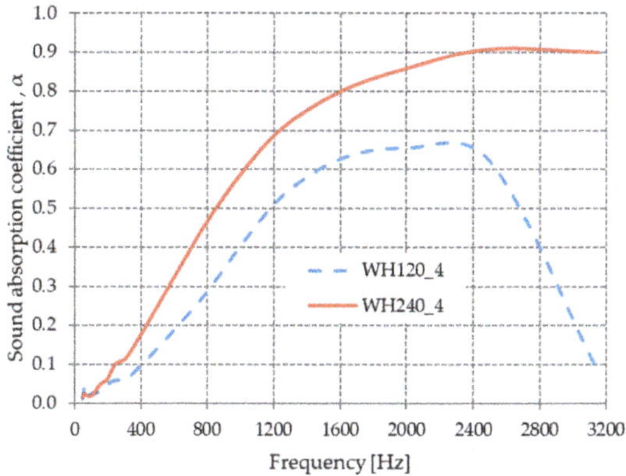

Figure 6. Variation in the acoustic absorption coefficient with the material thickness for WH120_4 and WH240_4.

From Figure 6, it can be observed that material WH240_4 with 25 mm thickness had an acoustic absorption coefficient greater than WH120_4, which had a smaller thickness (15 mm), over the entire analyzed frequency range.

The thickness of the compressed materials at 0.05 MPa was 35 mm for material WH120_0.05 and 50 mm for WH240_0.05.

The influence of compressed material thickness on the acoustic properties is shown in Figure 7. It can be observed that the material with the greatest thickness had the highest sound absorption coefficient values over the entire analyzed frequency range.

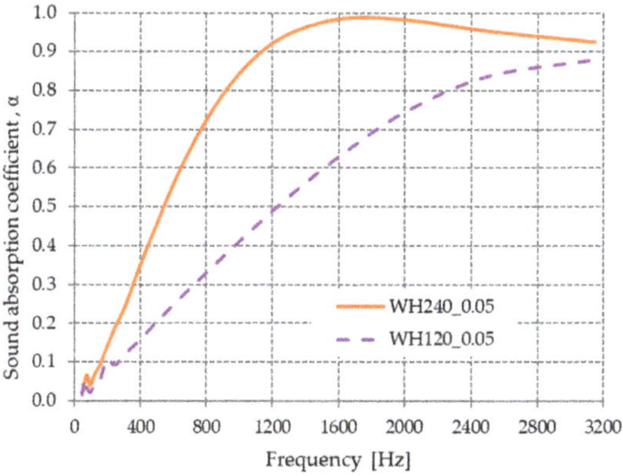

Figure 7. Variation in the acoustic absorption coefficient with the material thickness for WH120_0.05 and WH240_0.05.

The sound absorption coefficient improved by increasing the composite thickness, an aspect that has been demonstrated in the literature [39]. Experimental testing performed on materials such as fiber felts, glass wool, paddy straw, textile waste, rubber crumbs, and polyester have all shown an increase in sound absorption with an increase in material thickness, especially at lower frequencies [1,40–43].

An analysis of the sound absorption coefficient values of our hot-pressed, wool-based material (Figures 6 and 7) relative to other wool-based materials reported in the literature showed that the coefficient values were better at comparable thicknesses. The acoustic absorption coefficient values of the obtained materials at the frequency of 1000 Hz were 0.4 for WH120_4 (15 mm), 0.59 for WH240_4 (25 mm), and 0.84 for WH240_0.05 (50 mm). For other materials made from sheep's wool [21], the coefficient values for different material thicknesses were 0.331 for 20 mm, 0.415 for 30 mm, and 0.7 for 40 mm.

3.2.2. The Influence of Wool Compression on the Sound Absorption Coefficient

The studied materials were compressed at 0.05 and 4 MPa starting from an initial height of 240 mm and 120 mm, respectively, and the influence of the compression of wool fibers on the acoustic absorption coefficient is shown in Figure 8.

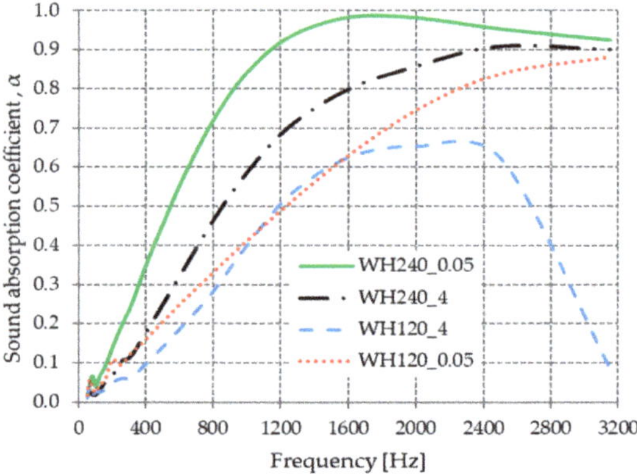

Figure 8. Variation in the acoustic absorption coefficient with the compaction pressure.

The compressed material WH240_0.05 had better sound-absorbing properties at 0.05 MPa with 0.01 g/cm^3 density than the compressed material WH240_4 at 4 MPa with 0.61 g/cm^3 density. For the materials marked as WH120_0.05 and WH120_4, the acoustic absorption coefficient had better values in the frequency range of 50 ÷ 1100 Hz and 1600 ÷ 3200 Hz compared to the compressed material at a lower pressure (WH120_0.05). This can be explained by the fact that the fibers within the material are brought closer to each other during compression. Thus, the material becomes more compact, the open porosity decreases, and the compression leads to a decrease in the thickness of the material [44].

The acoustic absorption properties of fibrous mat decrease during compression because the material thickness decreases during compression. Compression tests done on polyester fiber showed a drop in the absorption coefficient when the fibrous mat was compressed [45]. Fouladi et al. and Nor et al. they stated that compression affects the physical parameters of materials, including the flow resistivity, tortuosity, and porosity. These parameters define the link between the acoustic medium and the matrix [46,47].

3.2.3. The Influence of the Presence of Water on the Sound Absorption Coefficient

The variation in the sound absorption coefficient depending on the amount of water used to obtain WHW80_6_50 and WHW80_6_75 materials is shown in Figure 9. The sound-absorbing properties of the obtained material by wetting with 50 mL of water (WHW80_6_50) were better than that of the one obtained using 75 mL of water (WHW80_6_75).

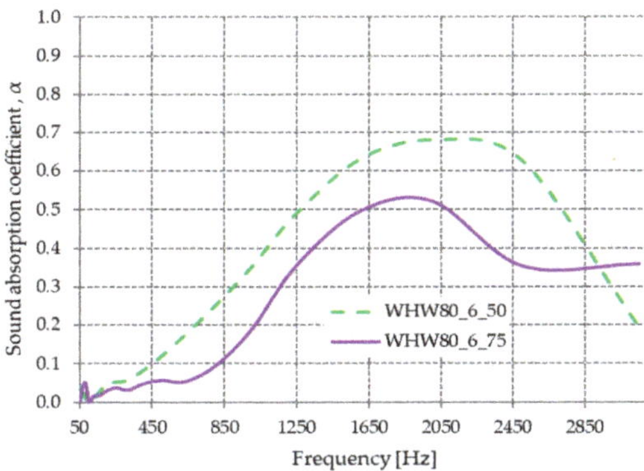

Figure 9. Variation in the acoustic absorption coefficient with the quantity of water.

The decrease in the absorption coefficient values at frequencies below 2850 Hz for material WHW80_6_75 was due to the change in the thermoplastic properties of wool in the presence of water at 80 °C. On the material's surface, plasticized and glossy areas appeared, which reflected the sound wave and did not allow it to penetrate the material for decreased sound intensity [31].

Polypropylene/jute webs with a thickness of 4.28 mm and a density of 0.65 g/cm^3 were found to have an acoustic absorption coefficient $\alpha < 0.2$ in the frequency range of 100 ÷ 1600 Hz [4], while WHW80_6_50 (2.5 mm) and WHW80_6_75 (3 mm) with a density of 0.7 g/cm^3 and 0.71 g/cm^3 had a sound absorption coefficient of $\alpha < 0.63$ in the frequency range of 100 ÷ 1600 Hz.

3.2.4. Influence of Cold/Hot Compression on the Sound Absorption Coefficient

The influence of the compression mode of the wool fibers (cold or hot) on the acoustic absorption coefficient is highlighted in Figure 10. It can be observed that the materials obtained by hot pressing (WH240_4 and WH120_4) had superior sound-absorbing properties compared to the material obtained by cold pressing (WC40).

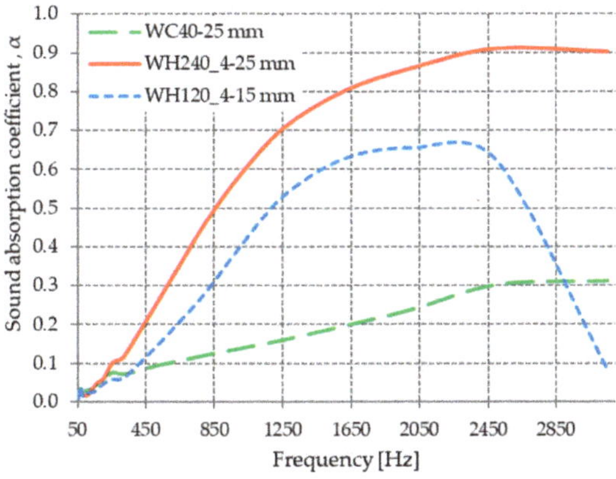

Figure 10. Variation in the acoustic absorption coefficient with frequency.

Considering the density of the hot-pressed materials (0.61 ÷ 0.65 g/cm^3) was higher than the density of the cold-pressed materials (WC40 0.02 g/cm^3), it can be said that the sound absorption coefficient at high and medium frequencies is higher for materials with higher density [48]. It can be seen from Figure 10 that sample WH240_4 with a density of 0.61 g/cm^3 and a thickness of 25 mm had the best acoustic absorption coefficient values for the entire frequency range analyzed, reaching a maximum of 0.91 at 2500 Hz. In comparison, sample WC40 with a density of 0.02 g/cm^3 and a thickness of 25 mm had much lower absorption coefficient values, barely reaching 0.3 in the frequency range of 2500 ÷ 3150 Hz.

3.2.5. Comparisons with Other Materials

In order to accentuate the sound-absorbing properties of the materials obtained in this research, a comparative study with other materials in the market (rigid polyurethane foam 40 mm, flexible polyurethane foam 40 mm, and mineral wool 50 mm) and a material from the literature (sheep wool mechanically fastened on cloth [21]) was carried out. The results obtained are shown in Figure 11. It can be observed that the obtained material WH240_0.05 (sheep wool hot pressed at 80 °C with 0.05 MPa) had the best sound-absorbing properties at frequencies below 2000 Hz, while it had values almost identical to mineral wool in the frequency range of 2000 ÷ 3200 Hz.

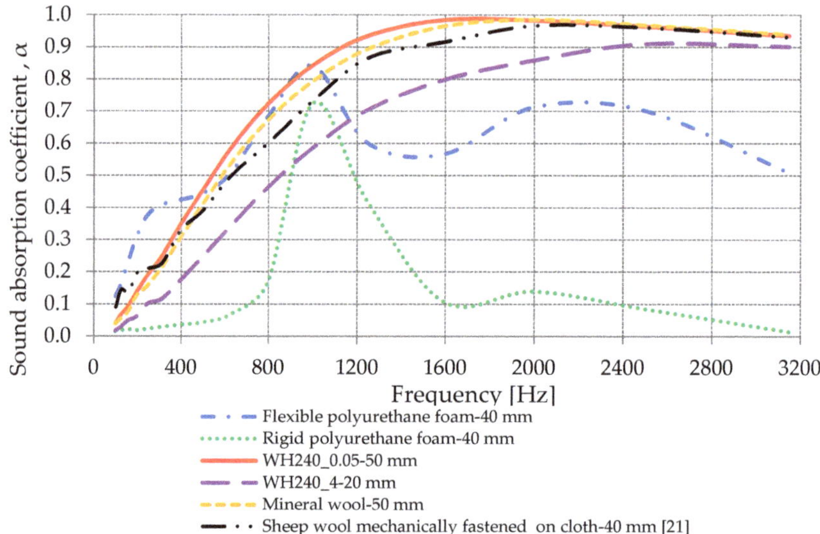

Figure 11. Variation in the acoustic absorption coefficient for WH240_0.05, WH240_4, flexible polyurethane foam, rigid polyurethane foam, and mineral wool.

Compared to the material obtained by Zach et al. [21], also from sheep wool, material WH240_0.05 obtained in this research had better acoustic absorption properties at frequencies higher than 315 Hz.

The material with 80% sheep wool fibers (40% first quality and 40% second quality) and 20% PET fibers (50 mm) with an acoustic absorption coefficient of more than 0.6 for the frequency range of 800 ÷ 3150 Hz [1] had lower acoustic properties compared to WH240_0.05 (50 mm) with an acoustic absorption coefficient of more than 0.72 for the frequency range of 800 ÷ 3150 Hz.

At frequencies greater than 1200 Hz, material WH240_4 with 20 mm thickness had an absorption coefficient value greater than the flexible polyurethane foam with 40 mm thickness. It should be mentioned that the flexible polyurethane foam maintained the best sound-absorbing properties at frequencies below 400 Hz.

4. Conclusions

Obtaining environmentally friendly materials with very good acoustic properties from natural and renewable raw materials, such as sheep wool without using any binder, is an important step in solving environmental problems and, at the same time, finding new methods of using wool. By simply hot pressing wool, a material that can be processed and manipulated can be obtained.

Hot-pressed materials have a much higher density than cold-pressed materials. The density of materials made from hot-pressed sheep wool increases with increasing pressure.

In this research, material WH240_0.05, which had a 240 mm layer of wool and 50 mm thickness and was hot-pressed at 80 °C and 0.05 MPa, had higher sound absorption coefficient values over the entire analyzed frequency range compared to WH120_0.05, which was obtained under the same conditions but with a smaller thickness and a 120 mm layer of wool.

Due to the thermoplastic properties of wool in the presence of water at a temperature of 80 °C, the sound absorption coefficient of material WHW80_6_75 had lower values at frequencies lower than 2850 Hz compared to the material with a lower water content.

WH240_0.05, which had 0.01 g/cm^3 density and was pressed at 0.05 MPa, had better sound-absorbing properties than WH240_4, which was pressed at 4 MPa and had 0.61 g/cm^3 density. During the compression, the fibers of materials come close, so the open porosity decreases and the compression increases.

The WH240_0.05 material obtained in this study had the best sound-absorbing properties at frequencies below 2000 Hz, while it had values almost identical to mineral wool in the frequency range of 2000 ÷ 3200 Hz. Thus, hot-pressed sheep wool has better or at least equal sound-absorbing properties as that of mineral wool, which is one of the most widely used sound-absorbing fibrous materials.

The field of use for the obtained materials is wide, but other characteristics will have to be determined.

Author Contributions: Conceptualization, S.I.B.M., A.-E.T. and O.N.; methodology, S.I.B.M.; validation, A.-E.T., H.V. and O.N.; investigation, S.I.B.M. and O.V.; resources, S.I.B.M, A.-E.T., H.V. and O.N.; writing—original draft preparation, S.I.B.M. and A.-E.T.; writing—review and editing, A.-E.T., O.N. and H.V.; supervision, O.N. and O.V. All authors have read and agreed to the published version of the manuscript.

Funding: This research received no external funding.

Conflicts of Interest: The authors declare no conflict of interest.

References

1. Del Rey, R.; Uris, A.; Alba, J.; Candelas, P. Characterization of sheep wool as a sustainable material for acoustic applications. *Materials* **2017**, *10*, 1277. [CrossRef]
2. Asdrubali, F.; D'Alessandro, F.; Schiavoni, S. A review of unconventional sustainable building insulation materials. *Sustain. Mater. Technol.* **2015**, *4*, 1. [CrossRef]
3. Schiavoni, S.; D'Alessandro, F.; Bianchi, B.; Asdrubali, F. Insulation materials for the building sector: A review and comparative analysis. *Renew. Sustain. Energy Rev.* **2016**, *62*, 988. [CrossRef]
4. Thilagavathi, G.; Pradeep, E.; Kannaian, T.; Sasikala, L. Development of natural fiber nonwovens for application as car interiors for noise control. *J. Ind. Text.* **2010**, *39*, 267–278. [CrossRef]
5. Asim, M.; Khalina Abdan, M.; Jawaid, M. A review on pineapple leaves fiber and its composites. *Int. J. Polym. Sci.* **2015**, 950567. [CrossRef]
6. Belakroum, R.; Gherfi, A.; Kadja, M.; Maalouf, C.; Lachi, M. Design and properties of a new sustainable construction material based on date palm fibers and lime. *Constr. Build. Mater.* **2018**, *184*, 330–343. [CrossRef]
7. Berardi, U.; Gino, I. Acoustic characterization of natural fibers for sound absorption. *Build. Environ.* **2015**, *94*, 840–852. [CrossRef]
8. Elammaran, J.; Hamdan, S.; Ezhumalai, P. Investigation on dielectric and sound absorption properties of banana fibers reinforced epoxy composites. *J. Teknol.* **2016**, *78*, 97–103. [CrossRef]
9. Hamzé, K.; Maalouf, C.; Bliard, C.; Moussa, T.; El Wakil, N. Hygrothermal and acoustical performance of starch-beet pulp composites for building thermal insulation. *Materials* **2018**, *11*, 1622. [CrossRef]
10. Lim, Z.Y.; Putra, A.; Nor, M.J.M.; Yaakob, M.Y. Sound absorption performance of natural kenaf fibers. *Appl. Acoust.* **2018**, *130*, 107–114. [CrossRef]
11. Yian, Z.; Wang, J.; Zhu, Y.; Wang, A. Research and application of kapok fiber as an absorbing material: A mini review. *J. Environ. Sci.* **2015**, *27*, 21–32. [CrossRef]

12. Gagan, B.; Singh, V.K.; Gope, P.C.; Gupta, T. Application and properties of chicken feather fiber (CFF) a livestock waste in composite material development. *J. Graph. Era Univ.* **2017**, *5*, 16–24.
13. João, B.; Souza, J.; Lopes, J.B.; Sampaio, J. Characterization of thermal and acoustic insulation of chicken feather reinforced composites. *Procedia Eng.* **2017**, *200*, 472–479. [CrossRef]
14. Berardi, U.; Gino, I.; Di Gabriele, M. Characterization of sheep wool panels for room acoustic applications. In Proceedings of the Meetings on Acoustics 22ICA, Buenos Aires, Argentina, 5–9 September 2016. [CrossRef]
15. Dunne, R.; Desai, D.; Sadiku, R.; Jayaramudu, J. A review of natural fibres, their sustainability and automotive applications. *J. Reinf. Plast. Compos.* **2016**, *35*, 1041. [CrossRef]
16. Sen, T.; Reddy, H.N.J. Various industrial applications of hemp, kinaf, flax and ramie natural fibres. *Int. J. Innov. Manag. Technol.* **2011**, *2*, 192.
17. Spritzendorfer, J. *Der Dämmstoff Schafwolle, Energetische Bewertung-CO_2 Bilanz und Ökobilanz Beratungsagentur für zukunftsfähiges Bauen, Energies*; Presse—EGGBI Publikationen: Basel, Schweiz, 2015.
18. Korjenic, A.; Klarić, S.; Hadžić, A.; Korjenic, S. Sheep wool as a construction material for energy efficiency improvement. *Energies* **2015**, *8*, 5765–5781. [CrossRef]
19. Dunne, R.; Desai, D.; Sadiku, R. A Review of the Factors that Influence Sound Absorption and the Available Empirical Models for Fibrous Materials. *Acoust Aust.* **2017**, *45*, 45. [CrossRef]
20. Asdrubali, F. Survey on the acoustical properties of new sustainable materials for noise control. In Proceedings of the Euronoise 2006, Tampere, Finland, 27–31 May 2006.
21. Zach, J.; Korjenic, A.; Petránek, V.; Hroudová, J.; Bednar, T. Performance evaluation and research of alternative thermal insulations based on sheep wool. *Energy Build.* **2012**, *49*, 246–253. [CrossRef]
22. Asis, P.; Mvubu, M.; Muniyasamy, M.; Botha, A.; Anandjiwala, R.D. Thermal and sound insulation materials from waste wool and recycled polyester fibers and their biodegradation studies. *Energ. Build.* **2015**, *92*, 161–169. [CrossRef]
23. Pennacchio, R.; Savio, L.; Bosia, D.; Thiebat, F.; Piccablotto, G.; Patrucco, A. Fitness: Sheep-wool and hemp sustainable insulation panels. *Energy. Proc.* **2017**, *111*, 287–297. [CrossRef]
24. Del Rey, R.; Alba, J.; Ramis, J.; Sanchis, V. New absorbent acoustics materials from plastic bottle remnants. *Mater. Constr.* **2011**, *61*, 547–558. [CrossRef]
25. Ramis, J.; Alba, J.; Del Rey, R.; Escuder, E.; Sanchís, V. New absorbent material acoustic based on kenaf's fiber. *Mater. Constr.* **2010**, *60*, 133–143. [CrossRef]
26. Baxter, B.P.; Cottle, D.J. Fiber diameter distribution characteristics of midside (fleece) samples and their use in sheep breeding. *Wool Technol. Sheep Breed.* **1998**, *46*, 154–171.
27. Hassan, M.M.; Carr, C.M. A review of the sustainable methods in imparting shrink resistance to wool fabrics. *J. Adv. Res.* **2019**, *18*, 39. [CrossRef] [PubMed]
28. Simpson, W.S.; Crawshaw, G.H. *Wool: Science and Technology*; Woodhead Publishing: Cambridge, UK, 2002.
29. Ibrahim, M.A.; Melik, R.W. Physical parameters affecting acoustic absorption characteristics of fibrous materials. *Proc. Math. Phys. Soc. Egypt* **1978**, *46*, 125–130.
30. Gombos, A.M.; Nemes, O.; Soporan, V.F.; Vescan, A. Toward New Composite Materials Starting from Multi-Layer Wastes. *STUD U BABES-BOL CHE* **2008**, *LIII*, 81.
31. Millington, K.R.; Rippon, J.A. Wool as a high-performance fiber. In *Structure and Properties of High-Performance Fibers*; Woodhead Publishing: Cambridge, UK, 2017; p. 367.
32. *Cellular Plastics and Rubbers—Determination of Apparent Density*; Standard ISO 845:2006; ISO: Geneva, Switzerland, 2006.
33. *Determination of Sound Absorption Coefficient and Acoustic Impedance with the Interferometer*; Standard SR EN ISO 10534-2; Part 2. Transfer Function Method; ISO: Geneva, Switzerland, 2002.
34. Tiuc, A.E.; Vasile, O.; Vermesan, H.; Nemes, O.; Borlea Muresan, S.I. New Multilayered Composite for Sound Absorbing Applications. *Rom. J. Acoust. Vib.* **2018**, *15*, 115.
35. Koizumi, T.; Tsujiuchi, N.; Adachi, A. *The Development of Sound Absorbing Materials Using Natural Bamboo Fibers, High Performance*; WIT Press: Southampton, UK, 2002.
36. Tiuc, A.E.; Nemeş, O.; Vermeşan, H.; Toma, A.C. New sound absorbent composite materials based on sawdust and polyurethane foam. *Compos. Part B Eng.* **2019**, *165*, 120–130. [CrossRef]
37. Jimenez-Espadafor, F.J.; Villanueva, J.A.B.; Garcia, M.T.; Trujillo, E.C.; Blanco, A.M. Optimal design of acoustic material from tire fluff. *Mater. Des.* **2011**, *32*, 3608–3616. [CrossRef]

38. Kalauni, K.; Pawar, S.J. A review on the taxonomy, factors associated with sound absorption and theoretical modeling of porous sound absorbing materials. *J. Porous Mater.* **2019**, *26*, 1795. [CrossRef]
39. Ersoy, S.; Küçük, H. Investigation of industrial tea-leaf-fiber waste material for its sound absorption properties. *Appl. Acoust.* **2009**, *70*, 215–220. [CrossRef]
40. Abdullah, A. *Ecological and Economic Attributes of Jute and Natural Fiber for Sustainable Eco-Management*; Primeasia University: Dhaka, Bangladesh, 2014.
41. Tiuc, A.E.; Vasile, O.; Vermesan, H. The analysis of factors that influence the sound absorption coefficient of porous materials. *RJAV* **2014**, *11*, 105.
42. Asdrubali, F.; D'Alessandro, F.D.; Schiavoni, S. Sound absorbing properties of materials made of rubber crumbs. *J. Acoust. Soc. Am.* **2008**, 35–40. [CrossRef]
43. Nick, A.; Becker, U.; Thoma, W. Improved acoustic behavior of interior parts of renewable resources in the automotive industry. *J. Polym. Environ.* **2002**, *10*, 115. [CrossRef]
44. Coates, M.; Kierzkowski, M. Acoustic textiles—lighter, thinner and more absorbent. *Tech. Text. Int.* **2002**, *11*, 15.
45. Castagnede, B.; Aknine, A.; Brouard, B. Effects of compression on the sound absorption of fibrous materials. *Appl. Acoust.* **2000**, *61*, 173. [CrossRef]
46. Fouladi, M.H.; Nor, M.J.M.; Ayub, M.D. Enhancement of coir fiber normal incidence sound absorption coefficient. *J. Comput. Acoust.* **2012**, *20*, 1250003. [CrossRef]
47. Nor, M.J.M.; Ayub, M.; Zulkifli, R. Effect of compression on the acoustic absorption of coir fiber. *Am. J. Appl. Sci.* **2010**, *7*, 1285. [CrossRef]
48. Delany, M.E.; Bazley, E.N. Acoustic properties of fibrous absorbent material. *Appl. Acoust.* **1970**, *3*, 105–116. [CrossRef]

Article

Comparative Analysis of the Thermal Insulation of Multi-Layer Thermal Inserts in a Protective Jacket

Dubravko Rogale [1], Goran Majstorović [2] and Snježana Firšt Rogale [1,*]

[1] Faculty of Textile Technology, University of Zagreb, 10000 Zagreb, Croatia; dubravko.rogale@ttf.hr
[2] Weltex, 11000 Beograd, Serbia; goranmajstor77@gmail.com
* Correspondence: sfrogale@ttf.hr

Received: 30 April 2020; Accepted: 10 June 2020; Published: 12 June 2020

Abstract: This paper presents the measurement results of the thermal insulation of the outer shell, thermal inserts, and clothing systems, as well as a comparative analysis of the thermal insulation of multi-layer thermal inserts in a thermal jacket intended for professional services in cold weather. Detachable thermal inserts are made of double-faced, diamond-shaped quilted lining with different masses per unit area, and together with the jacket, they form clothing systems with different thermal properties. Tests of the thermal properties of clothing were performed on a thermal manikin. They showed that an increase in the mass of thermal insulation textile materials contributes to an increase in the thermal insulation properties of clothing and are insufficient for a complete analysis of the thermal properties of clothing. Therefore, for the first time, three new parameters of integration efficiency of the thermal insert, thermal insulation efficiency parameters, and efficiency parameters of the integration of the textile material integrated into the clothing system were introduced. Based on these parameters, it is possible to perform an effective and accurate comparative analysis of the thermal insulation of multi-layer thermal inserts in clothing. This makes it possible to apply exact scientific methods largely in the technical design of the thermal properties of integrated textile materials, instead of experience-based methods as in the past.

Keywords: textile materials; thermal insulation; clothing system; multi-layer thermal inserts; thermal manikin

---

1. Introduction

Exact measurements in the field of thermal insulation of clothing began in the 1940s when the US Army was developing the first thermal manikin [1], and more intensely in the 1980s when international research on cold protective clothing, clothing physiology, and thermal functions of clothing began. Thermal insulation is nowadays expressed in SI-units by m^2 K W^{-1}. Gagge et al. published a scientific paper in 1941 defining a warm business suit providing thermal insulation of approximately 0.155 m^2 K W^{-1} for the whole body, which was originally equal to 1 Clo unit [2], and refers to a person who feels thermal comfort when sitting in a ventilated room with an ambient temperature of 21 °C, an airflow of 0.1 m s^{-1}, and a relative humidity of less than 50%. The Clo, was defined in J. R. Mather's "Climatology: Fundamentals and Applications" as units measuring the thermal insulation value of clothing, as well. To achieve a more simple perception of these units, it should be pointed out that the naked human body has an insulation value of 0.0 Clo, and a value of 1.0 Clo refers to a person who is dressed in a typical business suit [3,4]. The Clo unit is easier to understand and more used in clothing engineering.

In his work, Eryuruk examined the thermal properties of fabrics and their combinations in a three-layer garment composite. Sixteen garment composites were examined, combining two types of fabrics for making outer shells, four types of moisture barrier fabrics with membranes, and two types of thermal barrier fabrics. It was found that thermal and moisture comfort properties were significantly affected by different fabric layers [5].

Matusiak tested the thermal insulation properties of single-layer and multi-layer textile materials on the Alambeta instrument. Different types of materials intended for the manufacture of winter clothing: Cotton woven fabrics, thermal textile materials, and sets of both materials were examined [6].

Matusiak and Sybilska conducted a study in which they analysed the relationship between the thermal insulation of nine fabrics of different raw material compositions on the Alambeta instrument and the thermal insulation of T-shirts made from the tested fabrics of the same design and size [7]. The authors concluded that it is possible to design garments based on the measurement of the thermal insulation of the material with regard to the necessary thermal protection.

The Alambeta instrument was also used by Gupta et al. in testing the thermal properties of single and double layer fabric assemblies. The effect of layering fabrics with and without air gaps between them has been assessed to simulate the effect of a multi-layered garment assembly. Results show that thermal insulation increases markedly when an inner layer is paired with an outer layer of fabric [8].

Konarska et al. [3] compared the measurement results of the three ensembles of disposable medical clothing obtained by measuring the thermal manikin and subjective assessment of the wearer. On the thermal manikin the values of the heat balance of the thermal insulation of the thermal insert were measured, as well as the thermal insulation of the thermal insert, temperature, and heat losses, while a subjective assessment was used to determine values of dry heat losses under the conditions of heat comfort. The results showed that the thermal insulation value of the tested clothing samples was 13% higher in the subjective assessment of the wearer compared to the thermal manikin tests. It has also been shown that measurements on the thermal manikin are more accurate with regard to the measurements of subjective assessment of the wearer (manikin's error 2%, subjects' error 12–18%).

Koranska et al. also investigated the impact as thermal environment parameters, how heating power is transferred to the manikin, and thus required to reach thermal balance during tests with thermal clothing insulation. Tests are carried out on three double sets of cold protective clothing intended for use in very low temperatures. Based on the results, they conclude that methods for controlling the transfer of heating power do not have a major impact on the thermal insulation of clothing, that air velocity decreases thermal insulation, and that the temperature in the air-conditioning chamber should be determined in accordance with the anticipated clothing insulation of the clothing ensemble being tested. The authors of the paper propose changes of the measurement conditions on the thermal manikin prescribed by ISO 15831 [9] in terms of decreasing the admissible range of air velocity to values from 0.3 to 0.7 m s^{-1}, as well as increasing the temperature range between the thermal manikin surface and the environmental temperature in proportion to the thermal insulation of clothing [10].

Moreover, Holand dealt with the comparison of the results obtained by measuring on a thermal manikin and subjective assessment of the wearer [11]. He conducted tests of seven different types of sleeping bags, (one military and six commercial sleeping bags). Sleeping bags were tested on a thermal manikin in a climatic chamber where the temperature was adjusted to achieve a total heat loss from the manikin between 40 and 80 W/m^2 (normally around 55 W/m^2) when the system thermally ended up in a steady state with the environment. Subsequently, subjective assessment measurements were carried out on test persons who stayed in the climatic chamber (between 11:00 pm and 6:00 am), whereby the ambient temperatures were compared to the expected lower limit of sleeping comfort for different sleeping bags. The author states that the results are fairly equal, and the lower thermal insulation values of individual sleeping bags obtained by the subjective assessment of subjects that justify such cold spots (e.g., cold feet) may be a significant reason for thermal discomfort and consequently poor sleep. In addition, the thermal insulation of certain garments in the men's clothing system (undershirts, boxer shorts, socks, shirts, jeans, jackets, so-called "peaked caps", and shoes) and the entire clothing system was examined. This research was carried out on a thermal manikin on different menswear systems according to the international standard ISO 15831. The authors concluded that dressing in layers increases the overall thermal insulation of the clothing system. Exact measurements on the manikin have shown that the use of different clothing combinations can influence the thermal insulation of the clothing system [12].

Oliveira et al. compared the thermal properties of a summer ensemble, a typical business suit, and a cold protective clothing ensemble. Using different methods on a thermal manikin with 16 body parts, they measured the thermal insulation of the thermal insert and concluded that differences between calculation methods can be significant and that greater discrepancies arise when the distribution of clothing becomes less uniform. Other authors reached similar conclusions [13,14].

It is possible to implement new modern structures in the clothes, such as Smart Textile Materials with Shape Memory Alloys, whose task will be to preserve body heat or thermal protection insurance that were not previously used in conventional technical garment construction [15,16].

This brief overview of the use and benefits of measurement systems for testing the thermal properties of clothing, in particular of thermal manikins, shows that they can be very useful in the technical design of the thermal properties of clothing. However, they are still used on a small scale and are largely restricted for scientific research. When designing garments in the garment industry, visual design predominates from the point of view of aesthetics, and technical design, especially from the point of view of the thermal properties of the integrated textile materials, is not applied, but rather experience-based methods are used. Therefore, measurements of the resistance to heat transfer from the human body through the clothing to the environment should be introduced in the process of engineering clothing design. In this way, it is possible to accurately determine the success of designing the thermal properties of clothing and the usefulness of textile materials intended for the thermal insulation of clothing.

The paper deals with the design and practical performance of a professional protective jacket for cold conditions, where the outer shell consists of the base fabric and lining, and there is the possibility of inserting a detachable thermal insert into the inside of the jacket. The measurement results of the thermal insulation of the outer shell, thermal inserts, and clothing systems as a combination of outer shell and different multi-layer thermal inserts are presented.

Based on the measurement results of the thermal manikin, a new concept for the integration of the thermal insert has been presented in this work. Furthermore, a new method and procedure for evaluating thermal insulation efficiency is presented, which indicates how much mass of textile material had to be incorporated into a garment to achieve a thermal insulation effect of 1 Clo (thermal insulation of 0.155 m^2 K W^{-1}). The method and procedure for evaluating the efficiency of the incorporated mass of the textile insulation materials incorporated into the clothing system of the professional protective jacket and the efficiency of the mass of the textile insulation materials integrated into the thermal inserts are also presented. The efficiency of the integrated mass of textile insulation material indicates the value of the achieved thermal insulation when using 1 kg of textile insulation material. The measurements, the obtained results, and the adopted method presented in this paper confirm the thesis that it is necessary to abandon the current practice of the empirically determined selection of types and mas per unit area of textile materials intended for thermal insulation of clothing, and that modern measurement technology and scientific methods should be used in the engineering design of thermal properties of clothing. In assessing the characteristics of thermal parameters of protective clothing and uniforms for special services, when, following an invitation for tenders, garments are delivered by several different manufacturers using different garment cuts, materials, raw material compositions, and combinations of built-in composites (base fabrics, reinforcements, adhesive interlinings, and linings), the exact possibility of assessing the garment from the point of view of the precisely measured degree of thermal protection is not used. The paper presents a new method of technical design and construction of protective clothing from the point of view of thermal insulation, which enables a more reliable, faster, and technically accurate design and construction. This confirms the hypothesis that not only previous experience-based methods of clothing design can be used, but also more complex methods of technical clothing design. The tests of the thermal properties of clothing were carried out with a thermal manikin using the series method in a static mode according to ISO 15381.

2. Materials and Methods

The professional protective jacket (clothing systems) consists of an outer shell made of outer water repellent fabric (designation M01), lining fabric (designation M02), and five types of different thermal inserts (designations TI1 to TI5) as detachable thermal inserts with different masses of textile thermal insulation materials (Figure 1). Thermal inserts are made in five combinations of double-faced, diamond-shaped quilted lining (M03 to M07) with different masses per unit area. The four combinations consist of five layers: Two layers of cover fabrics, two layers of lining and one layer of padding, and a combination of nine layers: Four layers of cover fabrics, four layers of lining, and one layer of padding (Figure 2). The four five-layer combinations differ in weight from padding amounting to 0.385 g m^{-2} (designation M03), 0.790 g m^{-2} (designation M04), 0.114 g m^{-2} (designation M05), and 0.145 g m^{-2} (designation M06) (Table 1). Five clothing systems (designations CS1 to CS5) with professional jackets made for cold conditions, are shown in Table 2. The fifth combination of thermal inserts consists of two layers of a double-sided, diamond-shaped quilted lining with the designation M03. Table 2 shows the designations of the materials used to make the outer shell, thermal inserts, and clothing systems, and gives the values of their masses.

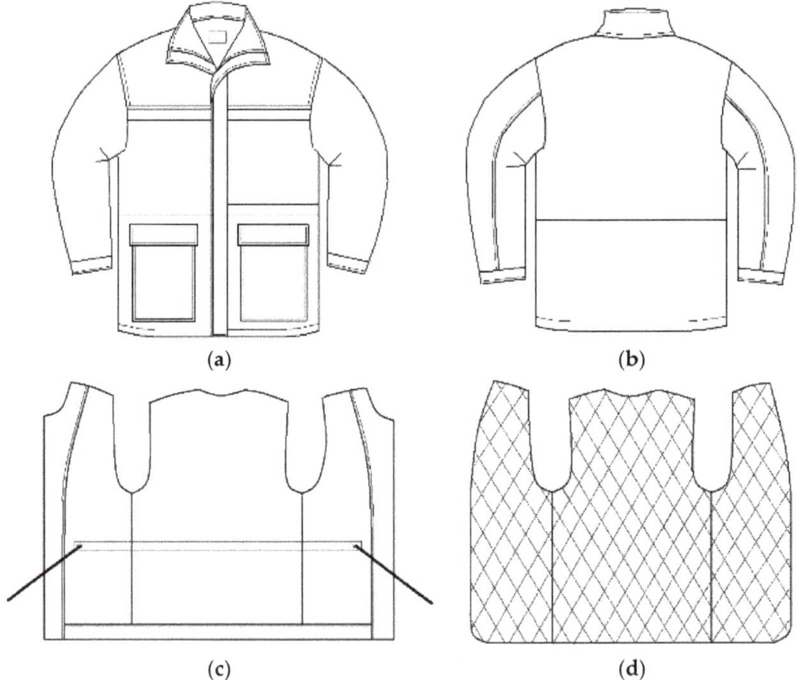

Figure 1. Basic construction features of the test garment. (**a**) The front of the outer shell; (**b**) the back of the outer shell; (**c**) lining; (**d**) detachable thermal insert.

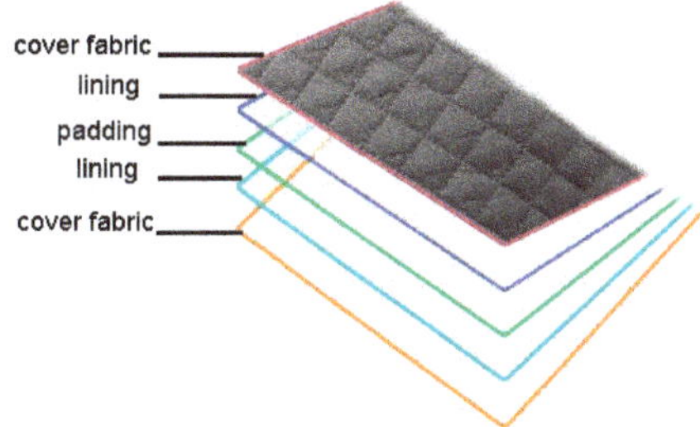

Figure 2. Double-faced, diamond-shaped quilted lining.

Table 1. Overview of the analysed technical characteristics of the sample of the integration material.

	Outer Shell		**Detachable Thermal Inserts**				
	M01	**M02**	**M03**	**M04**	**M05**	**M06**	**M07**
Raw material composition	Cotton 54.5% Polyester 41.5% Metallized fibres 4%	Polyester 100%	Double-faced, diamond-shaped quilted lining: Cover fabrics: Polyester 100% Lining: Polypropylene 100% Padding: Polyester 100%				
Mass per unit area: Cover fabrics (kg m^{-2})	-	-	0.475 × 2				0.475 × 4
Mass per unit area: lining (kg m^{-2})	-	-	0.132 × 2				0.132 × 4
Mass per unit area: padding (kg m^{-2})	-	-	0.385	0.790	0.114	0.145	0.770
Mass per unit area: in total (kg m^{-2})	0.162	0.768	0.160	0.200	0.235	0.263	0.320

Remarks: M01: Inner layer of outer shell; M02: Outer layer of outer shell; M03: Material of thermal insert (TI1); M04: Material of thermal insert (TI2); M05: Material of thermal insert (TI3); M06: Material of thermal insert (TI4); M07: Material of thermal insert (TI5).

Table 2. Summary of designations and masses of outer shells, detachable thermal inserts, and clothing systems.

Outer Shell				**Detachable Thermal Inserts**			**Clothing System**	
Designation of the Outer Shell Material		**Designation**	**Mass, kg**	**Designation of the Thermal Insert Material**	**Designation**	**Mass, kg**	**Designation**	**Mass, kg**
Outer Layer	**Inner Layer**							
M01	M02	OS1	0.985	M03	TI1	0.176	CS1	1.161
				M04	TI2	0.220	CS2	1.205
				M05	TI3	0.259	CS3	1.244
				M06	TI4	0.289	CS4	1.274
				M07	TI5	0.704	CS5	1.689

Remarks: OS1: Outer shell of clothing systems; TI1: Thermal insert of clothing system CS1; TI2: Thermal insert of clothing system (CS2); TI3: Thermal insert of clothing system (CS3); TI4: Thermal insert of clothing system (CS4); TI5: Thermal insert of clothing system (CS5).

The thermal insulation properties of a protective jacket for professional services without integrated thermal inserts were tested, then for thermal inserts made of double-sided, diamond-shaped quilted lining with different surface masses per unit area, and finally for five garment systems consisting of an outer shell and thermal inserts. Tests of thermal properties were performed on the thermal manikin.

Testing of thermal properties on a thermal manikin was performed using the serial method in static mode according to ISO 15381, Figure 3, implemented, installed, calibrated, and patented at the Faculty of Textile Technology of the University of Zagreb [17].

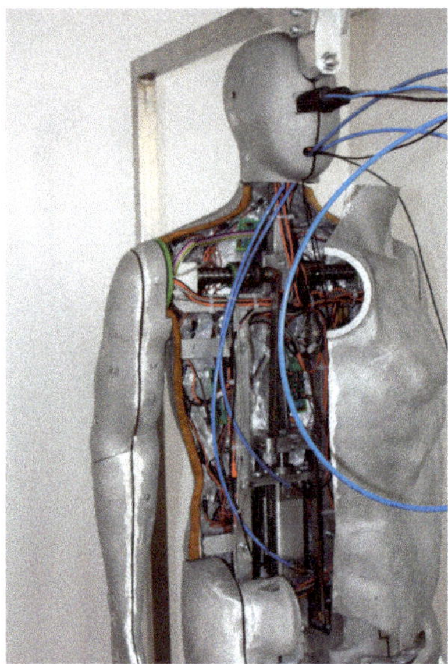

Figure 3. Thermal manikin.

After establishing the specified environmental conditions in accordance with ISO 15381 for the control and measurement of ambient air parameters (temperature, relative humidity, and flow velocity), the thermal insulation of the empty surface of the manikin together with the air boundary layer along the surface of the manikin R_{ct0} is determined in the measuring system.

$$R_{ct0} = \frac{(T_s - T_a) \cdot A}{H_0} \tag{1}$$

where R_{ct0} is the resultant total thermal insulation of the measuring device including the thermal insulation of the boundary air layer ($m^2\ kW^{-1}$); A is the surface area of the body segment i of the manikin (m^2); Ts is the mean skin surface temperature of the manikin (°C); Ta is the air temperature within the climatic chamber (°C), and H_0 is the total heating power supplied to the manikin (W).

The composite specimen to be tested is placed on a measuring unit of an electrically heated plate after the resultant total thermal insulation of the measuring device is determined and the heat flux through the test specimen is measured after new stable conditions have been reached. The assessment of thermal properties of clothing by means of the thermal manikin is conducted in such a way that the chosen garment or ensemble is placed around the body of the thermal manikin in the static or dynamic operational mode [18]. After thermal comfort has been established, which can be detected from the stabilization of parameter values, measurements are performed and thermal insulation is calculated from Equation (2):

$$R_{ct} = \frac{(T_s - T_a) \cdot A}{H_0} - R_{ct0} \tag{2}$$

In order to be able to compare the design results of different integration materials, thermal inserts, and clothing systems exactly and accurately, it is necessary to create a mathematical term for calculating the efficiency of thermal insulation. Based on the research conducted so far by a group of authors of this paper, a new term for the efficiency of thermal insulation, EI_{CS}, has been formulated, which represents the mass of thermal insulator required for a clothing system to have a thermal insulation of 1 Clo. Thermal insulation efficiency (EI_{CS}) is expressed by mathematical expression (3) [19].

$$EI_{CS} = \frac{m_{cs}}{Clo_{cs}}, \ (kg{\cdot}Clo^{-1}) \quad (3)$$

where Clo_{CS} is the thermal insulation of the clothing system (Clo); m_{CS} is the mass of the clothing system (kg).

Mathematical expressions were also determined for the calculation of the efficiency of the textile mass installed on the thermal insulation in the technical design of clothing systems. On the basis of the previous research, the original expression for the efficiency of the textile mass integrated into the clothing system was established, which represents the value of thermal insulation achieved with a mass of 1 kg of materials integrated into the clothing system. The efficiency of the textile mass integrated into the clothing system (EM_{CS}) is expressed by mathematical expression (4) [19].

$$EM_{CS} = \frac{Clo_{CS}}{m_{CS}}, \ (Clo{\cdot}kg^{-1}) \quad (4)$$

3. Results and Discussion

Testing the thermal insulation of an undressed thermal manikin in static mode was performed for outer shells and thermal inserts combined in different test clothing systems. Thermal insulation measurements of the test clothing systems on the thermal manikin were performed in the climatic chamber of the measuring system for testing the thermal properties of composites and clothing, in which the thermal insulation of the undressed thermal manikin was also performed. The measurements were performed at an ambient air temperature of 20 °C, a surface temperature of the thermal manikin of 34 °C, and an air flow velocity of 0.4 m s^{-1}. Relative humidity ranged from 47–50%. Measurements are made by measuring the temperature of the (heated) segments involved and the power of the heater, and the average values are plotted every minute. The measurement takes 20 min according to ISO 15831.

The measured total thermal insulation of the undressed manikin together with the air boundary layer along the surface (R_{ct0}) was 0.09683 m^2 K W^{-1}.

The measured value of thermal insulation of the outer shell (OS1) was 1.08 Clo (0.167 m^2 K W^{-1}).

The clothing system (CS1) consists of an outer shell (OS1) and a thermal insert (TI1). The value of the thermal insulation of the thermal insert (TI1) was 0.37 Clo (0.057 m^2 K W^{-1}). After these measurements, the outer shell (OS1) and the thermal insert (TI1) were assembled into a clothing system (CS1) and thermal insulation values of 1.72 Clo (0.266 m^2 K W^{-1}) were measured. Based on the measured data set, a graphical representation of the results was created (Figure 4a).

For the clothing system (CS2), a thermal insert (TI2) was selected in addition to the outer shell (OS1). The value of the thermal insulation of the thermal insert (TI2) was 0.39 Clo (0.060 m^2 K W^{-1}), and the measurement value of thermal insulation of the clothing system (CS2) was 1.84 Clo (0.285 m^2 K W^{-1}). Based on the measured data set, a graphical representation of the results was created (Figure 4b).

For the clothing system (CS3), a thermal insert (TI3) was selected in addition to the outer shell (OS1). The value of the thermal insulation of the thermal insert (TI3) was 0.43 Clo (0.067 m^2 K W^{-1}). After these measurements, the outer shell (OS1) and the thermal insert (TI1) were assembled into a clothing system (CS3) and thermal insulation of the new clothing system (CS3) were measured. The measurement value of thermal insulation of the clothing system (C3) was 2.02 Clo (0.304 m^2 K W^{-1}). Based on the measured data set, a graphical representation of the results was created (Figure 4c).

For the clothing system (CS4), a thermal insert (TI4) was selected in addition to the outer shell (OS1). The value of the thermal insulation of the thermal insert (TI4) was 0.45 Clo (0.070 m^2 K W^{-1}), and the measurement value of thermal insulation of the clothing system (CS4) was 2.08 Clo (0.323 m^2 K W^{-1}). Based on the measured data set, a graphical representation of the results was created (Figure 4d).

For the clothing system (CS5), a thermal insert (TI5) was selected in addition to the outer shell (OS1). The value of the thermal insulation of the thermal insert (TI5) was 0.5 Clo (0.078 m^2 K W^{-1}), and the measurement value of thermal insulation of the clothing system (CS5) (0.341 m^2 K W^{-1}) was measured. Based on the measured data set, a graphical representation of the results was created (Figure 4e).

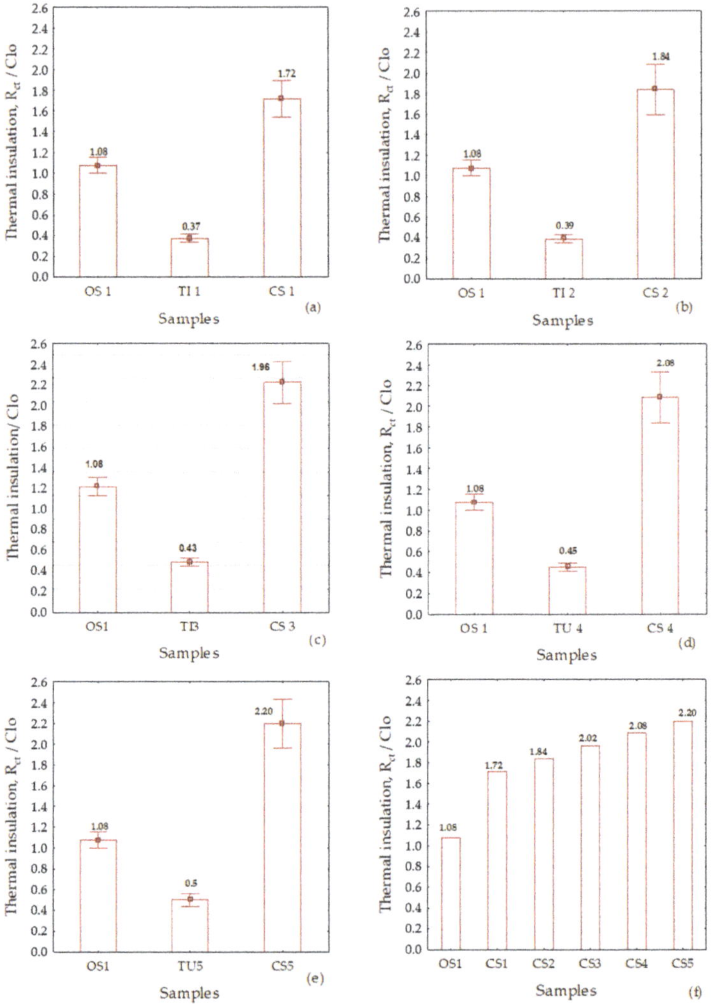

Figure 4. Graphical representations of the measurement results of the thermal insulation of the outer shell, thermal inserts, and clothing systems: (**a**) Clothing system (CS1); (**b**) clothing system (CS2); (**c**) clothing system (CS3); (**d**) clothing system (CS4); (**e**) clothing system (CS5); (**f**) increasing the thermal insulation of the clothing systems by installing thermal inserts of different masses per unit area.

On the basis of the measured values and the analysis of the measurement results on the manikin, the summarized results of the thermal insulation of all clothing systems are shown in Figure 5f. The measurement results for all clothing systems are presented according to the increasing value of the total mass of textile materials intended for thermal insulation. From Figure 4f, the basic conclusion can be drawn that the thermal insulation of the garment increases with the mass of the integrated textile material, which is evidence of an already known empirical fact. This is not enough for a scientific analysis, but it is necessary to approach the definition of new parameters and their comparative analysis to explain the effects of detachable multi-layer thermal inserts in the professional protective jacket for colder conditions.

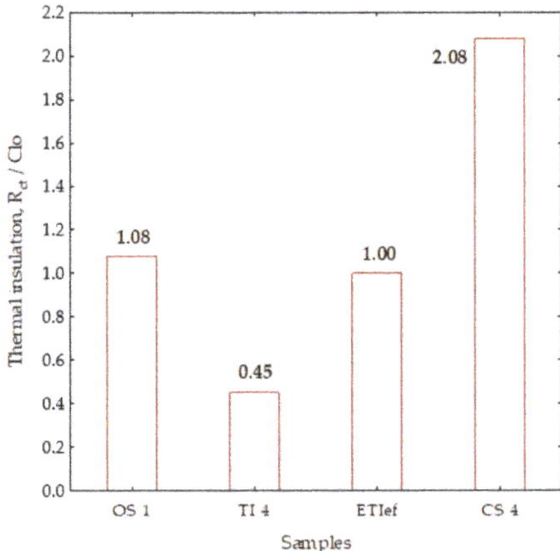

Figure 5. Presentation of the thermal insulation efficiency parameter (ETIef) calculated from the value of the thermal insulations of the clothing system (CS) and the outer shell (OS).

3.1. *Defining the Thermal Parameter of the Integration Efficiency of the Thermal Insert*

The analysis of the measured data on the thermal manikin for five different garment systems (CS) shown in Figure 4a–e reveals that the value of thermal insulation is greater than the algebraic sum of the measured thermal insulation of the outer shell (OS) and thermal insert (TI). It can be concluded from this that by inserting a thermal insert into the outer shell an additional air layer is formed, which additionally increases the overall thermal insulation properties, which is in any case a positive effect. This effect cannot be measured directly on the manikin, but, as a newly introduced analytical parameter of the integration efficiency of the thermal insert (ETI_{ef}), it can be calculated from the values of the thermal insulation of the clothing system (CS) and the outer shell (OS), as shown in Figure 5. The representation is based on the data in Figure 4d for the thermal insert (TI4). A column with the designation (ETI_{ef}) for the thermal insulation of the thermal insert and with a thermal insulation value of 1 Clo was determined and plotted as the difference between the total thermal insulation of the clothing system with 2.08 Clo and the thermal insulation of the outer shell (OS1) of 1.08 Clo. The introduction of the integration efficiency parameter (ETI_{ef}) defines the effects of the integration of this insert much better because, in addition to the insulation properties of the thermal insert measured on the thermal manikin, it also accurately evaluates the effects of the layer of air created by the integration of the thermal insert in the outer shell. The influence of the newly formed air layer can be determined for the case under consideration from the difference between the integration efficiency (1 Clo) and the value of the measured thermal insulation of the thermal insert on the manikin (0.45 Clo). The identified difference of 0.55 Clo can be attributed to the newly formed air layer between the outer shell and the thermal insert. Based on the defined parameter of integration efficiency of the thermal insert it is possible to calculate and accurately evaluate the thermal impact of the newly created airbag. The way for a comparative analysis of the thermal insulation of multi-layer, detachable thermal inserts is also paved.

The functional dependence on the thermal insulation value of clothing systems on the mass of the integrated textile material with thermal insulation properties is shown in Figure 6. It is obvious that the thermal insulation of clothing systems increases with the increase in mass of the clothing system, as a functional dependence $R_{ctCS} = f(m_{CS})$. The increase is stronger with smaller masses and then saturation occurs when the increase in the installed mass of thermal inserts does not significantly increase the thermal insulation. This observed effect can be attributed to the occurrence of chimney effect when, due to the increased cross-section of the insulation layer, heat convection caused by a slight air flow through the insulation material reduces the insulation properties. This is one possible reason why people do not wear one thick garment but thinner garments, which is called layered clothing.

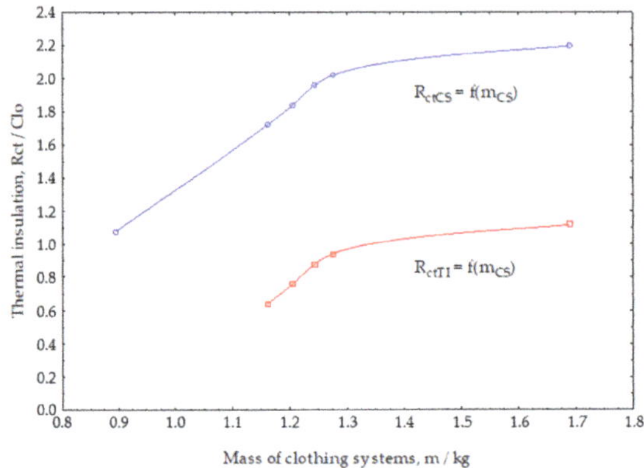

Figure 6. Functional dependence on the value of thermal insulations of clothing systems on the mass of the integrated textile material with thermal insulation properties.

Figure 6 also shows the functional dependence on the thermal insulation value of the thermal inserts as a function of the mass of the clothing system. The first functional dependence of R_{ctCS} is the cumulative effect of the outer shell and the clothing system (outer shells with thermal inserts). Functional dependence $R_{ctTI} = f\ (m_{CS})$ represents the change in thermal properties in the clothing system caused only by the change of mass of thermal inserts.

From both functional dependencies, it can be seen that above a certain mass of thermal inserts, i.e., the mass of the clothing system, there is no significant increase in the value of the thermal insulation. Thus, when technically designing the thermal properties of the garment at a certain limit, it is no longer necessary to increase the mass of the textile thermal insulation material. At this point, the increase in the mass of the garment will not lead to an increase in thermal insulation, and on the other hand, it will cause a heavier garment, increase the price of the garment due to the higher volume of the material that is integrated, while at the same time the bulkiness of the garment hinders the ease of movement.

3.2. Results and Discussion of Thermal Insulation Efficiency Parameters and Efficiency Parameter of the Textile Mass Integrated into the Clothing System

In order to determine the limit when a further increase in the mass of the textile thermal insulation material does not contribute to a significant increase in the thermal insulation properties of garments, it is necessary to define new parameters of thermal insulation efficiency and the textile mass integrated into the clothing system. These parameters were also defined for the first time for the purposes of this paper.

The functional dependence on changes in the values of the parameters of thermal insulation efficiency is shown in Figure 7 and is determined using expression (3).

The thermal insulation efficiency parameter shows how much mass of textile insulation material should be integrated into the garment to achieve a thermal insulation of 1 Clo. Therefore, this parameter is very important in the comparative analysis of the thermal insulation of garments, if technically exact, based on measured values on a manikin, the optimal multi-layer insulation effect can be accurately evaluated. The smaller the amount of textile insulation material required for a certain degree of thermal protection, the more successful the clothing project will be. Figure 7 shows that it is most difficult to achieve adequate thermal protection only by using an outer shell. The addition of detachable thermal inserts into the protective jacket, which is intended for cooler conditions, resulted in a great decrease in the required mass of thermal insulation textile materials.

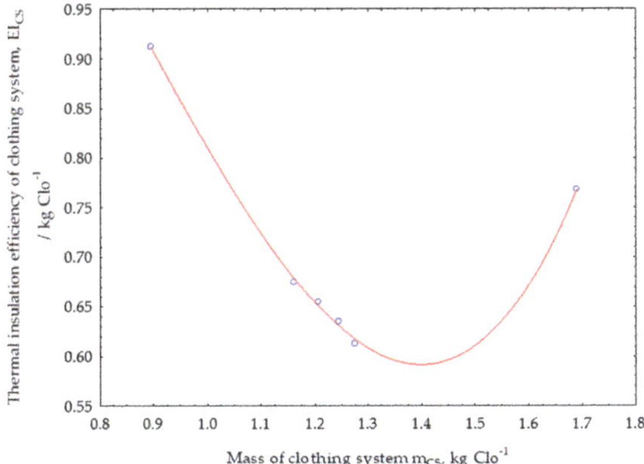

Figure 7. Functional dependence on changes in the values of the parameters of thermal insulation efficiency of clothing systems.

The curve has a pronounced minimum where the optimum of thermal insulation properties is achieved with a minimum of the integrated mass. Increasing the thermal insulation mass no longer achieves optimum effect, so the minimum of the function shown is the limit beyond which there is no rational reason to increase the mass of textile thermal insulation materials to increase the thermal properties of clothing.

The functional dependence on changes in the values of the efficiency parameters of the textile mass integrated into the clothing system is shown in Figure 8 and is determined by the expression (4). The efficiency parameter of the textile mass integrated into the thermal insulation system shows how much thermal insulation, expressed in Clo units, 1 kg of the textile thermal insulation material integrated into the clothing system can achieve. This parameter is also very important in the comparative analysis of the thermal insulation of clothing systems when it is necessary to determine the optimum efficiency of the textile mass integrated into the clothing system.

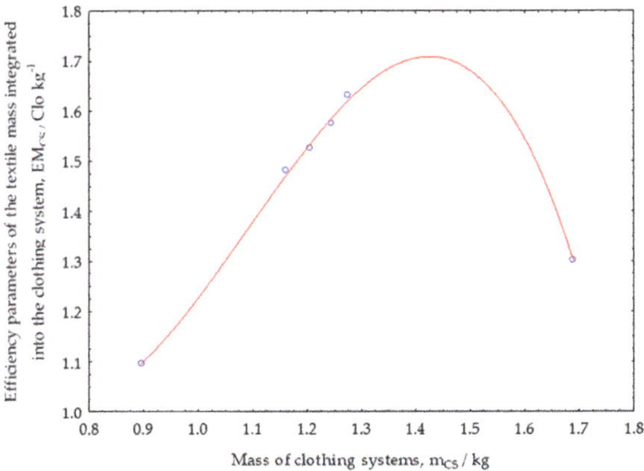

Figure 8. Functional dependence on changes in the values of the efficiency parameters of the textile mass integrated into the clothing systems.

The higher the thermal insulation by using 1 kg of the integrated textile thermal insulation material, the more successful was the technical design of the thermal properties of the clothing system. Figure 8 shows that increasing the total mass of the garment system increases the efficiency of the designed system and reaches a maximum of about 1.25 kg of the total mass of the clothing system.

Thereby, 1 kg of the textile material achieves the efficiency of the integrated textile mass of 1.7 Clo kg^{-1}. After maximum function dependence, any further increase in the mass of the integrated textile material leads to a decrease in the efficiency of the integrated textile mass. By using this parameter, it is also possible to determine the success of the technical design of the thermal properties of the clothing, to optimize the material consumption by quantity and price.

By the described measuring method on the thermal manikin, the evaluation of the measurement results and the introduction of three new parameters of the integration efficiency of the thermal insert, the efficiency parameter of the thermal insulation and the efficiency parameter of the textile mass integrated into the clothing system, it is possible to perform an efficient and accurate comparative analysis of the thermal insulation of multi-layer thermal inserts in the clothing designed for cold conditions. The described design method should completely replace the experience-based methods.

4. Conclusions

This research provides a comparative analysis of the detachable thermal insulation of multi-layer thermal inserts in the clothing system. A protective jacket for cold conditions, used by professional services, and five multi-layer thermal insulation inserts were designed for this purpose. Basic measurements were made on the thermal manikin with regard to thermal insulation, and it can be concluded that an increase in the mass of thermal insulation textile materials contributes to an increase in the thermal insulation properties of the clothing and that no further comprehensive analysis is possible. Therefore, the introduction of three new parameters for a more successful way of assessing the performance of the engineering design of thermal properties of clothing is proposed. These are the parameters of the integration efficiency of the thermal insert, the thermal insulation efficiency parameter, and the efficiency parameter of the textile mass integrated into the clothing system. Based on the above three parameters, it is possible to accurately determine the efficiency of the technical design of the thermal properties of clothing and optimal material consumption, resulting in an acceptable cost of clothing and through the ease of movement.

5. Patents

Dubravko Rogale, Gojko Nikolić, G.: Measuring System for Determination of Static and Dynamic Thermal Properties of Composite and Clothing. State Intellectual Property Office of the Republic of Croatia, Patent No. PK20130350 [17].

Author Contributions: Conceptualization, D.R., G.M. and S.F.R.; methodology: D.R. and G.M.; software, S.F.R. and D.R.; validation, D.R., G.M. and S.F.R.; formal analysis, D.R. and S.F.R.; investigation, D.R., G.M. and S.F.R.; resources, D.R. and G.M.; data curation, D.R.; writing—original draft preparation, D.R. and S.F.R.; writing—review and editing, D.R. and S.F.R.; visualization, D.R. and S.F.R.; supervision, D.R.; project administration: S.F.R.; funding acquisition, D.R. All authors have read and agreed to the published version of the manuscript.

Funding: This research was funded by the Croatian Science Foundation through the project IP-2018-01-6363 Development and thermal properties of intelligent clothing (ThermIC).

Conflicts of Interest: The authors declare no conflict of interest.

References

1. Holme'r, I. Thermal manikin history and applications. *Eur. J. Appl. Physiol.* **2004**, *92*, 614–618. [CrossRef] [PubMed]
2. Jussila, K. *Clothing Physiological Properties of Cold Protective-Clothing and Their Effects on Human Experience*; Tampere University of Technology: Tampere, Finland, 2016; p. 171.

3. Konarska, M.; Sołtynski, K.; Sudoł-Szopińska, I.; Chojnacka, A. Comparative Evaluation of Clothing Thermal Insulation Measured on a Thermal Manikin and on Volunteers. *Fibres Text. East. Eur.* **2017**, *15*, 73–79.
4. Mather, J.R. *Climatology, Fundamentals and Applications*; McGraw-Hill Inc.: New York, NY, USA, 1974.
5. Eryuruk, S.H. Effect of fabric layers on thermal comfort properties. *AUTEX Res. J.* **2018**, *19*, 271–278. [CrossRef]
6. Matusiak, M. Investigation of the thermal insulation properties of multilayer textiles. *Fibres Text. East. Eur.* **2006**, *14*, 98–102.
7. Matusia, M.; Sybilska, W. Thermal resistance of fabrics vs. thermal insulation of clothing made of the fabrics. *J. Text. Inst.* **2016**, *107*, 842–848. [CrossRef]
8. Gupta, D.; Srivastava, A.; Kale, S. Thermal properties of single and double layer fabric assemblies. *Indian J. Fibre Text. Res.* **2013**, *38*, 387–394.
9. ISO. *Clothing-Physiological Effects-Measurement of Thermal Insulation by Means of a Thermal Manikin*; ISO 15831: 1 Febuary 2004; ISO: Geneva, Switzerland, 2004.
10. Konarska, M.; Sołtyński, K.; Sudoł-Szopińska, I.; Młoźniak, D.; Chojnacka, A. Aspects of Standardisation in Measuring Thermal Clothing Insulation on a Thermal Manikin. *Fibres Text. East. Eur.* **2006**, *14*, 58–63.
11. Holand, B. Comfort temperatures for sleeping bags. In Proceedings of the Third International Meeting on Thermal Manikin Testing 3IMM at the National Institute for Working Life, Stockholm, Sweden, 12–13 October 1999; Nilsson, H., Holmér, I., Eds.; National Institute for Working Life: Stockholm, Sweden, 1999; pp. 25–33.
12. Rogale, S.F.; Benić, M.; Rogale, D. Investigation of resistance to the passage of heat for different men's clothing combinations. In Proceedings of the 10th Scientific–Professional Symposium Textile Science & Economy, Zagreb, Croatia, 10–16 October 2017.
13. AOliveira, V.M.; Branco, V.J.; Gaspar, A.; Quintela, D.A. Measuring Thermal Insulation of Clothing with Different Manikin Control Methods: Comparative Analysis of the Calculation Methods. In Proceedings of the 7th International Thermal Manikin and Modelling Meeting, Coimbra, Portugal, 3–5 September 2008.
14. Antonnen, H. Interlaboratory trial of thermal manikin. In Proceedings of the 3I3M-3rd International Meeting, Stockholm, Sweden, 12–13 October 1999.
15. Bartkowiak, G.; Dabrowska, A.; Greszta, A. Development of Smart Textile Materials with Shape Memory Alloys for Application in Protective Clothing. *Materials* **2020**, *13*, 689. [CrossRef] [PubMed]
16. He, J.; Yehu, L.; Wang, L.; Ma, N. On the Improvement of Thermal Protection for Temperature-Responsive Protective Clothing Incorporated with Shape Memory Alloy. *Materials* **2018**, *11*, 1932. [CrossRef] [PubMed]
17. Rogale, D.; Nikolić, G. *Measuring System for Determination of Static and Dynamic Thermal Properties of Composite and Clothing*; PK20130350; State Intellectual Property Office of the Republic of Croatia: Zagreb, Croatia, 2015.
18. Rogale, D.; Rogale, S.F.; Špelić, I. Development of the Measuring System for Analysing the Thermal Properties of Clothing. In *Book of 7th ITC&DC 2014*; Dragčević, Z., Ed.; Dubrovnik, Faculty of Textile Technology University of Zagreb: Zagreb, Croatia, 2014; pp. 322–327.
19. Majstorović, G. Determination of Thermal Properties of Special Purpose and Intelligent Clothing During Their Technical Design. Ph.D. Thesis, Faculty of Textile Technology University oh Zagreb, Zagreb, Croatia, 2015.

 materials

Article

Failure Analysis of Retrieved Osteosynthesis Implants

Mihai Nica [1,2], Bogdan Cretu [2], Dragos Ene [1,3,*], Iulian Antoniac [4], Daniela Gheorghita [4,*] and Razvan Ene [1,2]

1 University of Medicine and Pharmacy Carol Davila Bucharest, 050474 Bucharest, Romania; mikx99n@gmail.com (M.N.); razvan77ene@yahoo.com (R.E.)
2 Orthopedics and Traumatology Department, Bucharest Emergency University Hospital, 050098 Bucharest, Romania; jfrbogdan@yahoo.com
3 Surgery Department, Emergency Clinical Hospital of Bucharest, 050098 Bucharest, Romania
4 Politehnica University of Bucharest, 060042 Bucharest, Romania; antoniac.iulian@gmail.com
* Correspondence: dragoshene@yahoo.com (D.E.); daniela.mgm8@gmail.com (D.G.); Tel.: +4-074-0366082 (D.E.)

Received: 29 January 2020; Accepted: 4 March 2020; Published: 7 March 2020

Abstract: Failure of osteosynthesis implants is an intricate matter with challenging management that calls for efficient investigation and prevention. Using implant retrieval analysis combined with standard radiological examination, we evaluated the main causes for osteosynthesis implant breakdown and the relations among them for a series of cases. Twenty-one patients diagnosed with implant failure were assessed for this work. For metallurgical analysis, microscopy techniques such as scanning electron microscopy (SEM), stereomicroscopy, and optical microscopy were employed. The results showed that material structural deficiencies (nine patients) and faulty surgical techniques (eight patients) were the main causes for failure. An important number of patients presented with material structural deficiencies superimposed on an imperfect osteosynthesis technique (six patients). Consequently, the importance of failure retrieval analysis should not be overlooked, and in combination with other investigational techniques, must provide information for both implant manufacturing and design improvement, as well as osteosynthesis technique optimization.

Keywords: osteosynthesis; failure; analysis; implant; retrieval

1. Introduction

Failure analysis is of great importance not only for orthopedics [1–5], but also for other various medical areas such as general surgery [6–9], gynecology [10–12], cranioplasty [13,14], ophthalmology [15], and dentistry [16,17], as the investigation of retrieved implants offers insight into implant failure mechanisms and how to prevent such cases. Physiologic forces are transmitted to intact human bone under normal conditions without exceeding its ultimate strength. When non-physiologic forces are applied by trauma or normal bone strength is affected by various pathological entities, fractures ensue. There are different forces acting on the human skeleton, with natural physiologic loading comprising a combination of these. A fracture line is characterized by abnormal stress and strain which must be neutralized by stabilizing devices (external or internal) in order to provide favorable conditions for bone healing [18,19].

Internal fracture fixation, or osteosynthesis, uses metallic implants that have to meet specific requirements for orthopedic surgical use. These requirements refer to design, size, and material properties like strength, stiffness, resistance to fatigue, biocompatibility, and corrosion resistance [20–22]. Last but not least, pre-operative management plays an essential role and needs to be taken into account for favorable outcomes, as Table 1 presents. Implants used for osteosynthesis are manufactured using iron-based alloys such as surgical grade stainless-steel 316L, titanium-based alloys, cobalt-chromium-based composites, or shape-memory alloys [23–25]. Of these, stainless-steel 316L is the most commonly used, due to its cost-effectiveness and good integration of corrosion resistance and mechanical properties.

Table 1. Pre-operative planning.

Radiographic Images (Multiple Views)	Implant Length	Implant Diameter
Bone morphology	Contralateral bone x-rays (magnified)	IM canal isthmus (the narrowest portion of the canal)
Shape of the intramedullary (IM) canal	Traction radiographs	1.0 to 1.5 mm greater than anticipated IM nail diameter
Fracture pattern and comminution	Distance between palpable bony landmarks	

Successful clinical results are also dependent on adherence to sound osteosynthesis principles which achieve load sharing between the implant and bone during the healing period. The main implant failure mechanisms are mechanical or biomechanical [26]. A numerical analysis can also provide useful information regarding the loading forces that act on the bone and on the couple formed by a bone-osteosynthesis implant [27,28]. The tolerable mechanical stress level of an implant may be exceeded either through cyclic loading (fatigue) or under a single critical load (static or dynamic), combined with the corrosive effects of the internal biological environment. Especially for fatigue management, a good load-sharing level between bone and implant is crucial, which is contingent on good reduction of fracture and proper structural support, provided by adequately-reduced bone fragments [29,30]. Figure 1 presents the biomechanics of fractures and how forces act on the bone.

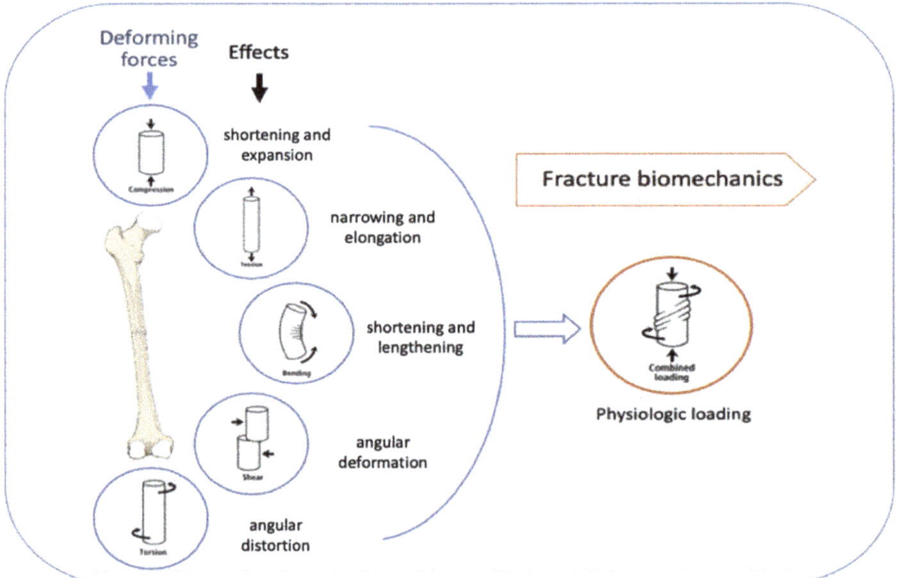

Figure 1. Deforming forces acting on long bones.

Therefore, both surgical technique and the level of bone comminution—which impacts structural support—become very important elements for reaching an equilibrium of forces needed for stable and reliable fixation until healing ensues. If this balance of forces is not established and maintained, areas of local stress concentration may appear on the implant surface and failure can occur. Biologic failure manifests as implant loosening caused by bone resorption at the contact interface with the implant, due to an inappropriate stabilizing device, excessive physiologic loading during injudicious rehabilitation, or prolonged bone healing (biologic factors play a key role) [31–33].

This paper presents the failure analysis of a series of stainless-steel osteosynthesis implants used for primary fixation and revised in the Orthopedics Department of Bucharest Emergency University Hospital Bucharest.

2. Materials and Methods

In one year, approximately eight hundred cases required osteosynthesis in the Orthopedics Department of the Bucharest Emergency University Hospital. The reported number of cases with osteosynthesis implant failure was just under 1%, which is consistent with reported rates in the literature. Twenty-one cases of osteosynthesis implant failure diagnosed and managed in our department during a three-year period were assessed for this work. All cases had the primary fixation surgery also done in our department, with a time-to-failure period ranging from four weeks to seven months. The majority of patients were male (19 cases), with an age span between 18 and 78 years. Table 2 presents the centralized data of patients with osteosynthesis implant failure.

Table 2. Statistics data on patients with osteosynthesis implant failure.

No.	Patient Age	Fracture Location (Bone Type)	Type of Implant	Failure Causes				
				External Factors (Traumatic Event)	Surgical Causes		Implant Defects	
					Inadequate Implant Size	Deficient Fracture Reduction	Materials Defects	Surface Defects
1	**36**	**Tibia**	**IMN**	(+)				
2	65	Femur	IMN			(+)		(+)
3	72	Femur	GN	(+)				
4	26	Humerus	PS				(+)	
5	44	Tibia	IMN	(+)	(+)	(+)		(+)
6	31	Tibia	PS	(+)		(+)	(+)	
7	71	Femur	GN	(+)				
8	67	Femur	DHS			(+)		(+)
9	39	Humerus	IMN	(+)	(+)			
10	78	Femur	DHS	(+)				
11	46	Tibia	IMN	(+)		(+)	(+)	
12	42	Humerus	PS	(+)				(+)
13	69	Femur	DHS			(+)		
14	18	Humerus	IMN	(+)				
15	68	Femur	DHS					
16	32	Humerus	PS	(+)				(+)
17	74	Femur	GN	(+)				
18	59	Femur	DHS	(+)				
19	46	Humerus	IMN			(+)		(+)
20	55	Tibia	IMN	(+)				
21	69	Femur	DHS			(+)		

IMN=Intramedullary Nail; GN=Gamma Nail; DHS=Dynamic Hip Screw; PS=Plate-Screw System

In terms of location, 10 were femoral implants, 5 tibia, and 6 humerus implants. The femoral implants used were 3 Gamma nails, one interlocking nail, and 6 dynamic hip screws. Cases with diagnosed infection and pathological fractures were not included in the study because of the confounding role they can have on research [34–36]. The osteosynthesis performed was by closed reduction and intramedullary nailing in 6 cases (3 femoral and 3 tibia fractures), one of the femoral fracture sites, and one tibia fracture requiring open reduction, followed by nailing. The humerus lesions were managed by open reduction and internal fixation with reconstruction or dynamic compression plates and screws (Auxein Medical, Haryana, India) in 3 cases or closed reduction and intramedullary nailing for the other 3. All implants (Auxein Medical, Haryana, India) used were made of austenitic stainless-steel type 316L, implanted using standard instrument sets and surgical techniques.

Diagnosis of implant failure was confirmed either by standard radiological examination during regular follow-up visits, or on admission for patients with inciting events and new symptoms generated by implant deterioration. Some radiographic images showing the aspect of deteriorated osteosynthesis implants are presented in Figure 2.

Images were assessed focusing on quality of reduction, implant type adequacy, size, and position. Retrieval of broken implants was performed with special care not to cause more damage, especially to the fracture surface.

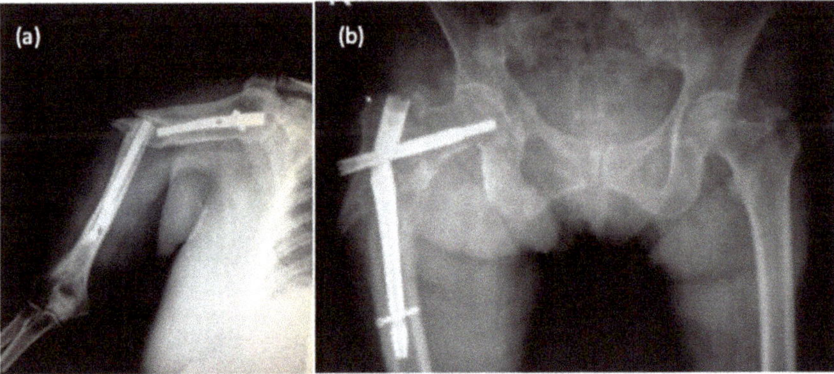

Figure 2. Radiological aspects of fractures with deterioration of osteosynthesis material: humerus fracture (**a**) and femoral fracture (**b**).

Implant failure investigation was conducted by means of macroscopic evaluation, optical microscopy, and scanning electron microscopy in concordance with recommended standards of retrieval analysis of failed internal fixation devices.

3. Results

Regarding external factors for osteosynthesis breakdown, we identified nine cases with critical loading during single traumatic events and five patients with injudicious rehabilitation and excessive loading. Subsequent radiological evaluation revealed that in eight cases (38%), fracture reduction was flawed with malrotation, which can add to the dynamic loading of implants or an interfragmentary gap of more than 2 mm. Only two cases (9.5%) had an unsuitably-sized implant, but always with an external factor added to the failure process of the fixation device.

3.1. Metallographic Analysis

Optical microscopy images obtained with an Olympus BX51 (Olympus Soft Imaging Solutions, Münster, Germany,) highlighted the metallographic structure of the sample, as seen in Figure 3. The presence of recrystallization slabs inside the crystalline grains is noted, and inhomogeneous grains of variable size are observed. The obtained images do not give reason to claim that the failure of the sample was due to technical errors, as no impurities and no gaps are present in the material.

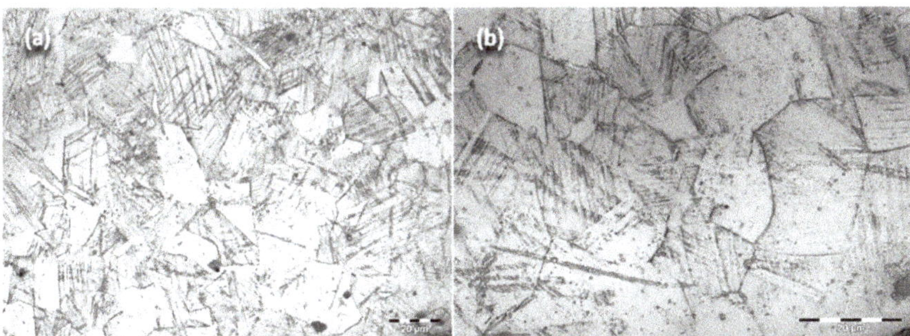

Figure 3. Optical microscopy images of experimental samples obtained from retrieved intramedullary nails made of stainless-steel type 316L: 500 × (**a**), 1000 × (**b**).

3.2. Stereomicroscopy

Following stereomicroscopy analysis performed on the fracture section, the image of the fracture zone can be observed (Figure 4c,d). Analysis was made using a Olympus SZX7 stereomicroscope (Olympus Soft Imaging Solutions, Münster, Germany). There are different areas with distinct aspects throughout the entire surface of the fracture. Some areas present an intergranular fracture with secondary cracks and a fibrous matte appearance, while other areas (glossy appearance) have a bright crystalline appearance and indicate the site of fracture initiation. Therefore, the fracture has a mixed character specific to a fatigue fracture with initiation, propagation, and sudden final fracture.

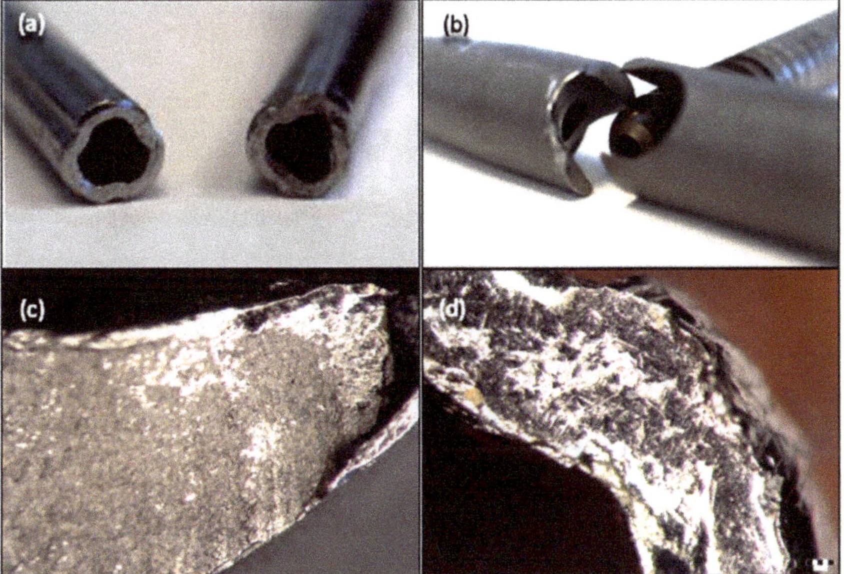

Figure 4. The macroscopic aspect (**a**,**b**) and stereomicroscopy images (**c**,**d**) of the fractured zone of some retrieved intramedullary nails.

3.3. Scanning Electron Microscopy

SEM (Philips model ESEM XL 30 TMP-FEI Company, Eindhoven, Netherlands) was used to characterize the surface morphology of the fracture site. Analysis of the images obtained with the scanning electron microscope (SEM) reinforces the results obtained from optical microscopy analysis. The fracture had a mixed character, and the propagation waves of the crack can be clearly observed (Figure 5).

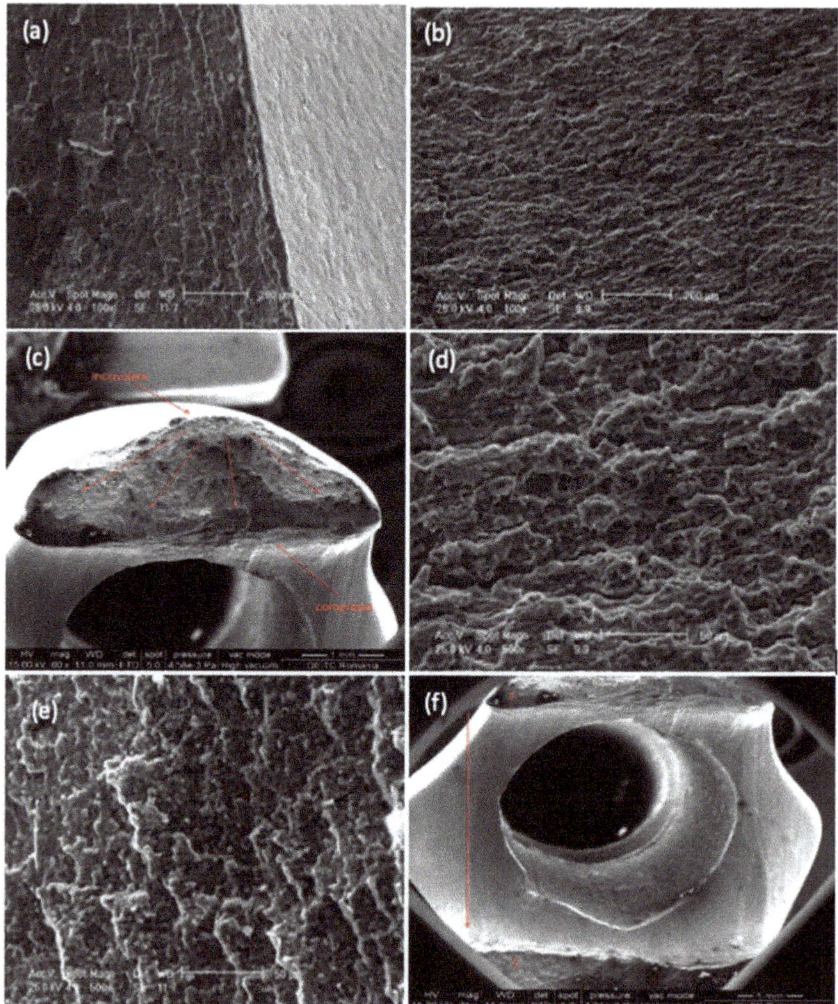

Figure 5. Scanning electron micrographs showing fracture surface morphology on failure region for retrieved intramedullary nails: 100 × (**a**,**b**), 60 × (**c**), 500 × (**d**,**e**), 45 × (**f**).

After implant failure analysis, the results showed that nine (42.85%) devices had structural abnormalities with either material inclusions inside the superficial layer, or surface defects with evidence of corrosion—all of which accounted for initiation points for material fracture (Table 2). The cause for surface defects could not be clearly attributed to either damage caused by a careless surgical implantation techniques or to engineering process errors. Material composition evaluation revealed elemental compositions with values compatible with the standard specifications for surgical grade stainless-steel 316L.

4. Discussion

Successful bone fracture treatment by means of osteosynthesis is dependent on a myriad of factors, such as correct indication and implant choice, surgical technique, appropriate rehabilitation, biological factors, and implant characteristics. Nevertheless, fixation devices are not intended for long-term load bearing in vivo, so the fracture healing process must restore the physiological equilibrium of forces and reclaim the stress-bearing role of the intact bony structure.

When mechanical failure of fixation devices occurs, it can be categorized as fatigue failure (under cyclic loading), plastic, or brittle. Fatigue and brittle breakdown are usually associated with flaws in the material or design of the implant. Fatigue combined with corrosion represents a fearsome and frequent failure cause, especially for stainless-steel internal fixation devices.

For intramedullary nails, it is well-recognized that the failure process is usually initiated in the locking holes, proximally or distally (case presented in Figure 6). Sliding screw hip plates are prone to significant wear and corrosion due to moving components [37,38]. For plates, the breakdown starting point is commonly situated around the holes where the cross-sectional area is reduced and where interaction with the screw heads generates wear and corrosion fatigue under localized stress concentration [39].

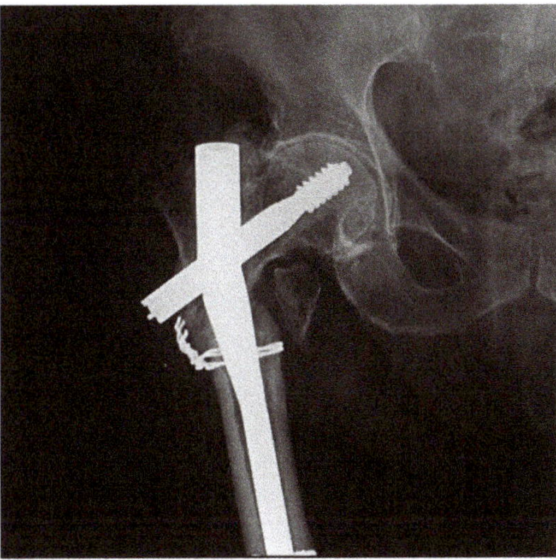

Figure 6. Radiography of a clinical case with failure of a Gamma Nail—72-year-old female patient.

Implant surface integrity is very important because it conventionally becomes the initiation point for implant fracture. Damage to the surface may be attributed to careless surgical implantation techniques, dynamic loading conditions (fatigue), wear at component junctions, or faulty manufacturing. Every type of surface defect, e.g., pitting, fretting, fatigue cracking, or crevicing, is in danger for corrosion attack in vivo, which increases even more the risk for failure [40].

Unevenly distributed mechanical load on the implant may be considered a crucial element of the failure process and commonly stems from improper device choice or faulty implantation technique with inadequate reduction of the bone fragments and insufficient structural support.

Therefore, we consider that the two main factors that generate the failure process of osteosynthesis implants are structural abnormalities, with focus on implant surface integrity (initiation point) and inappropriate surgical treatment algorithm (from implant choice to surgical implantation technique). The second may be easily identified in clinical practice, but failed implant analysis for identification of structural abnormalities does not represent a standard procedure. This may mean that many intrinsic structural flaws of the material or ones inflicted during implantation may be overlooked, and therefore failure attributed to other causes. We also consider that these two causes coexist for a large percent of implant failures.

As seen in our series, a considerable number of cases exhibit imperfect reduction, which calls for improvement of surgical technique, but we believe that failure should not be attributed only to this factor, with the metallurgical analysis results supporting our view. Of the eight failure cases with recognized fracture reduction shortcomings, six patients also showed structural abnormalities upon retrieved implant analysis.

Although a rare event, an implant breakdown before adequate bone healing requires a challenging management with repeated surgeries, added risks for complications, higher costs, and increased social and psychological burdens for the patient. Implant failure can manifest as various types of material fracture patterns, depending on the underlying mechanism. This is why legislation and protocols governing the manufacture, trade and use of osteosynthesis implants must be well- established and implemented, ensuring the best quality and reliability and that the relation between material structural abnormalities and surgical treatment flaws are better analyzed. Furthermore, retrieval analysis of failed devices used for internal fracture fixation, which provides useful data for manufacturing process optimization and surgical protocols development, must become a more prevalent tool for failure research and anticipation [41–44].

The results of our research show that the main two causes for implant failure are inadequate surgical technique and intrinsic material deficiencies, coupled with corrosion. All these can induce and promote failure initiation points and result in material breakdown.

5. Conclusions

Intramedullary nail failure represent a problematic matter with an even more demanding management. Albeit a rare occurrence, it brings significant human and financial costs, calling for continuous research and development of all factors involved in failure prevention.

Along with surgical implantation techniques and device manufacturing improvements, retrieval analysis of failed implants is an important tool that can provide data for development and implementation of better quality standards. This is supported by the results of our research, which show that structural material flaws account for a high percentage of the observed causes of failure. In order to identify the breaking mechanisms of osteosynthesis devices, explant analyses are needed. Macroscopic and microscopic investigations, including stereomicroscopy and scanning electron microscopy, are techniques that can help us identify the causes that lead to failure of implants.

Author Contributions: Conceptualization by M.N., I.A., R.E.; Methodology by M.N., D.E., I.A., R.E.; Software by D.E., I.A., D.G.; Validation by D.E., I.A., D.G., R.E.; Formal Analysis by B.C., D.G.; Investigation by M.N., B.C., D.G., R.E.; Resources by I.A., R.E.; Data Curation by M.N., I.A., D.G.; Writing-Original Draft Preparation by M.N., B.C., D.E., I.A., D.G., R.E.; Visualization by M.N., B.C., D.E., I.A., D.G., R.E.; Supervision by I.A., R.E.; Project Administration by M.N., I.A., R.E. All authors have read and agreed to the published version of the manuscript.

Funding: This research received no external funding.

Conflicts of Interest: The authors declare no conflict of interest

References

1. Marinescu, R.; Antoniac, V.I.; Stoia, D.I.; Lăptoiu, D.C. Clavicle anatomical osteosynthesis plate breakage-failure analysis report based on patient morphological parameters. *Romanian J. Morphol Embryol* **2017**, *58*, 593–598.

2. Atasiei, T.; Antoniac, I.; Laptoiu, D. Failure causes in hip resurfacing arthroplasty–retrieval analysis. *Int. J. Nano Biomater.* **2011**, *3*, 367–381. [CrossRef]
3. Bane, M.; Miculescu, F.; Blajan, A.I.; Dinu, M.; Antoniac, I. Failure analysis of some retrieved orthopedic implants based on materials characterization. *Solid State Phenom.* **2012**, *188*, 114–117. [CrossRef]
4. Grecu, D.; Antoniac, I.; Trante, O.; Niculescu, N.; Lupescu, O. Failure analysis of retrieved polyethylene insert in total knee replacement. *Mat. Plast.* **2016**, *53*, 776–780.
5. Antoniac, I.V.; Stoia, D.I.; Ghiban, B.; Tecu, C.; Miculescu, F.; Vigaru, C.; Saceleanu, V. Failure analysis of a humeral shaft locking compression plate—Surface investigation and simulation by finite element method. *Materials* **2019**, *12*, 1128. [CrossRef]
6. Mavrodin, C.I.; Pariza, G.; Ion, V.; Antoniac, V.I. Abdominal compartment syndrome—a major complication of large incisional hernia surgery. *Chirurgia* **2013**, *108*, 414–417.
7. Gradinaru, S.; Tabaras, D.; Gheorghe, D.; Gheorghita, D.; Zamfir, R.; Vasilescu, M.; Dobrescu, M.; Grigorescu, G.; Cristescu, I. Analysis of the Anisotropy for 3D Printed PLA Parts Usable in Medicine. *U.P.B. Sci. Bull. Series B Chem. Mater. Sci.* **2019**, *81*, 313–324.
8. Dumitrescu, D.; Savlovschi, C.; Borcan, R.; Pantu, H.; Serban, D.; Gradinaru, S.; Smarandache, G.; Trotea, T.; Branescu, C.; Musat, L.; et al. Clinical case-voluminous diaphragmatic hernia-surgically acute abdomen: Diagnostic and therapeutical challenges. *Chirurgia* **2011**, *106*, 657–660.
9. Cosmin, B.; Iulian, A.; Florin, M.; Marius, D.; Ionel, D. Investigation of a mechanical valve impairment after eight years of implantation. *Key Eng. Mater.* **2014**, *583*, 137–144.
10. Cirstoiu, M.M.; Antoniac, I.; Ples, L.; Bratila, E.; Munteanu, O. Adverse reactions due to use of two intrauterine devices with different action mechanism in a rare clinical case. *Mater. Plast.* **2016**, *53*, 666–669.
11. Cirstoiu, M.; Cirstoiu, C.; Antoniac, I.; Munteanu, O. Levonorgestrel-releasing intrauterine systems: Device design, biomaterials, mechanism of action and surgical technique. *Mater. Plast.* **2016**, *52*, 258–262.
12. Brătilă, E.; Comandasu, D.; Milea, C.; Berceanu, C.; Vasile, E.; Antoniac, I.; Mehedintu, C. Effect of the surface modification of the synthetic meshes used in the surgical treatment of pelvic organ prolapse on the tissue adhesion and clinical functionality. *J. Adhes. Sci. Technol.* **2017**, *31*, 2028–2043. [CrossRef]
13. Cavalu, S.; Antoniac, I.V.; Fritea, L.; Mates, I.M.; Milea, C.; Laslo, V.; Vicas, S.; Mohan, A. Surface modifications of the titanium mesh for cranioplasty using selenium nanoparticles coating. *J. Adhes. Sci. Technol.* **2018**, *32*, 2509–2522. [CrossRef]
14. Rivis, M.; Pricop, M.; Talpos, S.; Ciocoiu, R.; Antoniac, I.; Gheorghita, D.; Trante, O.; Moldovan, H.; Grigorescu, G.; Seceleanu, V.; et al. Influence of the bone cements processing on the mechanical properties in cranioplasty. *Rev. Chim.* **2018**, *69*, 990–993. [CrossRef]
15. Antoniac, I.V.; Burcea, M.; Ionescu, R.D.; Balta, F. IOL's opacificatiofn: A complex analysis based on the clinical aspects, biomaterials used and surface characterization of explanted IOL's. *Mater. Plast.* **2015**, *52*, 109–112.
16. Gabor, A.; Zaharia, C.; Todericiu, V.; Szuhanek, C.; Cojocariu, A.C.; Duma, V.F.; Sticlaru, C.; Negrutiu, M.L.; Antoniav, I.V.; Sinescu, C. Adhesion of scaffolds with implants to the mandibular bone with a defect. *Mater. Plast.* **2018**, *55*, 393–397. [CrossRef]
17. Corobea, M.S.; Albu, M.G.; Ion, R.; Cimpean, A.; Miculescu, F.; Antoniac, I.V.; Raditoiu, V.; Sirbu, I.; Stoenescu, M.; Voicu, S.; et al. Modification of titanium surface with collagen and doxycycline as a new approach in dental implants. *J. Adhes. Sci. Technol.* **2015**, *29*, 2537–2550. [CrossRef]
18. Ionescu, R.; Mardare, M.; Dorobantu, A.; Vermesan, S.; Marinescu, E.; Saban, R.; Antoniac, L.; Ciocan, D.N.; Ceausu, M. Correlation between materials, design and clinical issues in the case of associated use of different stainless steels as implant materials. *Key Eng. Mater.* **2014**, *583*, 41–44. [CrossRef]
19. Perren, S.M. Physical and biological aspects of fracture healing with special reference to internal fixation. *Clin. Orthop. Relat. Res.* **1979**, *138*, 175–196.
20. Panti, Z.; Cretu, B.; Panaitescu, C.; Nica, M.; Tecu, C.; Semenescu, A.; Ene, D.; Ene, R. Implant associated local recurrence in primary bone sarcoma. *Rev. Chim.* **2020**, *71*, 172–175. [CrossRef]
21. Iulian, A.; Dan, L.; Camelia, T.; Claudia, M.; Sebastian, G. Synthetic materials for osteochondral tissue engineering. *Osteochondral Tissue Eng.* **2018**, *1058*, 31–52.
22. Marinescu, R.; Antoniac, I.; Laptoiu, D.; Antoniac, A.; Grecu, D. Complications Related to Biocomposite Screw Fixation in ACL Reconstruction Based on Clinical Experience and Retrieval Analysis. *Mater. Plast.* **2015**, *52*, 340–344.

23. Stefanescu, T.; Antoniac, I.V.; Popovici, R.A.; Galuscan, A.; Tirca, T. Ni-Ti rotary instrument fracture analysis after clinical use. Structure changes in used instruments. *Environ. Eng. Manag. J.* **2016**, *15*, 981–988.
24. Wadood, A. Brief overview on nitinol as biomaterial. *Adv. Mater. Sci. Eng.* **2016**, *2016*, 1–9. [CrossRef]
25. Cirstoiu, C.; Ene, R.; Panti, Z.; Ene, P.; Cîrstoiu, M.-M. Particularities of shoulder recovery after arthroscopic bankart repair with bioabsorbable and metallic suture anchors. *Mater. Plast.* **2015**, *52*, 361–363.
26. Guazzo, R.; Gardin, C.; Bellin, G.; Sbricoli, L.; Ferroni, L.; Ludovichetti, F.S.; Piattelli, A.; Antoniac, I.; Bressan, E.; Zavan, B. Graphene-Based Nanomaterials for Tissue Engineering in the Dental Field. *Nanomaterials.* **2018**, *8*, 349. [CrossRef] [PubMed]
27. Pascoletti, G.; Cianetti, F.; Putame, G.; Terzini, M.; Zanetti, E.M. Numerical simulation of an intramedullary elastic nail: Expansion phase and load-bearing behavior. *Front. Bioeng. Biotechnol.* **2018**, *6*, 174. [CrossRef]
28. Putame, G.; Pascoletti, G.; Franceschini, G.; Dichio, G.; Terzini, M. Prosthetic hip ROM from multibody software simulation. In Proceedings of the 2019 41st Annual International Conference of the IEEE Engineering in Medicine and Biology Society (EMBC), Berlin, Germany, 23–27 July 2019; pp. 5386–5389.
29. Gheorghe, D.; Pencea, I.; Antoniac, I.V.; Turcu, R.-N. Investigation of the Microstructure, Hardness and Corrosion Resistance of a New 58Ag24Pd11Cu2Au2Zn1.5In1.5Sn Dental Alloy. *Materials* **2019**, *12*, 4199. [CrossRef]
30. Antoniac, I.; Negrusoiu, M.; Mardare, M.; Socoliuc, C.; Zazgyva, A.; Niculescu, M. Adverse local tissue reaction after 2 revision hip replacements for ceramic liner fracture: A case report. *Medicine* **2017**, *96*, e6687. [CrossRef]
31. Buzatu, M.; Geanta, V.; Stefanoiu, R.; Buţu, M.; Petrescu, M.-I.; Buzatu, M.; Ghica, S.-I.; Antoniac, I.; Iacob, G.; Niculescu, F.; et al. Mathematical Modeling for Correlation of the Resistance to Compression with the Parameters Md, Bo and E/A, for the Design of Titanium Beta Alloys Developed for Medical Applications. *U.P.B. Sci. Bull. Series B Chem. Mater. Sci.* **2019**, *81*, 183–192.
32. Aksakal, B.; Yildirim, Ö.S.; Gul, H. Metallurgical failure analysis of various implant materials used in orthopedic applications. *J. Fail. Anal. Prev.* **2004**, *4*, 17–23. [CrossRef]
33. Ene, R.; Panti, Z.; Nica, M.; Pleniceanu, M.; Ene, P.; Cirstoiu, M.; Antoniac, V.I.; Cirstoiu, C. Mechanical failure of angle locking plates in distal comminuted tibial fractures. *Key Eng. Mater.* **2016**, *695*, 118–122. [CrossRef]
34. Popescu, D.; Ene, R.; Popescu, A.; Cirstoiu, M.; Sinescu, R.; Cirstoiu, C. Total hip joint replacement in young male patient with osteoporosis, secondary to hypogonadotropic hypogonadism. *Acta Endocrinol. Buchar.* **2015**, *11*, 109–113. [CrossRef]
35. Popa, M.; Ene, R.; Streinu-Cercel, A.; Popa, V.; Pleniceanu, M.; Nica, M.; Panti, Z.; Cirstoiu, M.; Cirstoiu, C. Algic syndrome in osteoarticular infectious pathology; detection and rapid treatment of the causative agent using microcalorimetry. In Proceedings of the 14th National Congress of Urogynecology and the National Conference of the Romanian Association for the Study of Pain, Eforie, Romania, 7–9 September 2017; pp. 589–594.
36. Ene, R.; Panti, Z.A.; Nica, M.; Popa, M.G.; Cîrstoiu, M.M.; Munteanu, O.; Vasilescu, S.L.; Simion, G.; Vasilescu, A.; Daviţoiu, D.V.; et al. Chondrosarcoma of the pelvis–case report. *Rom. J. Morphol. Embryol.* **2018**, *59*, 927–931.
37. Álvarez, D.B.; Aparicio, J.P.; Fernandez, E.L.; Mugica, I.G.; Batalla, D.N.; Jiménez, J.P. Implant breakage, a rare complication with the Gamma nail. A review of 843 fractures of the proximal femur treated with a Gamma nail. *Acta Orthop. Belg.* **2004**, *70*, 435–443.
38. Shahgaldi, B.F.; Compson, J. Wear and corrosion of sliding counterparts of stainless-steel hip screw-plates. *Injury* **2000**, *31*, 85–92. [CrossRef]
39. Proverbio, E.; Bonaccorsi, L.M. Microstructural analysis of failure of a stainless steel bone plate implant. *Pract. Fail. Anal.* **2001**, *1*, 33–38. [CrossRef]
40. Walczak, J.; Shahgaldi, F.; Heatley, F. In vivo corrosion of 316L stainless-steel hip implants: Morphology and elemental compositions of corrosion products. *Biomaterials* **1998**, *19*, 229–237. [CrossRef]
41. *ISO 12891-1:1999(E)—Retrieval and analysis of surgical implants—Part 1: Retrieval and handling*; International Organization for Standardization: Geneva, Switzerland, 1999.
42. *ISO 12891-2:2000(E)—Retrieval and analysis of surgical implants—Part 2: Analysis of retrieved metallic surgical implants*; International Organization for Standardization: Geneva, Switzerland, 2000.

43. Azevedo, C.R.F.; Hippert Jr, E.D.U.A.R.D.O. Failure analysis of surgical implants in Brazil. *Eng. Fail. Anal.* **2002**, *9*, 621–633. [CrossRef]
44. Buzatu, M.; Geanta, V.; Stefanoiu, R.; Butu, M.; Petrescu, M.I.; Buzatu, M.; Antoniac, I.; Iacob, G.; Niculescu, F.; Ghica, S.I.; et al. Investigations into Ti-15Mo-W alloys developed for medical applications. *Materials* **2019**, *12*, 147. [CrossRef]

Article

Electrochemical Analysis and In Vitro Assay of Mg-0.5Ca-xY Biodegradable Alloys

Bogdan Istrate [1], Corneliu Munteanu [1,*], Stefan Lupescu [1,*], Romeu Chelariu [2], Maria Daniela Vlad [3] and Petrică Vizureanu [2,4]

1. Mechanical Engineering Department, Gheorghe Asachi University of Iasi, 6 D. Mangeron Blvd, 700050 Iasi, Romania; bogdan_istrate1@yahoo.com
2. Faculty of Material Science and Engineering Department, Gheorghe Asachi University of Iasi, 41 DimitrieMangeron str., 700050 Iasi, Romania; rchelariu@yahoo.com (R.C.); peviz2002@yahoo.com (P.V.)
3. Faculty of Medical Bioengineering, "Grigore T. Popa" University of Medicine and Pharmacy from Iasi, 9-13 Kogălniceanu Str, 700454 Iasi, Romania; maria.vlad@umfiasi.ro
4. Romanian Inventors Forum, Sf. P. Movila 3, 700089 Iasi, Romania

* Correspondence: cornelmun@gmail.com (C.M.); lupescustefan@ymail.com (S.L.); Tel.: +40-744-647-991 (C.M.); +40-753-867-926 (S.L.)

Received: 24 May 2020; Accepted: 4 July 2020; Published: 10 July 2020

Abstract: In recent years, biodegradable Mg-based materials have been increasingly studied to be used in the medical industry and beyond. A way to improve biodegradability rate in sync with the healing process of the natural human bone is to alloy Mg with other biocompatible elements. The aim of this research was to improve biodegradability rate and biocompatibility of Mg-0.5Ca alloy through addition of Y in 0.5/1.0/1.5/2.0/3.0wt.%. To characterize the chemical composition and microstructure of experimental Mg alloys, scanning electron microscopy (SEM), energy-dispersive spectroscopy (EDS), light microscopy (LM), and X-ray diffraction (XRD) were used. The linear polarization resistance (LPR) method was used to calculate corrosion rate as a measure of biodegradability rate. The cytocompatibility was evaluated by MTT assay (3-(4,5-dimethylthiazole-2-yl)-2,5-diphenyltetrazolium bromide) and fluorescence microscopy. Depending on chemical composition, the dendritic α-Mg solid solution, as well as lamellar Mg_2Ca and $Mg_{24}Y_5$ intermetallic compounds were found. The lower biodegradability rates were found for Mg-0.5Ca-2.0Y and Mg-0.5Ca-3.0Y which have correlated with values of cell viability. The addition of 2–3 wt.%Y in the Mg-0.5Ca alloy improved both the biodegradability rate and cytocompatibility behavior.

Keywords: Mg-Ca-Y alloys; microstructure; electrochemical evaluation; in vitro test

1. Introduction

Nowadays, because of an increasing number of humans having a diversity of diseases or traumas of skeletal system, there has been an increasing demand for osteosynthesis devices (plates, screws, prostheses, rods, implants, etc.) [1–3]. Some of these devices are dedicated to maintaining structural stability and aligning bone fragments for a finite time during the healing process of fractures, so that they became temporary devices [4]. Traditionally, such temporary osteosynthesis devices are made from inert metallic materials, but this approach requires a second surgical procedure to extract the device at the end of the healing process [1–7]. Rahim et al. [2] set some important requirements for degradable orthopedic implants to allow the healing of broken bones, like high biocompatibility, tissue friendly self-degrading, and minimal stress-shielding effects. Furthermore, recent research mentioned that the interference screws are made of an MgYREZr-alloy which has been introduced to the market (Milagro; DePuyMitek, Leeds, UK). Banerjee et al. [3] highlights the prospects of Mg alloy implants and various coatings due to their fast degradation in physiological fluids. Another approach in the fabrication of temporary osteosynthesis devices appeared with the development of body-absorbable polymers [1,5]. To eliminate the main drawback of body-absorbable polymers, the unacceptable decreasing of mechanical properties during the healing process [3], in the last two decades Mg-, Fe-, and Zn-based materials were intensively investigated as body-degradable materials [8], mainly Mg-based materials [2,3,5].

Mg is an important mineral in the human body, it has an important role for many physiological functions, and it is an essential mineral for bone formation [5,9,10]. The density, elastic modulus, compressive yield stress, and fracture toughness are closer to those of natural bone in comparison with those of inert metallic or ceramic biomaterials [5]. Mg corrodes in aqueous environments, which makes it a body-degradable material [5], but it has the drawback of degrading with a corrosion rate much higher than that required in temporary osteosynthesis applications, and this process led, among others, to the hydrogen evolution in a significant volume [2,3,5,11]. It is known that other elements such as impurities (Fe, Ni, Cu, etc.) are identified in Mg-based alloys. Their chemical concentration significantly influences the corrosion process. Thus, the studies performed by Atrens et al. [12] and Liu et al. [13] highlighted the major effects that metallic impurities have on Mg-based alloys. These elements form secondary phases and influence by at least an order of magnitude the degradation rate in specific solutions.

In order to correlate the corrosion rate to the osteosynthesis timeframe of Mg, in a quantity that can be considered to estimate body-degradability rate, some have used alloys with different metals, which can be a beneficial approach for mechanical properties as well [5,11,14]. Mg-rare earths (REs) alloys from binary, ternary, and quaternary system alloys were investigated regarding the mechanical properties and the corrosion behavior, as well as from the biocompatibility point of view [15]. Although an improvement of both the mechanical properties and the corrosion behavior for an appreciable number of alloys was attained, the biocompatibility did not follow the same trend [15], controversial experimental data being reported about effects and toxicity of REs [15–19]. Because REs are mixtures of some lanthanides in various chemical compositions, a way to better control the effects of such elements on the biocompatibility of Mg-REs alloys is to use only one of the elements as alloying element in biodegradable Mg alloys [16]. From REs group besides lanthanides there are Y and Sc [16] which can have similar effects on the microstructure and properties of Mg alloys with those of lanthanide mixtures [20–26]. Adding of Y significantly improves the mechanical properties in monolithic Mg, while Sc increases corrosion resistance. Liu et al. [27] divided the rare earth elements into two main categories: I (Sc, Nd, Sm, Eu, Gd, Tb, Tm, and Yb) with applications in the cardiovascular field (stents) and II (Y, La, Ce, and Pr) for orthopedic applications (biodegradable implants). Y presents a maximum solubility of 4.7 at.%(15.28 wt.%) in Mg and it strengthens the solid solution. Similar aspects of increasing mechanical strength and ductility were observed for Mg-2.0Y and Mg-3.0Y [20]. Tekumalla et al. also reported that Y highlights a negative aspect on Mg corrosion resistance and Mg-1Y presented very low toxicity to osteoblast cells.

Ca has a critical role for a broad range of physiological functions and it is the most abundant mineral in the human body [28–31]. Ca in Mg alloys has the role of grain refiner that leads to an improvement of ultimate tensile strength and creep resistance [32]. Binary Mg-Ca alloys have been investigated for orthopedic applications and it has been found that Mg-Ca alloys with a Ca content up to 1 wt.% are suitable to make degradable implants [32–50], the higher contents affecting the castability [32], the corrosion behavior, and mechanical properties due to the higher volume fraction of Mg_2Ca phase [33–35,38,39], and the biocompatibility [35,36].

The cellular viability evaluation using L-929 cells presented by Li et al. [35] showed that Mg-1Ca alloy did not induce toxicity to cells. A 1 wt.% Ca in Mg alloy presented high activity of osteoblasts and osteocytes due to the implanted pin into the femoral shafts of a rabbit, respectively, for 1, 2 and 3 months. Also, Erdmann et al. [41] and Zeng et al. [48] showed that Mg-0.8Ca and Mg–0.79Ca alloys have good cellular response and high mechanical resistance (hardness, ultimate tensile strength, and yielding strength) for up to six weeks during in vivo implantation. After this period, a gradual degradation was observed.

The properties of the Mg-Ca alloys are influenced by mechanical or thermo-mechanical processing [39,41–43,46–50]. However, e.g., Mg0.8Ca alloy tested in vivo has shown an insufficient initial strength and a fast degradation [40]. Thus, alloying the Mg-Ca alloys (Ca wt.% < 1) with a third element, e.g., from REs group, can be a way to improve the properties and biocompatibility.

In this paper, a Mg-0.5Ca alloy was alloyed witha content of 0.5, 1.0, 1.5, 2.0, and 3.0 wt.%Y to improve the degradability. Additionally, the biocompatibility of the studied alloys was tested in vitro.

2. Materials and Methods

2.1. Synthesis of Mg-Ca-Y Alloys, Morphological and Structural Analysis

The master alloys used for sample manufacturing were purchased from Hunan China Co., Hunan, China [51,52] and they have the chemical composition presented in Table 1. The casting of Mg alloys was performed in an induction furnace (Inductro S.A., Bucharest, Romania) with an inert Ar5.0 protective atmosphere at a temperature of 680–690 °C for 30 min using rectangular bars as raw materials. The resulting mini-ingots were cut into spherical samples having different concentrations as is presented in Table 2, with a diameter of 20 mm and a thickness of 2 mm. The amount of metallic charge for the experimental samples is presented in Table 2 and is about 23 g/ingot. The samples were grounded with abrasive discs with granulation between 340–2000MPi, polished with alumina suspension (1 μm–6 μm), cleaned with alcohol, and then ultrasonically cleaned in ethyl alcohol for 10 min. The experimental samples were etched with $Mg(CH_3COO)_2{*}4H_2O$ acetate solution for microstructural analysis. Surface morphology was investigated by light microscopy (Leica DMI 5000 microscope, Wetzlar, Germany) and scanning electron microscopy (SEM FEI Quanta 200 3D, dual beam, equipped with energy dispersive X-Ray spectroscopy analysis unit—Xflash Bruker, Harvard, MA, USA). X-Ray diffractions (XRD, Panalytical, Almelo, The Netherlands) were performed using a Xpert PRO MPD 3060 facility from Panalytical (Almelo, The Netherlands), with a Cu X-ray tube (Kα = 1.54051Å), 2 theta: 30°–100°, step size: 0.13°, time/step: 51 s, and a scan speed of 0.065651°/s.

Table 1. Chemical composition of Mg master alloy—Hunan China Co. [51,52].

Alloys	Mg/Ca/Y (wt.%)	Fe (wt.%)	Ni (wt.%)	Cu (wt.%)	Si (wt.%)	Al (wt.%)
Pure Mg	Mg (99 wt.%)	0.15–0.2	0.17–0.2	0.14–0.2	0.15–0.2	0.16–0.2
Mg15Ca	Ca (15.29 wt.%)	0.004	0.001	0.003	0.013	0.011
Mg30Y	Y (28.05 wt.%)	0.010	0.001	0.001	0.006	0.011

Table 2. Correspondence of codes of studied materials and the amount of metallic charge for the experimental alloys.

Sample Code	Mg (g)	Mg-15Ca (g)	Mg-30Y (g)
Mg-0.5Ca-0.5Y	21.82	0.77	0.41
Mg-0.5Ca-1.0Y	21.42	0.77	0.82
Mg-0.5Ca-1.5Y	21.00	0.77	1.23
Mg-0.5Ca-2.0Y	20.59	0.77	1.64
Mg-0.5Ca-3.0Y	19.77	0.77	2.46

2.2. Electrochemical Analysis

Mg alloys were tested in a simulated body fluid (SBF) with the chemical composition presented in Table 3, according to the procedure described in [53]. Potential measurements were performed using a VoltaLab 21 potentiostat (Radiometer Analytical SAS, Lyon, France). Data acquisition and processing was performed with Volta Master 4 software. A three-electrode cell was used: a platinum auxiliary electrode, a calomel saturated electrode, and a work electrode, with the specification that the working electrodes (samples) were removed from the resin and polished in parallelepiped forms. The experimental samples were mounted in Teflon support to allow connection to the electrochemical system electrode. The representation of linear polarization curves in the following coordinates: current density (mA/cm^2) versus potential (V), allowed the highlighting of the corrosion potential (E_{cor}), and the corrosion currents (J_{cor}). Measurements were made at 25 °C and the electrolyte was naturally aerated, the linear polarization curves were recorded at a scanning potential of 1mV of the electrode and the cyclic polarization curves were performed at a scanning speed of 10mV.

The instantaneous corrosion rate, V_{cor} (mm/y), was determined from the corrosion current density, J_{cor} (mA/cm^2) [22]:

$$V_{cor} = 22.85 \times J_{cor}, \tag{1}$$

Table 3. Chemical composition of immersion solution.

Chemical Composition (Ions) ($mmol/dm^3$)	**Na^+**	**K^+**	**Mg^{2+}**	**Ca^{2+}**	**Cl^-**	**HCO_3^-**	**HPO_4^{2-}**	**SO_4^{2-}**
Simulated body fluid	142	5	1.5	2.5	147.8	4.2	1	0.5
Human blood plasma	142	5	1.5	2.5	103	27	1	0.5

2.3. *Cytocompatibility Testing*

2.3.1. Alloy Sample Preparation

Samples in the form of flat metal pieces (weighing between 0.68 g and 0.98 g) were cleaned with acetone–ethanol by sonication and exposed for 30 min on each side to the ultraviolet (UV) action for sterilization, and subsequently placed in hanging cell culture inserts (with a pore size of 0.4 µm) used in a 24-well plates to coincubate the alloy samples with the cells for the cell viability study. In addition, the studied Mg-0.5Ca-xY alloy samples were immersed in complete culture media, at 37 °C and 5% CO_2, over a period of 5 days, for evaluating the pH variation during the co-incubation times (1, 3, and 5 days).

2.3.2. Cell Culture

Albino rabbit dermal primary fibroblasts at passage 3 were selected for the viability study. The cells were cultured in DMEM-F12 Ham culture medium (Dulbecco's modified Eagle medium/Nutrient F-12 Ham) supplemented with 10% fetal bovine serum (FBS) and 1% antibiotic (penicillin–streptomycin–neomycin) in humidified atmosphere of 95% air, 5% CO_2 at 37 °C. The culture medium was changed every 2 days.

Cultures of 90% confluent cells were rinsed with prewarmed phosphate buffered saline (PBS) and harvested by incubating with trypsin/EDTA (Sigma Chemical Co., Saint Louis, MO, USA). Afterwards, the cells were detached, suspended in fresh media again, counted using a Neubauer counting chamber, seeded in 24 well culture plates at a density of 1×10^4 cells/well/0.5 mL complete DMEM-F12, and incubated under the above mentioned conditions for 24 h to facilitate cell adhesion. Then, the medium was removed by aspiration, the cells were washed with PBS three times to eliminate unattached or dead cells and the inserts containing the studied alloy samples were placed on the assigned wells containing the cells cultured for 24 h. In addition, wells containing cell culture without alloy samples were used as controls (control-wells, i.e., negative control). It should be mentioned that the alloys' interaction/reaction with the cell culture media took place during the cell culture test. Coincubation of the metal samples (3 samples for each alloy) with the cells was performed for 1, 3, and 5 days respectively for both the cytocompatibility testing and cell morphology study.

2.3.3. Cell Viability

Cell viability was tested by the MTT assay (3-(4,5-dimethylthiazole-2-yl)- 2,5-diphenyltetrazolium bromide), which allows quantification of a metabolic activity causally related to the live cell [54]. For this purpose, following the incubation period, inserts were removed from the wells, the cells were rinsed with PBS and then MTT dye solution dissolved in fresh medium was added to the cells for 3 h at 37 °C in order to assure formation of the intracellular formazan crystals [55,56]. Subsequently dark-blue insoluble product formed inside the viable cells was solubilized with isopropyl alcohol under continuous agitation (Environmental Shaker-Incubator ES-20, Biosan, Riga, Latvia) for 15 min. The liquid of each sample was removed for the assay, which was performed in a 96-well plate, on a microplate reader (Tecan Sunrise, with Magelan V.7.1 soft for data acquisition, Tecan Group Ltd., Männedorf, Switzerland) at a wavelength of 570 nm.

Cell viability was expressed as a percentage related to the control wells, according to the formula CV = 100 × (ODs-ODb)/(ODc-ODb) where ODs represents the optical density for the sample wells (i.e., wells containing cell culture coincubated with alloy samples), ODb—optical density of the wells without cells or medium (blank), and ODc—optical density for the control-wells (i.e., wells containing cell culture without alloy samples).

For statistical analysis of cell viability results, the ANOVA one-way test was used and the data were compared using Tukey's method, the statistically significant difference being accepted for $p < 0.05$.

2.3.4. Cell Morphology

The cell morphology study was performed, at specified time periods of 1 day and 5 days, by fluorescence microscopy. For this, the cells were washed with Hanks' Balanced Salt solution (HBSS; H8264, Sigma–Aldrich, Taufkirchen, Germany) without red phenol, and 200 µL of a 1:1000 calcein solution (Calcein AM; C1359, Sigma–Aldrich, Taufkirchen, Germany) in HBSS was added to each well and the plate was incubated in the dark for 30 min at 37 °C. Subsequently, cellular morphology was assessed by an inverted microscope (Leica DMIL LED equipped with camera Leica DFC450C and soft Leica Application Suite, Version 7.4.1 for image acquisition, Leica Microsystems, Wetzlar, Germany).

3. Results and Discussions

The microstructure of the synthesized alloys was presented through the experimental results. Through microscopy both the microstructure of new Mg-based alloys and the state of the surface before and after the tests of corrosion resistance were analyzed. The chemical compounds formed on the surface were determined and were followed by the cellular viability testing by MTT assay.

3.1. Structural Characterization

The structural characterizations of the five Mg alloys by microscopy are shown in Figure 1. The structure presents specific morphology of as-cast metallic materials, Mg_2Ca lamellar intermetallic compounds and relatively uniform presence of Y-containing particles. The addition of Y led to the formation of globular particles with a segregating tendency of relatively uniform color [26]. The morphological aspect of the experimental samples is presented in Figure 2, where the presence of an intermetallic phase is observed. Y-containing particles are typically of about 15 µm and appear in a brighter contrast than Mg_2Ca lamellar intermetallic compounds. The surface morphological investigations of the present study add new information to some previous researches conducted by Istrate et al. [57] in the case of Mg-based alloys having a Y concentration between 1.0 wt.% and 3.0 wt.%.

The chemical composition of the Mg-0.5Ca-xY alloys was investigated by Energy-Dispersive X-ray Spectroscopy analysis. Transverse sections of the samples were used for the examination in ten different areas and the average results are presented in Table 4.

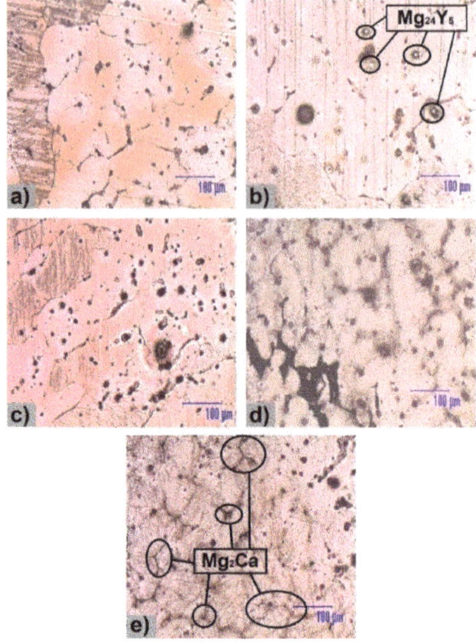

Figure 1. Light microscopy analysis of the Mg-0.5Ca-xY: (**a**) Mg-0.5Ca-0.5Y; (**b**) Mg-0.5Ca-1.0Y [57]; (**c**) Mg-0.5Ca-1.5Y; (**d**) Mg-0.5Ca-2.0Y [57]; (**e**) Mg-0.5Ca-3.0Y [57].

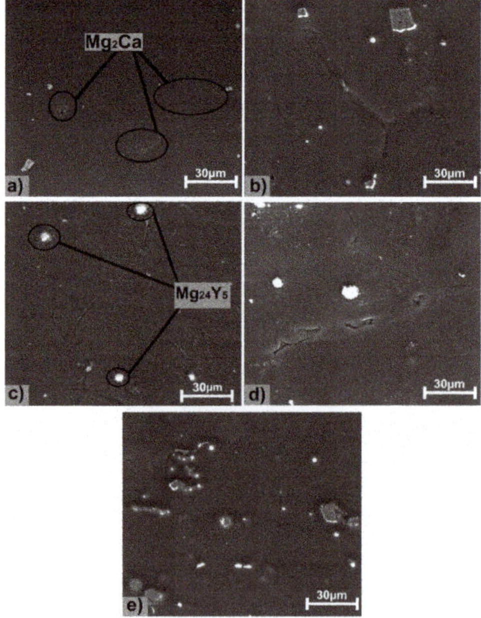

Figure 2. Surface SEM images of the Mg-0.5Ca-xY: (**a**) Mg-0.5Ca-0.5Y; (**b**) Mg-0.5Ca-1.0Y [57]; (**c**) Mg-0.5Ca-1.5Y; (**d**) Mg-0.5Ca-2.0Y [57]; (**e**) Mg-0.5Ca-3.0Y [57].

Table 4. Average elemental compositions obtained by Energy-Dispersive X-ray Spectroscopy analysis.

Alloy		Mg (wt.%)	Ca (wt.%)	Y (wt.%)	Si (wt.%)	Fe (wt.%)	Ni (wt.%)	Cu (wt.%)
Mg-0.5Ca-0.5Y	Average	97.8	0.7	0.5	0.2	0.3	0.2	0.3
	Stdev	±0.4	±0.1	±0.1	±0.1	±0.1	±0.1	±0.1
Mg-0.5Ca-1.0Y	Average	97.3	0.6	0.9	0.2	0.3	0.3	0.3
	Stdev	±0.5	±0.1	±0.3	±0.1	±0.1	±0.1	±0.2
Mg-0.5Ca-1.5Y	Average	96.6	0.7	1.3	0.3	0.3	0.3	0.2
	Stdev	±0.2	±0.1	±0.1	±0.1	±0.1	±0.1	±0.1
Mg-0.5Ca-2.0Y	Average	96.2	0.7	1.9	0.2	0.3	0.4	0.4
	Stdev	±0.5	±0.1	±0.3	±0.1	±0.1	±0.2	±0.2
Mg-0.5Ca-3.0Y	Average	95.6	0.7	2.8	0.1	0.2	0.2	0.3
	Stdev	±0.4	±0.1	±0.1	±0.1	±0.1	±0.1	±0.2

Average elemental composition with calculated standard deviation on 10 analyzed surfaces per sample.

Microstructural analysis highlights homogeneous structures with the formation of specific phases, namely α-Mg-based solid solution (α-Mg), Mg_2Ca, and $Mg_{24}Y_5$. The Mg_2Ca compound is located at the boundaries of the α-Mg grains, forming an eutectic compound with α-Mg. The XRD patterns for the Mg-0.5Ca-xY are presented in Figure 3. α-Mg (ICDD-PDF: 01-071-9399) has been identified at 2θ = 36.54°, 47.69°, 57.28°, 68.49°, and 90.21°, as predominant phase having a hexagonal crystalline structure. Furthermore, the presence of Mg_2Ca (ICDD-PDF: 01-073-5122) was revealed at 2θ = 34.11°, 52.35°, and 69.12° and $Mg_{24}Y_5$ (ICDD-PDF: 01-071-9618) in cubic form at 2θ = 37.48°, 57.02° and 75.78°.

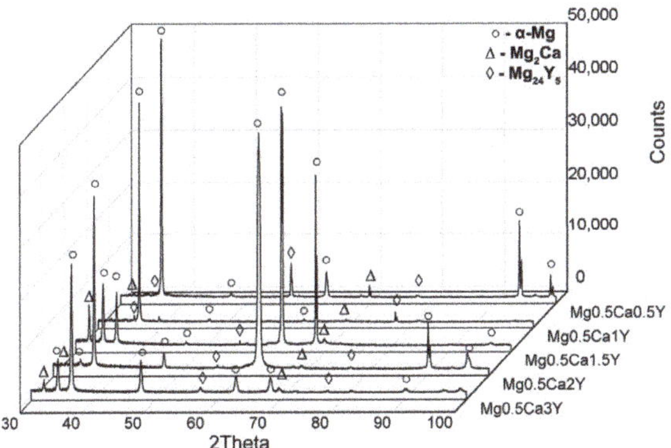

Figure 3. XRD analysis of Mg-0.5Ca-xY experimental alloys [57].

3.2. Electrochemical Evaluation

The main parameters of the corrosion process performed in SBF solution are presented in Table 5. Since Mg alloys eliminate high amounts of gas on the surface of the samples, gas bubbles are constantly formed. These gas bubbles were removed by using a magnetic stirrer that operated at a relatively slow rate of agitation of the electrolyte solution. At each test, the fresh electrolyte solution was used. The corrosion current (J_{cor}) shows the degradation degree of the experimental alloys. It revealed different values of corrosion density between 0.3663 mA/cm^2 and 2.9812 mA/cm^2 for all the samples.

Table 5. Parameters obtained from the electro-corrosion resistance tests of the experimental alloys Mg0.5Cax (x = 0.5; 1.0; 1.5; 2.0; and 3.0 wt.%) Y.

Sample	E_0	b_a (mV)	b_c (mV)	R_p (ohm/cm^2)	J_{cor} (mA/cm^2)	V_{cor} (mm/y)
Mg-0.5Ca-0.5Y	−3562.1	73.7	−40.9	49.99	2.8753	65.70
Mg-0.5Ca-1.0Y	−3801.9	479.8	−237.2	13.70	2.9812	68.12
Mg-0.5Ca-1.5Y	−3697.5	112.1	−102.8	12.95	1.5805	36.11
Mg-0.5Ca-2.0Y	−3470.8	269.3	−205.6	160.31	0.3663	8.37
Mg-0.5Ca-3.0Y	−2969.2	63.4	−156.6	276.38	0.4463	10.20

To confirm the effect of the polarization (Figure 4) and to understand the electrochemical corrosion mechanism of the Mg-0.5Ca-xY alloys, surface morphology was investigated by SEM (Figure 5). The purity of Mg is 99 wt.%, the rest of the elements being accompanying elements, used in the synthesizing process, which led to the obtaining of experimental alloys with some impurities (Fe, Ni, Cu, Si). The alloying elements contribute to the formation of additional phases in the α-Mg matrix (Mg_2Ca and $Mg_{24}Y_5$).

The analysis of the experimental results reveals the effect of the alloying element Y on the electro-corrosion resistance properties, namely a reduction of the corrosion rate of up to 6.68 times with the addition of 3.0 wt.% of Y or 8.14 or for the addition of a percentage of at least 2.0 wt.%Y. From the point of view of corrosion resistance, the percentage of Y significantly influences the behavior of the alloy to values higher than 1.5 wt.%Y. It is known that the both intermetallic compounds significantly influence the corrosion behavior of Mg alloys [23,24]. Mg_2Ca suffers of dissolution at high rate [24], and when the fraction of $Mg_{24}Y_5$ increases, the corrosion current density increases [23]. Also, although Y has a limited solubility in Mg, up to a certain content of Y dissolved in Mg improve corrosion behavior of Mg matrix. Having in view those previously mentioned, at low content of Y (up to 1.0 wt.%) the Mg_2Ca and Mg matrix lead to high values of J_{cor}. When content of Y increases (up to 2.0 wt.%) the dissolution rate of Mg matrix decreases. A supplementary increase of Y content (up to 3.0 wt.%) causes an increase of the $Mg_{24}Y_5$ fraction which leads to a slight increasing of J_{cor} values. Also, the impurities could either form specific compounds or segregate at grain boundary, what can significantly accelerate corrosion by micro-galvanic corrosion. This can lead to higher degradation rates [58], but, as the diffraction patterns show, probably these phases are in extremely low fraction because all detectable peaks of the XRD patterns were assigned to α−Mg, Mg_2Ca, and $Mg_{24}Y_5$.

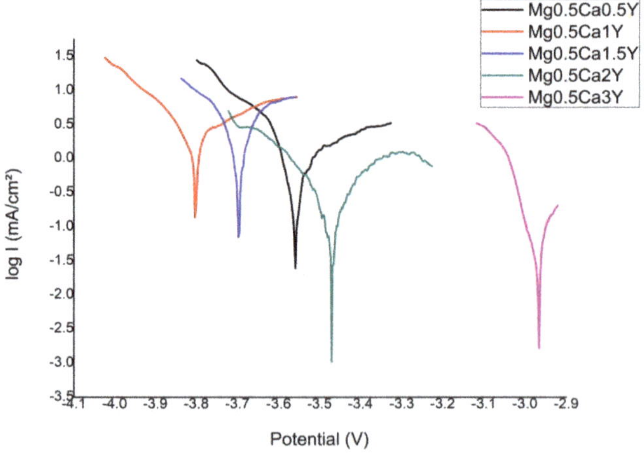

Figure 4. Tafel diagrams of Mg-0.5Ca-xY-based experimental alloys.

All samples exhibit a corroded surface, especially due to the potentiodynamic polarization test, the corrosion being of a generalized type, and in all cases the formation of surface compounds occurs by the interaction of the alloy and the electrolyte solution. The corrosion products formed on the surface of Mg alloys exhibit numerous cracks and have different morphological aspects.

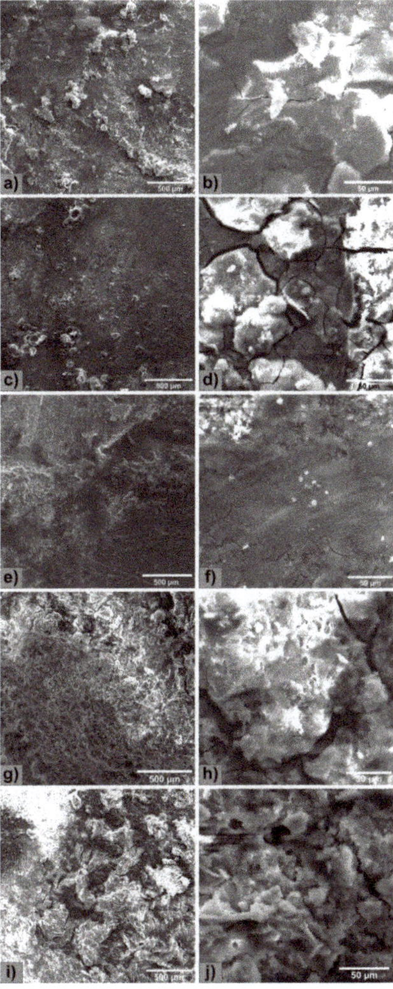

Figure 5. SEM images of experimental surfaces after electrochemical tests: (**a**,**b**) Mg-0.5Ca-0.5Y; (**c**,**d**) Mg-0.5Ca-1.0Y; (**e**,**f**) Mg-0.5Ca-1.5Y; (**g**,**h**) Mg-0.5Ca-2.0Y; and (**i**,**j**) Mg-0.5Ca-3.0Y.

The quantitative determination of the elements existing on the surface of the alloy led to the results shown in Table 6. It can be considered that due to the alloying elements there is availability for the formation of oxides and hydroxides by the reaction with the electrolytic SBF medium.It can be observed that the chemical composition of the sample surface is influenced by the behavior of the Y element on the Mg alloy which is close to that of the original material. The difference is being attributed to the standard EDS detector's error and also to the partial oxidation of the Y-based compounds. The distribution of the elements identified on the Mg-0.5Ca-xY alloy surface after the electrochemical corrosion test was investigated by SEM-EDS (Figure 6). A higher percentage of Y over 1.5 wt.% is a success in terms of corrosion resistance, an affirmation confirmed by the linear polarization results. According to references [22] and [37], Atrens et al. and Li et al. showed the formation of Mg-Y and Mg-Ca compounds after the electrocorrosion process. The results of the present research highlight that the entire surface of the samples is covered by O and Mg compounds. The Y and Ca elements exhibit a similar behavior to the other experimental alloys [22,37], by forming rich-Y precipitates ($Mg_{24}Y_5$) and Mg_2Ca compound at the grains boundary.

In the case of the 2.0 wt.%Y alloy, the presence of O, Na, C, and Cl elements was identified on the surface by the interaction of the experimental alloy with the SBF electrolyte solution. The presence of chlorine-based salts is also noted. On the surface of the Mg-0.5Ca-3.0Y alloy, in addition to the basic elements of the experimental alloy, the elements: O, K, and Na were also identified, mainly due to the interaction with the SBF.

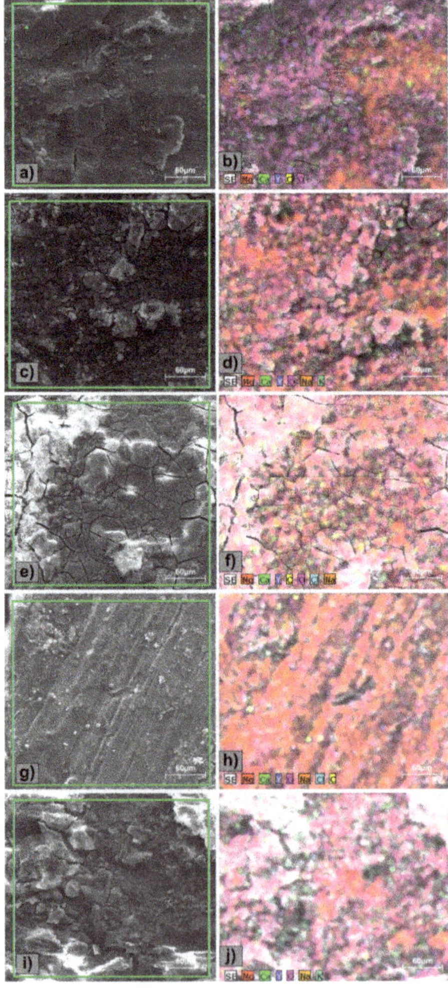

Figure 6. Distribution of the identified elements on the surface of the Mg-0.5Ca-xY alloys after electrochemical tests: (**a**,**b**) Mg-0.5Ca-0.5Y; (**c**,**d**) Mg-0.5Ca-1.0Y; (**e**,**f**) Mg-0.5Ca-1.5Y; (**g**,**h**) Mg-0.5Ca-2.0Y; and (**i**,**j**) Mg-0.5Ca-3.0Y.

Table 6. Chemical composition of Mg-0.5Ca-xY alloy surface after electrochemical test.

	Chemical Elements	Mg wt.%	Ca wt.%	Y wt.%	O wt.%	Cl wt.%	Na wt.%	K wt.%
Mg-0.5Ca-0.5Y	Surface with oxides	61.6	0.4	0.4	37.6	-	-	-
	Surface without oxides	96.7	1.3	2.1	-	-	-	-
Mg-0.5Ca-1.0Y	Surface with oxides	58.4	0.7	0.5	33.2	4.0	2.4	-
	Surface without oxides	95.6	2.3	2.1	-	-	-	-
Mg-0.5Ca-1.5Y	Surface with oxides	45.9	0.7	0.9	46.3	-	5.4	0.9
	Surface without oxides	90.2	3.0	6.8	-	-	-	-
Mg-0.5Ca-2.0Y	Surface with oxides	44.1	0.8	0.9	50.1	0.4	3.8	-
	Surface without oxides	87.8	4.0	8.2	-	-	-	-
Mg-0.5Ca-3.0Y	Surface with oxides	47.7	0.5	2.8	44.8	-	1.8	2.9
	Surface without oxides	83.6	3.1	13.4	-	-	-	-

3.3. Cytocompatibility Study

Figure 7 shows the results of the MTT assay for testing the cytocompatibility of the studied Mg-0.5Ca-xY experimental alloys, expressed as percentages of the control-wells' viability (i.e., negative control). After 1 day of coincubation of the cells with the studied samples, cell viability profile was significantly higher in the case of alloys containing 1.5 wt.%Y and 2.0 wt.%Y respectively compared to the other studied alloys ($p < 0.05$). Cell viability profile after 3 days of coincubation was significantly lower in the case of alloys containing 0.5 wt.%Y and 1.0 wt.%Y respectively, compared to the other studied alloys ($p < 0.05$), while cell viability was not significantly different ($p > 0.05$) for alloys with a Y content of 1.5 wt.%, 2.0 wt.%, and 3.0 wt.%, respectively. In addition, it was observed that the cells' viability profile decreased significantly with the time, after 5 days being lower than for the other time periods ($p < 0.05$), and appearing to be inversely proportional to the amount of Y used for alloying. However, cell viability after 5 days was not significantly different ($p > 0.05$) for alloys with Y amount of 1.5 wt.%, 2.0 wt.%, and 3.0 wt.%, respectively. In the case of the alloy obtained by alloying with 3.0 wt.% Y the cell viability level was not significantly different ($p > 0.05$) after 5 days comparing with those obtained after 1 day of incubation.

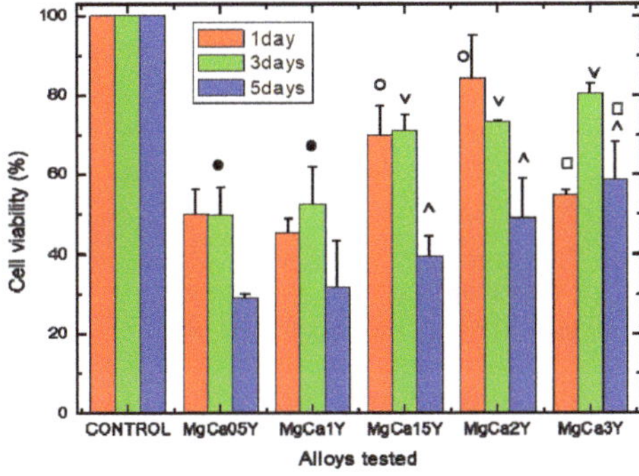

Figure 7. Cell viability profile (%) evaluated by 3-(4,5-dimethylthiazole-2-yl)- 2,5-diphenyltetrazolium bromide - (MTT)-assay: Effect of Mg-0.5Ca-xY experimental alloys on cell viability after 1, 3, and 5 days of co-incubation. Data expressed as percent related to the negative control. (∧; o; •; ∨; □) No significant differences on cell viability ($p > 0.05$); see details in the text.

The results of the cytocompatibility study seem to suggest that by increasing the amount of Y used for alloying, cytocompatibility was improved and this was probably due tothe increase of the experimental alloys stability in humid environment having a complex composition (such as culture medium or biological fluids). The lower viability level compared with the control-wells (i.e., negative control) could be attributed to alloy reactivity at the contact with the culture medium [59], resulting in H_2 release and immediate pH changes (i.e., subsequent alkalinization) of the cell culture medium (see Figure 8). It is known that the environment has a considerable influence on the degradation behavior of degradable materials [12], and consequently, the degradation of the Mg-alloys (accompanied by ions and degradation products release) is responsible for pH and osmolarity increasing, and this might adversely affect cells' metabolic activity. In this sense, the increase of pH value after Mg-alloy immersion in aqueous environment is mainly the result of an anodic reaction of Mg^{2+} by reduction of H_2O (to H_2 gas and OH^-) and production of $Mg(OH)_2$ corrosion product.

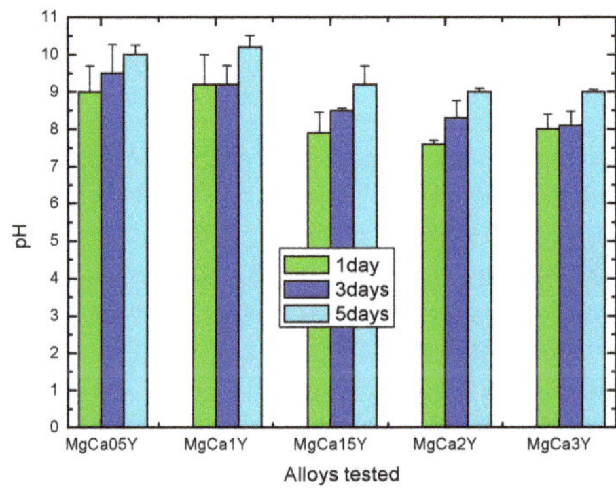

Figure 8. Variation of the pH in the DMEM-F12 complete media (Dulbecco's modified Eagle medium/Nutrient F-12 Ham) during co-incubation with the studied Mg-0.5Ca-xY alloys samples (at 37 °C, 5% CO_2), over a period of 5 days.

However, the viability level both at 1 day and 3 days (≥70%) for alloys with Y content of 1.5 wt.%, 2.0 wt.% allows us to appreciate (based on ISO 10993-5, [60]) that the two alloys don't have a cytotoxic effect, and this is so because of the decision to carry out the cytotoxicity tests by direct co-incubation of the studied alloys with the cells using hanging inserts (by which the alloys were suspended in the wells, in order to avoid cell damage by mechanical effect as a result of alloys samples' disintegration). Consequently, by this way, the cells were continuously exposed (up to 5 days) to an 100% extract and, furthermore, continuously exposed to the sum of the phenomena that take place during Mg-alloys degradation, like hydrogen, ions, and degradation products release.

In this sense, a non-cytotoxic character could also be attributed to the alloy containing 3.0 wt.%Y, excepting the lower viability for this alloy at one day, possibly due to a higher reactivity, being a potential signal that by alloying with Y, an optimum is reached for 1.5 and 2.0 wt.%Y amounts. Indeed, this observation is sustained by other complementary data, such as corrosion resistance tests (see the above section).

Cell viability after 5 days decreased at values around 50%, especially for alloys with Y content of 1.5, 2.0, and 3.0 wt.% (and below 50% for other alloys), and these findings could be attributed to many factors such as modification of the pH values (as is shown in Figure 8) that may affect cell metabolic activity, ionic massive release from alloys, and precipitation of salts with inhibitory or toxic effect, the increase of osmolarity as a consequence of reactivity that may lead to hyperosmotic shock [61]. It is true that any inhibitory or cytotoxic effect is of concern regarding the potential in vivo cytotoxicity, but only the in vitro cytotoxicity data not essentially discard the biomaterials' potential use in clinical application [60], because of different real corrosion/degradation scenario of Mg-alloys under biomedical applications (i.e., both composition and flow of biological fluids) that may affect the local surface chemistry of the implants.

The cytocompatibility data were correlated with the cell density and morphology results (Figure 9) obtained for the studied alloys after 1 day and 5 days of coincubation. In this sense, cell' density after 1 day was lower than that of the controlwells, which agrees with the cell viability data and this observation is maintained after 5 days of coincubation.

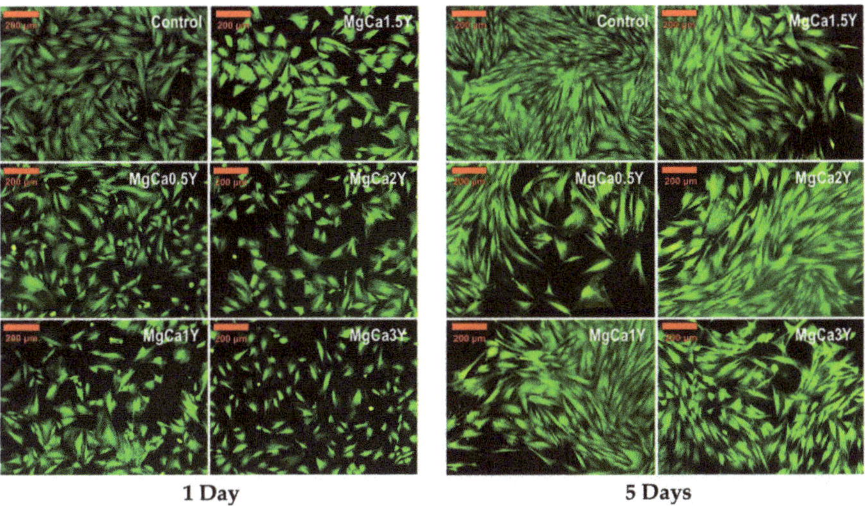

Figure 9. Fluorescence microscopy images of fibroblastic cells' morphology after 1 and 5 days of coincubation with the experimental alloys. Viable cells stained in green because of calcein fluorescent dye presence inside the cells. Bar: 200 μm.

Figure 9 shows the observation performed by fluorescence microscopy on fibroblast cells density and morphology, coincubated with the studied alloys, for 1 day and 5 days. It is observed that after 1 day the cell density evidenced for the alloys is lower than that of the control-wells (i.e., negative control, containing only cells without metallic samples). Moreover, different cell morphology is observed, in the sense of an elongated aspect with bipolar morphology in areas with greater cell density, and a polygonal morphology with wide lamellipodia cytoplasmic processes in areas with low cell density. Furthermore, polygonal morphology was observed towards the center of the well and in the immediate surrounding area of the insert membrane where cell density was lower. Different cell morphology can also be attributed to different concentrations of Ca, Mg, and Y ions in the well microenvironment, knowing that Ca is associated with a structural role in the reorganization of cytoskeleton elements [62], whereas the accumulation of Mg ions over a certain limit may have an inhibitory effect [63]. However, it should be noted that all of the physico-chemical processes taking place during the degradation process of Mg alloys depend on the microenvironment conditions [64] and, from this point of view, under amicroenvironment governed by active transport processes (as is the case of the in vivo degradation), these physico-chemical phenomena may be very well balanced and the results of in vivo biodegradation and osteointegration are somehow optimal. Accordingly, in order to prove the biocompatible behaviour of the studied Mg-alloys a pending in vivo study is expected to clarify this concern.

4. Conclusions

In the present study, five differently prepared biodegradable Mg alloys, Mg-0.5Ca-0.5Y, Mg-0.5Ca-1.0Y, Mg-0.5Ca-1.5Y, Mg-0.5Ca-2.0Y, and Mg-0.5Ca-3.0Y, were investigated. The results can be concluded as follows:

(1) Addition of Y in the experimental alloys refines the microstructure, resulting in the $Mg_{24}Y_5$ cubic structure compound. Y compounds have a white spherical form in the metallic matrix andtypically a size of 15 µm. Also, Ca forms an eutectic compound—Mg_2Ca, founded at the Mg grainsboundary.

(2) The corrosion resistance was performed in SBF solution and presented a generalized type with very few areas not affected by corrosion. Addition of Y leads to an increase in electro-corrosion resistance, especially at alloying percentages greater than 1.0 wt.%. Increasing the content of Y, the immersion and electro-chemical tests show an improved degradation rate as following: 65.7 mm/y (0.5 wt.%) and 10.20 mm/y (3.0 wt.%).

(3) The Mg-0.5Ca-xY alloys have a cytocompatible behavior, i.e., the viability level at 3 days for Mg-Ca-Y alloys with a Y amount of 1.5, 2.0 and 3.0 wt.% was above 70%. The decrease of cell viability level after 5 days at values around 50%, especially in the case of alloys with 1.5 wt.%, 2.0 wt.% and 3.0 wt.% Y, should be attributed to the following factors: change of pH value, ion release from alloys, increasing of osmolarity, and salt precipitation with toxic or inhibitory effect. However, the increase of the Y alloying amount seems to increase the cytocompatibility of these alloys and open the way for future studies concerning alloys with a content higher than 3.0 wt.% Y.

Author Contributions: Conceptualization, B.I., C.M., and S.L.; methodology, C.M. and P.V.; software, B.I.; validation, C.M., R.C., and M.D.V.; formal analysis, S.L.; investigation, B.I., R.C., M.D.V; resources, C.M., P.V.; data curation, B.I.; writing—original draft preparation, B.I.; writing—review and editing, C.M., R.C., M.D.V.; visualization, P.V.; supervision, C.M., S.L.; project administration, C.M.; funding acquisition, C.M. All authors have read and agreed to the published version of the manuscript.

Funding: This work was supported by a grant of the Romanian Ministry of Research and Innovation, CCCDI–UEFISCDI, project number PN-III-P1-1.2-PCCDI-2017-0239/60PCCDI 2018, within PNCDI III.

Acknowledgments: The authors would like to thank Cimpoesu Nicanor, from "Gheorghe Asachi" Technical University of Iasi for his contributions to the research work.

Conflicts of Interest: The authors declare no conflict of interest.

References

1. Schumann, P.; Lindhorst, D.; Wagner, M.; Schramm, A.; Gellrich, N.C.; Rucker, M. Perspectives on ResorbableOsteosynthesis Materials in Cranio maxilla facial Surgery. *Pathobiology* **2013**, *80*, 211–217. [CrossRef]
2. Rahim, M.I.; Ullah, S.; Mueller, P. Advances and Challenges of Biodegradable Implant Materials with a Focus on Magnesium-Alloys and Bacterial Infections. *Metals* **2018**, *8*, 532. [CrossRef]
3. Banerjee, P.C.; Al-Saadi, S.; Choudhary, L.; Harandi, S.E.; Singh, R. Magnesium Implants: Prospects and Challenges. *Materials* **2019**, *12*, 136. [CrossRef]
4. Tune, D. Body-Absorbable Osteosynthesis Devices. *Clin. Mater.* **1991**, *8*, 119–123. [CrossRef]
5. Staiger, M.P.; Pietak, A.M.; Huadmai, J.; Dias, G. Magnesium and its alloys as orthopedic biomaterials: A review. *Biomaterials* **2006**, *27*, 1728–1734. [CrossRef] [PubMed]
6. Sandu, A.V.; Baltatu, M.S.; Nabialek, M.; Savin, A.; Vizureanu, P. Characterization and Mechanical Proprieties of New TiMo Alloys Used for Medical Applications. *Materials* **2019**, *12*, 2973. [CrossRef]
7. Baltatu, M.S.; Tugui, C.A.; Perju, M.C.; Benchea, M.; Spataru, M.C.; Sandu, A.V.; Vizureanu, P. Biocompatible Titanium Alloys used in Medical Applications. *Rev. Chim.* **2019**, *70*, 1302–1306. [CrossRef]
8. Han, H.S.; Loffredo, S.; Jun, I.; Edwards, J.; Kim, Y.C.; Seok, H.K.; Witte, F.; Mantovani, D.; Glyn-Jones, S. Current status and outlook on the clinical translation of biodegradable metals. *Mater. Today* **2019**, *23*, 57–71. [CrossRef]

9. Jahnen-Dechent, W.; Ketteler, M. Magnesium basics. *Clin. Kidney J.* **2012**, *5* (Suppl. 1), 3–14. [CrossRef]
10. Al Alawi, A.M.; Majoni, S.W.; Falhammar, H. Magnesium and Human Health: Perspectives and Research Directions. *Int. J. Endocrinol.* **2018**, *2018*. [CrossRef]
11. Ding, W. Opportunities and challenges for the biodegradable magnesium alloys as next-generation biomaterials. *Regen. Biomater.* **2016**, *3*, 79–86. [CrossRef]
12. Atrens, A.; Johnston, S.; Shi, Z.; Dargusch, M.S. Viewpoint—Understanding Mg corrosion in the body for biodegradable medical implants. *Scr. Mater.* **2018**, *154*, 92–100. [CrossRef]
13. Liu, M.; Uggowitzer, P.J.; Nagasekhar, A.V.; Schmutz, P.; Easton, M.; Song, G.L.; Atrens, A. Calculated phase diagrams and the corrosion of die-cast Mg–Al alloys. *Corros. Sci.* **2009**, *51*, 602–619. [CrossRef]
14. Riaz, U.; Shabib, I.; Haider, W. The current trends of Mg alloys in biomedical applications—A review. *J. Biomed. Mater. Res. B Appl. Biomater.* **2019**, *107*, 1970–1996. [CrossRef]
15. Liu, D.; Yang, D.; Li, X.; Hu, S. Mechanical properties, corrosion resistance and biocompatibilities of degradable Mg-RE alloys: A review. *J. Mater. Res. Technol.* **2019**, *8*, 1538–1549. [CrossRef]
16. Angrisani, N.; Reifenrath, J.; Seitz, J.M.; Meyer-Lindenberg, A. Rare Earth Metals as Alloying Components in Magnesium Implants for Orthopaedic Applications. In *New Features on Magnesium Alloys*; Monteiro, W.A., Ed.; IntechOpen: Rijeka, Croatia, 2012. [CrossRef]
17. Hirano, S.; Suzuki, K. Exposure, Metabolism, and Toxicity of Rare Earths and Related Compounds. *Environ. Health Perspect.* **1996**, *104*, 85–95. [CrossRef]
18. Rim, K.T.; Koo, K.H.; Park, J.S. Toxicological Evaluations of Rare Earths and Their Health Impacts to Workers: A Literature Review. *Saf. Health Work* **2013**, *4*, 12–26. [CrossRef]
19. Pagano, G.; Guida, M.; Tommasi, F.; Oral, R. Health effects and toxicity mechanisms of rare earth elements—Knowledge gaps and research prospects. *Ecotoxicol. Environ. Saf.* **2015**, *115*, 40–48. [CrossRef]
20. Tekumalla, S.; Seetharaman, S.; Almajid, A.; Gupta, M. Mechanical Properties of Magnesium-Rare Earth Alloy Systems: A Review. *Metals* **2015**, *5*, 1–39. [CrossRef]
21. Peng, Q.; Huang, Y.; Zhou, L.; Hort, N.; Kainer, K.U. Preparation and properties of high purity Mg–Y biomaterials. *Biomaterials* **2010**, *31*, 398–403. [CrossRef] [PubMed]
22. Liu, M.; Schmutz, P.; Uggowitzer, P.; Song, G.; Atrens, A. The influence of Y (Y) on the corrosion of Mg–Y binary alloys. *Corros. Sci.* **2010**, *52*, 3687–3701. [CrossRef]
23. Sudholz, A.D.; Gusieva, K.; Chen, X.B.; Muddle, B.; Gibson, M.; Birbilis, N. Electrochemical behaviour and corrosion of Mg–Y alloys. *Corros. Sci.* **2011**, *53*, 2277–2282. [CrossRef]
24. Südholz, A.D.; Kirkland, N.T.; Buchheit, R.G.; Birbilis, N. Electrochemical Properties of Intermetallic Phases and Common Impurity Elements in Magnesium Alloys. *Electrochem. Solid-State Lett.* **2011**, *14*, C5–C7. [CrossRef]
25. Lupescu, S.; Munteanu, C.; Istrate, B.; Stanciu, S.; Cimpoesu, N.; Oprisan, B. Microstructural Investigations on Alloy Mg-2Ca-0.2Mn-0.5Zr-1Y. *IOP Conf. Ser. Mater. Sci. Eng.* **2017**, *209*, 012018. [CrossRef]
26. Ben-Hamu, G.; Eliezer, D.; Shin, K.S.; Cohen, S. The relation between microstructure and corrosion behavior of Mg–Y–RE–Zr alloys. *J. Alloys Compounds* **2007**, *431*, 269–276. [CrossRef]
27. Liu, J.; Bian, D.; Zheng, Y.; Chu, X.; Lin, Y.; Wang, M.; Lin, Z.; Li, M.; Zhang, Y.; Guan, S. Comparative in vitro study on binary Mg-RE (Sc, Y, La, Ce, Pr, Nd, Sm, Eu, Gd, Tb, Dy, Ho, Er, Tm, Yb and Lu) alloy systems. *Acta Biomater.* **2020**, *102*, 508–528. [CrossRef]
28. Pravina, P.; Sayaji, D.; Avinash, M. Calcium and its Role in Human Body. *Int. J. Res. Pharm. Biomed. Sci.* **2013**, *4*, 659–668.
29. Pang, X.; Lin, L.; Tang, B. Unraveling the role of Calcium ions in the mechanical properties of individual collagen fibrils. *Sci. Rep.* **2017**, *7*, 46042. [CrossRef] [PubMed]
30. Pu, F.; Chen, N.; Xue, S. Calcium intake, calcium homeostasis and health. *Food Sci. Hum. Wellness* **2016**, *5*, 8–16. [CrossRef]
31. Peron, M.; Torgersen, J.; Berto, F. Mg and Its Alloys for Biomedical Applications: Exploring Corrosion and Its Interplay with Mechanical Failure. *Metals* **2017**, *7*, 252. [CrossRef]
32. Witte, F.; Hort, N.; Vogt, C.; Cohen, S.; Kainer, K.U.; Willumeit, R.; Feyerabend, F. Degradable biomaterials based on magnesium corrosion. *Curr. Opin. Solid State Mater. Sci.* **2008**, *12*, 63–72. [CrossRef]
33. Kim, W.C.; Kim, J.G.; Lee, J.Y.; Seok, H.K. Influence of Ca on the corrosion properties of magnesium for biomaterials. *Mater. Lett.* **2008**, *62*, 4146–4148. [CrossRef]

34. Wan, Y.; Xiong, G.; Luo, H.; He, F.; Huang, Y.; Zhou, X. Preparation and characterization of a new biomedical magnesium–calcium alloy. *Mater. Des.* **2008**, *29*, 2034–2037. [CrossRef]
35. Li, Z.; Gu, X.; Lou, S.; Zheng, Y. The development of binary Mg-Ca alloys for use as biodegradable materials within bone. *Biomaterials* **2008**, *29*, 1329–1344. [CrossRef]
36. Von Der Hoh, N.; Bormann, D.; Lucas, A.; Denkena, B.; Hackenbroich, C.; Meyer-Lindenberg, A. Influence of different surface machining treatments of magnesium-based resorbable implants on the degradation behavior in rabbits. *Adv. Eng. Mater.* **2009**, *11*, B47–B54. [CrossRef]
37. Li, Y.; Li, M.; Hu, W.; Hodgson, P.; Wen, C. Biodegradable Mg-Ca and Mg-Ca-Y alloys for Regenerative Medicine. *Mater. Sci. Forum* **2010**, *654–656*, 2192–2195. [CrossRef]
38. Liu, C.L.; Wang, Y.J.; Zeng, R.C.; Zhang, X.M.; Huang, W.J.; Chu, P.K. In vitro corrosion degradation behaviour of Mg–Ca alloy in the presence of albumin. *Corros. Sci.* **2010**, *52*, 3341–3347. [CrossRef]
39. Kirkland, N.; Birbilis, N.; Walker, J.; Woodfield, T.; Dias, G.; Staiger, M. In-vitro dissolution of magnesium–calcium binary alloys: Clarifying the unique role of calcium additions in bioresorbable magnesium implant alloys. *J. Biomed. Mater. Res. Part B Appl. Biomater.* **2010**, *95B*, 91–100. [CrossRef]
40. Krause, A.; von der Hoh, N.; Bormann, D.; Krause, C.; Bach, F.W.; Windhagen, H.; Meyer-Lindenberg, A. Degradation behaviour and mechanical properties of magnesium implants in rabbit tibiae. *J. Mater. Sci.* **2010**, *45*, 624–632. [CrossRef]
41. Erdmann, N.; Angrisani, N.; Reifenrath, J.; Lucas, A.; Thorey, F.; Bormann, D.; Meyer-Lindenberg, A. Biomechanical testing and degradation analysis of MgCa0.8 alloy screws: A comparative in vivo study in rabbits. *Acta Biomater.* **2011**, *7*, 1421–1428. [CrossRef] [PubMed]
42. Harandi, S.E.; Idris, M.H.; Jafari, H. Effect of forging process on microstructure, mechanical and corrosion properties of biodegradable Mg–1Ca alloy. *Mater. Des.* **2011**, *32*, 2596–2603. [CrossRef]
43. Salahshoor, M.; Guo, Y. Biodegradable Orthopedic Magnesium-Calcium (MgCa) Alloys, Processing, and Corrosion Performance. *Materials* **2012**, *5*, 135–155. [CrossRef] [PubMed]
44. Rad, H.R.B.; Idris, M.H.; Kadir, M.R.A.; Farahany, S. Microstructure analysis and corrosion behavior of biodegradable Mg–Ca implant alloys. *Mater. Des.* **2012**, *33*, 88–97. [CrossRef]
45. Li, N.; Zheng, Y. Novel Magnesium Alloys Developed for Biomedical Application: A Review. *J. Mater. Sci. Technol.* **2013**, *29*, 489–502. [CrossRef]
46. Jeong, Y.S.; Kim, W.J. Enhancement of mechanical properties and corrosion resistance of Mg–Ca alloys through microstructural refinement by indirect extrusion. *Corros. Sci.* **2014**, *82*, 392–403. [CrossRef]
47. Jeong, Y.S.; Kim, W.J. Development of biodegradable Mg–Ca alloy sheets with enhanced strength and corrosion properties through the refinement and uniform dispersion of the Mg_2Ca phase by high-ratio differential speed rolling. *Acta Biomater.* **2015**, *11*, 531–542. [CrossRef]
48. Zeng, R.C.; Qi, W.C.; Cui, H.Z.; Zhang, F.; Li, S.Q.; Han, E.H. In vitro corrosion of as-extruded Mg–Ca alloys—The influence of Ca concentration. *Corros. Sci.* **2015**, *96*, 23–31. [CrossRef]
49. Bita, A.I.; Antoniac, A.; Cotrut, C.; Vasile, E.; Ciuca, I.; Niculescu, M.; Antoniac, I. In Vitro Degradation and Corrosion Evaluation of Mg-Ca Alloys for Biomedical Applications. *J. Optoelectron. Adv. Mater.* **2016**, *18*, 394–398. Available online: https://joam.inoe.ro/articles/in-vitro-degradation-and-corrosion-evaluation-of-mg-ca-alloys-forbiomedical-applications (accessed on 15 April 2020).
50. Rau, J.V.; Antoniac, I.; Fosca, M.; De Bonis, A.; Blajan, A.I.; Cotrut, C.; Graziani, V.; Curcio, M.; Cricenti, A.; Niculescu, M.; et al. Glass-ceramic coated Mg-Ca alloys for biomedical implant applications. *Mater. Sci. Eng. C* **2016**, *64*, 362–369. [CrossRef]
51. Hunan High Broad New Material, Co.Ltd. Available online: http://www.hbnewmaterial.com/supplier-129192-master-alloy (accessed on 10 June 2019).
52. Lupescu, S.; Istrate, B.; Munteanu, C.; Minciuna, M.G.; Focsaneanu, S.; Earar, K. Characterization of Some Master Mg-X System (Ca, Mn, Zr, Y) Alloys Used in Medical Applications. *Rev. Chim.* **2017**, *68*, 1408–1413. [CrossRef]
53. Mansfield, F.; Bertocci, U. Electrochemical Corrosion Testing. *ASTM STP* **1981**, *727*, 1981–2015. Available online: https://www.astm.org/digital_library/stp/source_pages/stp727.htm (accessed on 15 April 2020).
54. Mosmann, T. Rapid colorimetric assay for cellular growth and survival: Application to proliferation and cytotoxicity assays. *J. Immunol. Methods* **1983**, *65*, 55–63. [CrossRef]

55. Vlad, M.D.; Valle, L.J.; Poeată, I.; Barracó, M.; López, J.; Torres, R.; Fernández, E. Injectable iron-modified apatitic bone cement intended for kyphoplasty: Cytocompatibility study. *J. Mater. Sci. Mater. Med.* **2008**, *19*, 3575–3583. [CrossRef] [PubMed]
56. Vlad, M.D.; Valle, L.J.; Poeată, I.; López, J.; Torres, R.; Barracó, M.; Fernández, E. Biphasic calcium sulfate dihydrate/iron-modified alpha-tricalcium phosphate bone cement for spinal applications: In vitro study. *Biomed. Mater.* **2010**, *5*, 025006. [CrossRef] [PubMed]
57. Istrate, B.; Munteanu, C.; Lupescu, S.; Antoniac, V.I.; Sindilar, E. Structural Characterization of Mg-0.5Ca-xY Biodegradable Alloys. *Key Eng. Mater.* **2018**, *782*, 129–135. [CrossRef]
58. Atrens, A.; Song, G.L.; Cao, F.; Shi, Z.; Bowen, P.K. Advances in Mg corrosion and research suggestions. *J. Magnes. Alloy.* **2013**, *1*, 177–200. [CrossRef]
59. Li, Z.; Sun, S.; Chen, M.; Fahlman, B.D.; Liu, D.; Bi, H. In vitro and in vivo corrosion, mechanical properties and biocompatibility evaluation of MgF2-coated Mg-Zn-Zr alloy as cancellous screws. *Mater. Sci. Eng. C* **2017**, *75*, 1268–1280. [CrossRef]
60. ISO 10993-5:2009—Biological Evaluation of Medical Devices—Part 5: Tests for In Vitro Cytotoxicity. Available online: http://nhiso.com/wp-content/uploads/2018/05/ISO-10993-5-2009.pdf (accessed on 15 April 2020).
61. Gilles, R.; Belkhir, M.; Compere, P.; Libioulle, C.; Thiry, M. Effect of high osmolarity acclimation on tolerance to hyperosmotic shocks in L929 cultured cells. *Tissue Cell* **1995**, *27*, 679–687. [CrossRef]
62. Mbele, G.O.; Deloulme, J.C.; Gentil, B.J.; Delphin, C.; Ferro, M.; Garin, J.; Takahashi, M.; Baudier, J. The zinc and calcium-binding S100B interacts and co-localizes with IQGAP1 during dynamic rearrangement of cell membranes. *J. Biol. Chem.* **2002**, *277*, 49998–50007. [CrossRef] [PubMed]
63. Yang, L.; Hort, N.; Laipple, D.; Höche, D.; Huang, Y.; Kainer, K.U.; Willumeit, R.; Feyerabend, F. Element distribution in the corrosion layer and cytotoxicity of alloy Mg–10Dy during in vitro biodegradation. *Acta Biomater.* **2013**, *9*, 8475–8487. [CrossRef] [PubMed]
64. Esmaily, M.; Svensson, J.E.; Fajardo, S.; Birbilis, N.; Frankel, G.S.; Virtanen, S.; Arrabal, R.; Thomas, S.; Johansson, L.G. Fundamentals and advances in magnesium alloy corrosion. *Prog. Mater. Sci.* **2017**, *89*, 92–193. [CrossRef]

Article

Third-Generation Cephalosporin-Loaded Chitosan Used to Limit Microorganisms Resistance

Letiția Doina Duceac [1,2], Gabriela Calin [1,*], Lucian Eva [1,2,*], Constantin Marcu [3,4,*], Elena Roxana Bogdan Goroftei [3,5,*], Marius Gabriel Dabija [2,6], Geta Mitrea [3,7], Alina Costina Luca [6,8], Elena Hanganu [6,8,9], Cristian Gutu [3,10], Liviu Stafie [1,11], Elena Ariela Banu [3,5], Carmen Grierosu [1,12] and Alin Constantin Iordache [2,6]

1 Faculty of Dental Medicine, "Apollonia" University of Iasi, 11 Pacurari Str., 700511 Iasi, Romania; letimedr@yahoo.com (L.D.D.); dr_liviustafie@yahoo.com (L.S.); grierosucarmen@yahoo.com (C.G.)
2 Nicolae Oblu Neurosurgery Hospital of Iasi, 2 Ateneului, 700309 Iasi, Romania; mariusdabija.md@gmail.com (M.G.D.); alinciordache@gmail.com (A.C.I.)
3 Faculty of Medicine and Pharmacy, University Dunarea de Jos, 47 Domneasca Str., 800008 Galati, Romania getamitrea@yahoo.com (G.M.); dr_cgutu@gmail.com (C.G.); banuariela@yahoo.com (E.A.B.)
4 Saarbrucken-Caritasklink St.Theresia University Hospital, 66113 Saarbrücken, Germany
5 Sf. Ioan, Emergency Clinical Hospital, 2 Gheorghe Asachi Str., 800494 Galati, Romania
6 "Grigore T. Popa", University of Medicine and Pharmacy of Iasi, 16 Universitatii Str., 700115 Iasi, Romania; acluca@yahoo.com (A.C.L.); dr.elenahanganu@gmail.com (E.H.)
7 Sf. Ap. Andrei Emergency Clinical Hospital, 177 Brailei Str., 800578 Galati, Romania
8 Sf. Maria Emergency Clinical Hospital for Children of Iasi, 62 Vasile Lupu Str., 700309 Iasi, Romania
9 Discipline of Pediatric Surgery and Orthopedics, Faculty of Medicine, "Grigore T. Popa" University of Medicine and Pharmacy of Iasi, 16 Universitatii Str., 700115 Iasi, Romania
10 Emergency Military Hospital, 199 Traian Str., 800150 Galati, Romania
11 Public Health Directorate of Iasi, 2-4 Vasile Conta, 7001016 Iasi, Romania
12 Orthopaedic Trauma Surgery Clinic, Clinical Rehabilitation Hospital, 14 Pantelimon Halipa Str., 700661 Iasi, Romania
* Correspondence: m_gabriela2004@yahoo.com (G.C.); lucianeva74@yahoo.com (L.E.); marcu_saar@yahoo.de (C.M.); elenamed84@yahoo.com (E.R.B.G.)

Received: 18 August 2020; Accepted: 22 October 2020; Published: 27 October 2020

Abstract: From their discovery, antibiotics have significantly improved clinical treatments of infections, thus leading to diminishing morbidity and mortality in critical care patients, as well as surgical, transplant and other types of medical procedures. In contemporary medicine, a significant debate regarding the development of multi-drug resistance involves all types of pathogens, especially in acute care hospitals due to suboptimal or inappropriate therapy. The possibility of nanotechnology using nanoparticles as matrices to encapsulate a lot of active molecules should increase drug efficacy, limit adverse effects and be an alternative helping to combat antibiotic resistance. The major aim of this study was to obtain and to analyze physico-chemical features of chitosan used as a drug-delivery system in order to stop the antibiotic resistance of different pathogens. It is well known that World Health Organization stated that multidrug resistance is one of the most important health threats worldwide. In last few years, nano-medicine emerged as an improved therapy to combat antibiotic-resistant infections agents. This work relies on enhancement of the antimicrobial efficiency of ceftriaxone against gram(+) and gram(−) bacteria by antibiotic encapsulation into chitosan nanoparticles. Physicochemical features of ceftriaxone-loaded polymer nanoparticles were investigated by particle size distribution and zeta potential, Fourier-transform infrared spectroscopy (FTIR), Thermal Gravimetric Analysis (TG/TGA), Scanning Electron Microscopy (SEM) characteristics techniques. The obtained results revealed an average particle size of 250 nm and a zeta potential value of 38.5 mV. The release profile indicates an incipient drug deliverance of almost 15%, after 2 h of approximately 83%, followed by a slowed drug release up to 24 h. Characteristics peaks of chitosan were confirmed by FTIR spectra indicating a similar structure in the case of ceftriaxone-loaded chitosan nanoparticles. A good encapsulation of the antibiotic into chitosan nanoparticles was also provided by thermo-gravimetric analysis. Morphological characteristics shown by SEM micrographs exhibit spherical nanoparticles of 30–250 nm in size with agglomerated architectures. Chitosan, a natural polymer which is used to load different drugs, provides sustained and prolonged release of antibiotics at a specific target by possessing antimicrobial activity against gram(+) and gram(−) bacteria. In this research, ceftriaxone-loaded chitosan nanoparticles were investigated as a carrier in antibiotic delivery.

Keywords: chitosan; cardiology; pediatrics; epidemiology; neurosurgery; pediatric surgery; pulmonology; obstetrics/gynecology; orthopedics

---

1. Introduction

Nowadays, many researchers are focused on the evolution mechanism of microbial resistance to broad-spectrum antibiotics in order to develop novel formulations of antimicrobial agents that reveal increased activity against serious multidrug-resistant pathogen agents. Reduced bacteria cellular penetration limits the efficiency of many antibiotic current therapies, thus decreasing their activity against various microbial infections. It is obvious that bacteria have adapted survival mechanisms over time, which allow them to occur in any environmental condition, representing an important challenge to the medical field and healthcare procedures [1,2]. Strategies on infection management take the prolonged contact time of antibiotics with pathogens into account, and so identified antimicrobial materials based on polymers. Therefore, chitosan, a natural polysaccharide, is a highly biocompatible nanomaterial possessing antimicrobial activity, non-toxicity, non-antigenicity and biodegradability. The mechanism of its action consists of connection to the bacterial cell wall followed by alteration in the cell membrane and reduced permeability, and ending with binding to bacteria deoxyribonucleic acid (DNA), thereby inhibiting its replication. It has been demonstrated that chitosan acts as an antimicrobial agent against both gram (−) and gram (+) bacteria. Considering all these issues, drug potency against pathogens could be enhanced by loading chitosan nanoparticles with active biomolecules, inducing synergistic action against microorganisms [3–8]. Ceftriaxone is a broad-spectrum, third-generation cephalosporin, launched in 1988 in the community for several infections' treatment or multi-resistant strain infections. It is a β-lactamase-resistant antibiotic with an exceptional serum half-life, up to ten times longer than other antibiotics belonging to this class. Ceftriaxone is administered to newborns, children and adults to treat various infections such as bacterial meningitis, acquired community pneumonia, hospital-acquired pneumonia, acute media otitis, intra-abdominal infections, complicated urinary tract infections (including pyelonephritis), gonorrhea, pox, infections of bones and joints, complicated skin and soft tissue infections, bacterial endocarditis, for the treatment of acute exacerbations of chronic obstructive pulmonary disease in adults, for the treatment of disseminated Lyme borreliosis (early stage and advanced disease stage) in adults and children, including infants over 15 days of age, for preoperative prophylaxis of local infections associated with surgery, in the control of neutropenia in patients with fever, which is suspected to be caused by a bacterial infection and for the treatment of patients with bacteremia associated with, or suspected of being associated with, any of the infections mentioned above. For this reason, ceftriaxone is used in various medical fields such as cardiology, pediatrics, epidemiology, neurosurgery, pediatric surgery, pulmonology, obstetrics/gynecology, orthopedics [9–14].

Figure 1 presents a schematic view of chitosan and ceftriaxone. An encouraging strategy to ensure efficacy in drug therapy is the evolution of suitable drug delivery systems. There are studies reporting the use of polymer nanoparticles as a self-assembled vehicle to deliver the drug to the target site due to the unique properties such as biodegradability, biocompatibility, antimicrobial qualities and non-toxicity [15–17]. Their capacity to incorporate various active molecules offers the possibility of enhancing the bactericidal and bacteriostatic efficacy of the antibiotic, limiting, at the same time, the adverse effect [18–20].

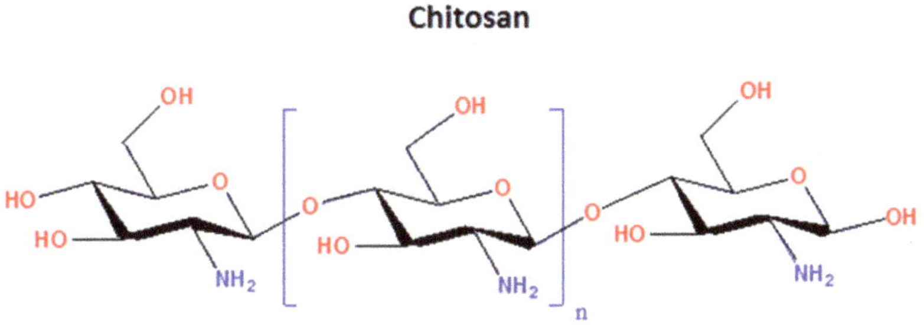

(a)

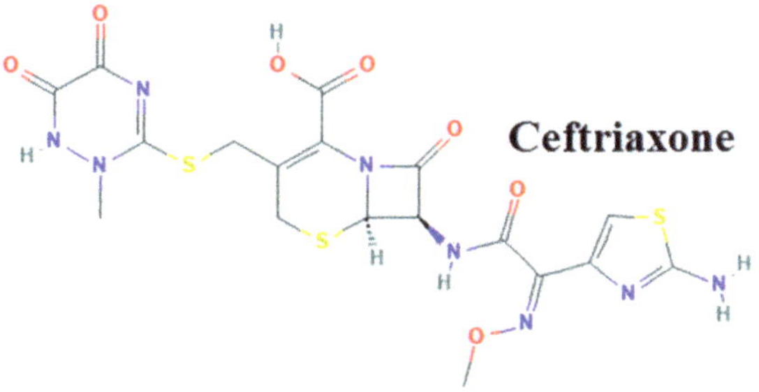

(b)

Figure 1. Schematic representation of chitosan (**a**) and ceftriaxone (**b**).

In order to limit ceftriaxone resistance, drug carriers based on chitosan nanoparticles may increase the therapeutic efficiency of the antibiotic by prolonging its release as well as minimizing its side effects [21–27].

These delivery systems improve the usage of novel formulations by perfecting preparation methods and bringing together natural or synthetic polymer nanoparticles, which were extensively engaged in pharmaceutical and medical areas, and active molecules used in different treatments. Thereby, these approaches include a wide variety of anti-infection agents as, currently, the medical field worldwide is confronted with multidrug-resistant microorganisms. The prolonged and controlled release of the encapsulated drug consists of its higher concentration in the circulatory system, which obtains the average concentration by keeping away the side effects.

In this study, we prepared ceftriaxone-loaded chitosan nanoparticles and analyzed their physico-chemical features.

2. Materials and Methods

2.1. Material Synthesis

Samples containing ceftriaxone-intercalated chitosan nanoparticles were obtained using chitosan as a polymer and trisodium-polyphosphate (TPP) as a crosslinking agent by the coacervation method. A pre-established amount of chitosan was dissolved in 10 mL acetic acid (1% v/v) and the pH of the solution was maintained at 5.0 by adding 1 M NaOH solution under sonication at room temperature. The degree of deacetylation of the chitosan is 80.3% according to the producer's certificate. Three amounts of ceftriaxone, 25, 50 and 100 mg, were dissolved in ultrapure water before being added to the as-prepared polymer and continuously stirred for 60 min. Then, 1 mg/mL TPP was added to the antibiotic–polymer solution under stirring at 800 rpm for 5 h in order to obtain ceftriaxone-loaded chitosan nanoparticles. Finally, the fresh prepared nanostructures were centrifuged and freeze-dried for 24 h (Figure 2).

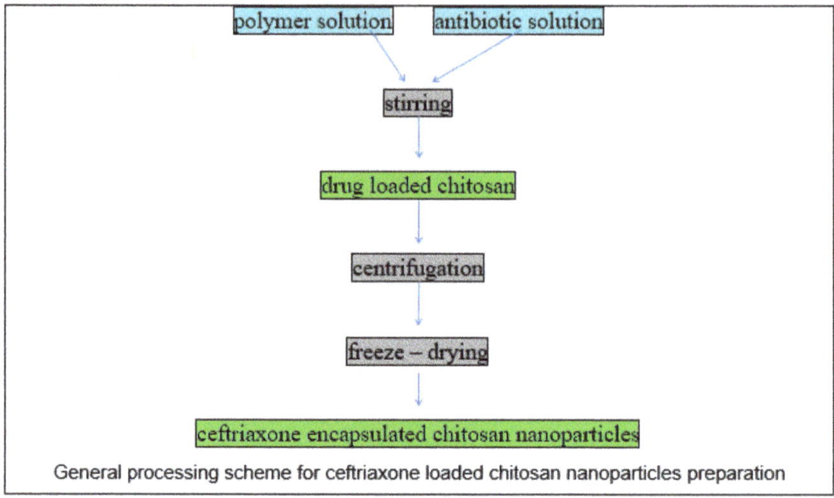

Figure 2. General processing scheme for ceftriaxone-loaded chitosan nanoparticles preparation.

The obtained nano-scaled formulations were stored at 5 °C and further analyzed.

2.2. Characterization Methods

Simultaneous thermal analysis STA 449 F1 Jupiter by NETZSCH (Selb, Germany) was used to investigate the compositional analysis of multi-component materials or blends; thermal stabilities; oxidative stabilities; estimation of product lifetimes; decomposition kinetics; effects of reactive atmospheres on materials; filler content of materials; moisture and volatile content.

Particle size distribution was evaluated using a ZetaSizer Nano ZS analyzer (Malvern, UK), for the measurements of the particles size in the range of 1–8000 nm (by dynamic light scattering at an angle of 90°, using a He-Ne laser at $\lambda = 633$ nm) and zeta potential of the nanoparticles.

Scanning Electron Microscope (Thermo Fisher Scientific, Waltham, MA, USA) equipped with an energy dispersive spectrometer (EDS, EDAX Octane Elite) allows for complete, high-resolution morphological investigations of this type of material.

FTIR spectrometer, Model Vertex 70 by Bruker (Berlin, Germany) was used to determine the structure and molecular composition of the samples in spectral domain: mid IR (5000 ÷ 400 cm^{-1}), far IR (400 ÷ 50 cm^{-1}).

3. Results

Chitosan nanoparticles were prepared by the ion gelation method, and the loading process of ceftriax-one-chitosan nanoparticles was accomplished by varying the concentration of antibiotic active molecule, as shown in Table 1.

Table 1. Formulation model for the preparation of ceftriaxone loaded chitosan nanoparticles.

Formulation Code	Amount of Antibiotic (mg/mL)	Amount of Chitosan (mg)	Amount of TPP (mg/mL)	Drug–Polymer Ratio
S1	10	10	0.25	1:1
S2	10	30	0.25	1:3
S3	10	50	0.25	1:5

Particle size distribution and zeta potential are two parameters considered important in drug delivery (Figure 3). Ceftriaxone-encapsulated chitosan nanoparticles reveal an average particle size of 250 nm and a zeta potential value of 38.5 mV, indicating the stability of nanoparticles and high electric surface charge on antibiotic-loaded chitosan nanoparticles.

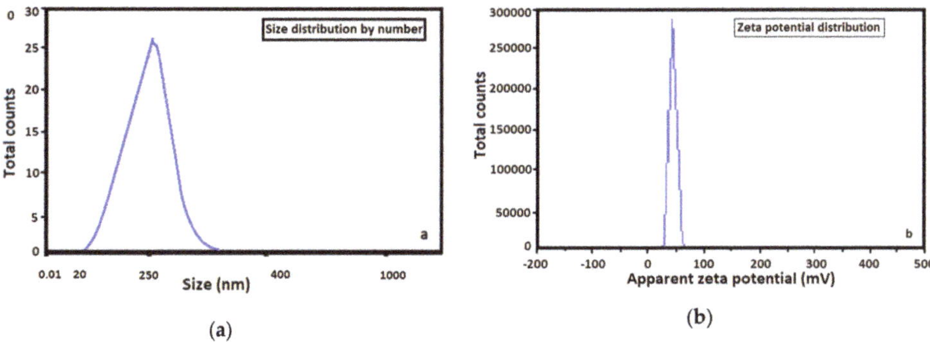

Figure 3. (**a**) Particle size distribution of ceftriaxone-loaded chitosan nanoparticles and (**b**) zeta potential of ceftriaxone loaded chitosan nanoparticles.

The drug release profile (Figures 4 and 5) was performed at a pH value of 7.4 and 37 °C and occurred by several operations such as surface erosion, disintegration, diffusion and desorption. There was a continual mixing indicating a rapid release of ~20% over the first 2 h, a steadier release from 2 to 12 h and from 12 to 24 h there were minor ceftriaxone releases of from 5 to 10% for each complex.

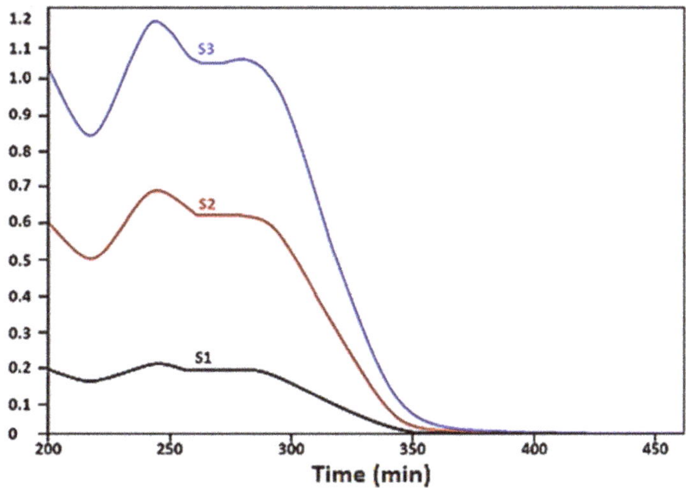

Figure 4. UV–VIS spectrum of S1, S2, S3.

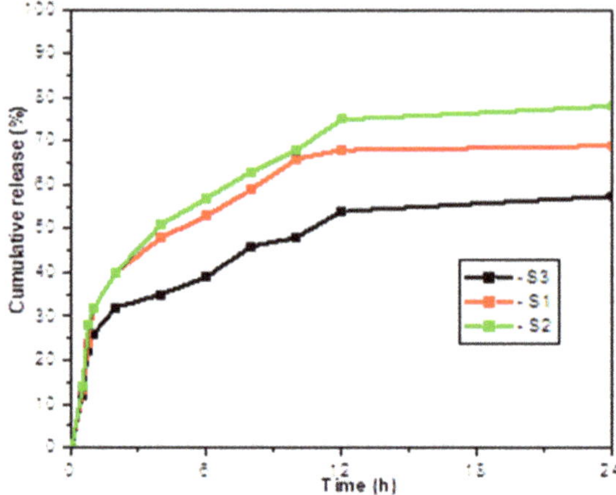

Figure 5. In vitro drug release profile of ceftriaxone-loaded chitosan nanoparticles.

The FTIR spectrum of ceftriaxone encapsulated chitosan nanoparticles presented in Figure 6 revealed characteristic peaks at around 3430 cm^{-1} of the N-H stretching vibration, at around 1740 cm^{-1} for the C=O stretching vibration and at 1590 cm^{-1} for the C=N stretching vibration, indicating that IR spectra matches with antibiotic spectra.

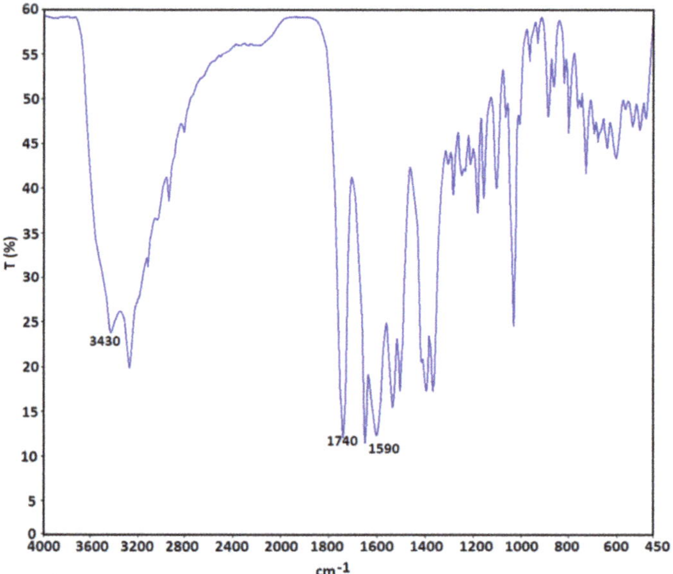

Figure 6. FTIR spectrum of ceftriaxone loaded chitosan nanoparticles.

Furthermore, characteristic peaks of chitosan were noticed, and there was no modification of characteristic peaks, denoting the lack of chemical interaction between ceftriaxone and chitosan nanoparticles.

Thermo-gravimetric analysis (TGA) is used to evaluate the thermal stability of ceftriaxone-loaded chitosan nanoparticles and weight loss at various temperatures, confirming the successful encapsulation of the antibiotic within polymer nanoparticles. Four steps of decomposition patterns on drug-loaded chitosan nanoparticles were noticed, as shown in Figure 7. At the beginning, almost 40% weight loss was remarked at 40 °C due to the water molecule release. The next step refers to 75% weight loss at 290 °C, denoting the decomposition of chitosan.

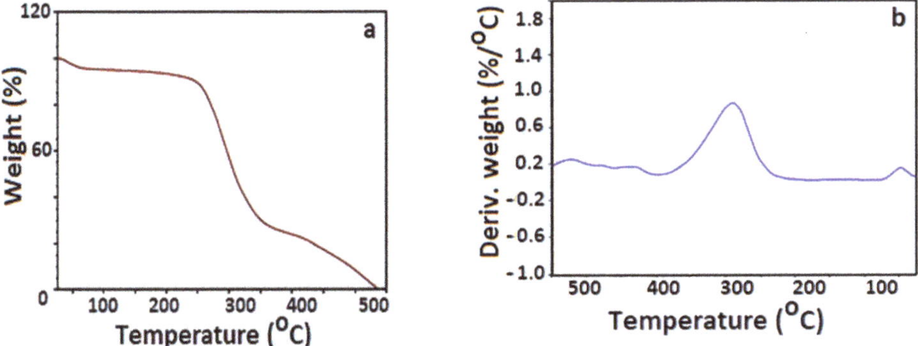

Figure 7. Thermo-gravimetric analysis of ceftriaxone encapsulated chitosan nanoparticles: (**a**) TG; (**b**) DTA.

The third step was noticed at 450 °C, with 19% weight loss, possibly due to the decomposition of chitosan and ceftriaxone, respectively. The last step of decomposition was eventually observed at 540 °C of about 10%, due to the decomposition of pure antibiotic.

SEM micrographs presented in Figure 8 exhibit the morphological features of ceftriaxone loaded chitosan nanoparticles as being spherical in shape with aggregated particles with approximately 30–250 nm in size.

Figure 8. SEM image of ceftriaxone loaded chitosan nanoparticles.

Small-size particles tend to aggregate due to the raised surface energy. Increased particle size is possibly due to the encapsulation of an antibiotic into chitosan nanoparticles.

4. Discussion

In order to design novel formulation-type drug delivery approaches, some specific demands must be respected, especially the increase in therapeutic efficacy and efficiency, minimizing of adverse effects and, most of all, physical and chemical properties, structural and morphological features, biocompatibility and usage in the medical field. The as-prepared carriers based on natural polymers were characterized to check their potential to encapsulate active molecules. Polymer nanoparticles' capacity to load various antibiotics and then to release the drugs in a controlled and prolonged manner could improve the current therapies. Ceftriaxone-loaded chitosan nano-sized particles present an average particle size of 250 nm and a zeta potential value of 38.5 mV, pointing to the stability of nanoparticles and the high electric surface charge on drug-loaded chitosan nanoparticles. Samples containing an antibiotic-loaded polymer revealed an incipient ceftriaxone release of approximately 15%, and over 2 h of approximately 83%, due to the filling property of chitosan, followed by a decelerated antibiotic release up to 24 h. FTIR spectra of antibiotic-loaded polymer revealed characteristic peaks of chitosan, indicating that it was no alteration of chitosan structure denoting the absence of chemical interaction between ceftriaxone and chitosan nanoparticles. Thermo-gravimetric analysis (TGA) used to analyze the thermal stability of ceftriaxone-loaded chitosan nanoparticles and their weight loss at various temperatures evidenced the successful encapsulation of the antibiotic within polymer nanoparticles. SEM micrographs indicate the morphological characteristics of ceftriaxone-loaded chitosan nanoparticles as being spherical in shape, with agglomerated nano-architectures of approximately 30–250 nm in size. The major aim of this study was to prepare a new nanocomposite based on chitosan and ceftriaxone molecules for targeted drug transport followed by structural and morphological analysis of the obtained nanostructures. The obtained results confirmed that the antibiotic was successfully loaded in the polymer matrix.

A first step focused on preparation of ceftriaxone encapsulated chitosan nanoparticles and characterization of the obtained samples, then, preceding in vivo evaluation, the antimicrobial effect against multidrug-resistant bacteria and other issues that were not included in this study.

Further studies concern the antimicrobial activity assay that validates the limitation of bacterial multiplication, the improving therapy and the limitation of antibiotic side effects.

5. Conclusions

Multidrug-resistant pathogens represent a major concern worldwide because of the inefficient conventional therapies. In this study, ceftriaxone encapsulated chitosan nanoparticles were prepared and structurally and morphologically tested for possible use in limiting bacterial resistance and improving antibiotic efficacy against some dangerous pathogens. Due to their small size, the obtained nano-hybrids are able to better penetrate the bacterial cell, thus increasing antimicrobial activity by inhibiting gram (+) and gram (−) microorganisms' multiplication. Our work proves that chitosan nanoparticles could be successfully used in the medical area as a vehicle for antimicrobials, due to their biocompatibility and antibacterial activity, respectively, acting as drug-delivery systems for prolonged and sustained release of the drug. Therefore, nano-antibiotics are novel formulations applied to defeat multi-drug resistant mechanism by inhibiting biofilm formation as well as by enhancing drug efficiency.

Author Contributions: All authors have equal contribution to the work. Conceptualization, L.D.D.; project administration, G.C.; data curation, L.E.; formal analysis, C.M., E.R.B.G.; investigation, M.G.D., G.M.; resources, A.C.L., L.S.; visualization, E.H.; validation, C.G. (Carmen Geriosu); investigation and writing—review and editing, E.A.B., C.G. (Cristian Gutu); methodology, A.C.I. All authors have read and agreed to the published version of the manuscript.

Funding: This research received no external funding

Conflicts of Interest: The authors declare no conflict of interest.

References

1. Raffi, M.; Mehrwan, S.; Bhatti, T.M.; Akhter, J.I.; Hameed, A.; Yawar, W.; Hasan, M.M.U. Investigations into the antibacterial behavior of copper nanoparticles against Escherichia coli. *Ann. Microbiol.* **2010**, *60*, 75–80. [CrossRef]
2. Zaki, N.M.; Hafez, M.M. Enhanced Antibacterial Effect of Ceftriaxone Sodium-Loaded Chitosan Nanoparticles Against Intracellular Salmonella typhimurium. *AAPS Pharm. Sci. Tech.* **2012**, *13*, 411–421. [CrossRef] [PubMed]
3. Kong, M.; Chen, X.; Xing, K.; Park, H.J. Antimicrobial properties of chitosan and mode of action: A state of the art review. *Int. J. Food Microbiol.* **2010**, *144*, 51–63. [CrossRef] [PubMed]
4. Rao, K.S.V.K.; Reddy, P.R.; Lee, Y.-I.; Kim, C. Synthesis and characterization of chitosan–PEG–Ag nanocomposites for antimicrobial application. *Carbohydr. Polym.* **2012**, *87*, 920–925. [CrossRef]
5. Li, L.-H.; Deng, J.-C.; Deng, H.-R.; Liu, Z.-L.; Li, X.-L. Preparation, characterization and antimicrobial activities of chitosan/Ag/ZnO blend films. *Chem. Eng. J.* **2010**, *160*, 378–382. [CrossRef]
6. Wei, D.; Sun, W.; Qian, W.; Ye, Y.; Ma, X. The synthesis of chitosan-based silver nanoparticles and their antibacterial activity. *Carbohydr. Res.* **2009**, *344*, 2375–2382. [CrossRef]
7. Cho, S.-W.; Lee, J.S.; Choi, S.-H. Enhanced Oral Bioavailability of Poorly Absorbed Drugs. I. Screening of Absorption Carrier for the Ceftriaxone Complex. *J. Pharm. Sci.* **2004**, *93*, 612–620. [CrossRef]
8. Gan, Q.; Wang, T.; Cochrane, C.; McCarron, P. Modulation of surface charge, particle size and morphological properties of chitosan–TPP nanoparticles intended for gene delivery. *Colloids Surf. B Biointerfaces* **2005**, *44*, 65–73. [CrossRef]
9. Arlet, G.; Pors, M.J.S.-L.; Rouveau, M.; Fournier, G.; Marie, O.; Schlemmer, B.; Philippon, A. Outbreak of nosocomial infections due toKlebsiella pneumoniae producing SHV-4 beta-lactamase. *Eur. J. Clin. Microbiol. Infect. Dis.* **1990**, *9*, 797–803. [CrossRef]
10. Brun-Buisson, C.; Philippon, A.; Ansquer, M.; Legrand, P.; Montravers, F.; Duval, J. Transferable enzymatic resistance to third-generation cephalosporins during nosocomial outbreak of multiresistant klebsiella pneumoniae. *Lancet* **1987**, *330*, 302–306. [CrossRef]
11. Goldstein, F.W.; Péan, Y.; Rosato, A.E.; Gertner, J.; Gutmann, L. Characterization of ceftriaxone-resistant Enterobacteriaceae: a multicentre study in 26 French hospitals. *J. Antimicrob. Chemother.* **1993**, *32*, 595–603. [CrossRef]
12. Jacoby, G.A.; Medeiros, A.A. More extended-spectrum beta-lactamases. *Antimicrob. Agents Chemother.* **1991**, *35*, 1697–1704. [CrossRef] [PubMed]
13. Sirot, D.; Sirot, J.; Labia, R.; Morand, A.; Courvalin, P.; Darfeuille-Michaud, A.; Perroux, R.; Cluzel, R. Transferable resistance to third-generation cephalosporins in clinical isolates of Klebsiella pneumoniae: identification of CTX-1, a novel β-lactamase. *J. Antimicrob. Chemother.* **1987**, *20*, 323–334. [CrossRef] [PubMed]
14. A Lipinski, C.; Lombardo, F.; Dominy, B.W.; Feeney, P.J. Experimental and computational approaches to estimate solubility and permeability in drug discovery and development settings 1PII of original article: S0169-409X(96)00423-1. The article was originally published in Advanced Drug Delivery Reviews 23 (1997) 3–25. 1. *Adv. Drug Deliv. Rev.* **2001**, *46*, 3–26. [CrossRef] [PubMed]
15. Vikulina, A.S.; Skirtach, A.G.; Volodkin, D. Hybrids of Polymer Multilayers, Lipids, and Nanoparticles: Mimicking the Cellular Microenvironment. *Langmuir* **2019**, *35*, 8565–8573. [CrossRef]
16. Leng, D.; Thanki, K.; Fattal, E.; Foged, C.; Yang, M. Engineering of budesonide-loaded lipid-polymer hybrid nanoparticles using a quality-by-design approach. *Int. J. Pharm.* **2018**, *548*, 740–746. [CrossRef]
17. Wang, T.; Hu, Q.; Lee, J.-Y.; Luo, Y. Solid Lipid–Polymer Hybrid Nanoparticles by In Situ Conjugation for Oral Delivery of Astaxanthin. *J. Agric. Food Chem.* **2018**, *66*, 9473–9480. [CrossRef]
18. Alshubaily, F.A.; Al-Zahrani, M.H. Appliance of fungal chitosan/ceftriaxone nano-composite to strengthen and sustain their antimicrobial potentiality against drug resistant bacteria. *Int. J. Biol. Macromol.* **2019**, *135*, 1246–1251. [CrossRef]
19. Kravanja, G.; Primožič, M.; Knez, Ž.; Leitgeb, M. Chitosan-Based (Nano)Materials for Novel Biomedical Applications. *Molecules* **2019**, *10*, 1960. [CrossRef]

20. Glaser, T.K.; Plohl, O.; Vesel, A.; Ajdnik, U.; Ulrih, N.P.; Hrnčič, M.K.; Bren, U.; Zemljič, L.F. Functionalization of Polyethylene (PE) and Polypropylene (PP) Material Using Chitosan Nanoparticles with Incorporated Resveratrol as Potential Active Packaging. *Materials* **2019**, *12*, 2118. [CrossRef]
21. Azhdarzadeh, M.; Lotfipour, F.; Zakeri-Milani, P.; Mohammadi, G.; Valizadeh, H. Anti-bacterial performance of azithromycin nanoparticles as colloidal drug delivery system against different gram-negative and gram-positive bacteria. *Adv. Pharm. Bull.* **2012**, *2*, 17–24. [PubMed]
22. Luca, A.C.; Eva, L.; Duceac, L.D.; Mitrea, G.; Marcu, C.; Stafie, L.; Ciuhodaru, M.I.; Ciomaga, I.M.; Goroftei, E.R.B.; Hanganu, E.; et al. Drug Encapsulated Nanomaterials as Carriers Used in Cardiology Field. *Rev. Chim.* **2020**, *71*, 413–417. [CrossRef]
23. Lee, S.; Kim, S.K.; Lee, D.Y.; Park, K.; Kumar, T.S.; Chae, S.Y.; Byun, Y. Cationic Analog of Deoxycholate as an Oral Delivery Carrier for Ceftriaxone. *J. Pharm. Sci.* **2005**, *94*, 2541–2548. [CrossRef]
24. Duceac, L.D.; Mitrea, G.; Banu, E.A.; Ciuhodaru, M.I.; Ciomaga, I.M.; Ichim, D.L.; Constantin, M.; Luca, A.C. Synthesis and Characterization of Carbapenem Based Nanohybrids as Antimicrobial Agents for Multidrug Resistant Bacteria. *Mater. Plast.* **2019**, *56*, 388–391. [CrossRef]
25. Shanmugarathinam, A.; Puratchikody, A. Formulation and characterisation of ritonavir loaded ethylcellulose buoyant microspheres. *J. Pharm. Sci. Res.* **2014**, *8*, 274–277.
26. Duceac, L.D.; Marcu, C.; Ichim, D.L.; Ciomaga, I.M.; Tarca, E.; Iordache, A.C.; Ciuhodaru, M.I.; Florescu, L.; Tutunaru, D.; Luca, A.C.; et al. Antibiotic Molecules Involved in Increasing Microbial Resistance. *Rev. Chim.* **2019**, *70*, 2622–2626. [CrossRef]
27. Ichim, D.L.; Duceac, L.D.; Marcu, C.; Iordache, A.C.; Ciomaga, I.M.; Luca, A.C.; Goroftei, E.R.B.; Mitrea, G.; Damir, D.; Stafie, L. Synthesis and Characterization of Colistin Loaded Nanoparticles Used to Combat Multi-drug Resistant Microorganisms. *Rev. Chim.* **2019**, *70*, 3734–3737. [CrossRef]

Publisher's Note: MDPI stays neutral with regard to jurisdictional claims in published maps and institutional affiliations.

Article

High Temperature Behavior of RuAl Thin Films on Piezoelectric CTGS and LGS Substrates

Marietta Seifert

Leibniz IFW Dresden, Helmholtzstraße 20, 01069 Dresden, Germany; marietta.seifert@ifw-dresden.de; Tel.: +49-351-4659-639

Received: 9 March 2020; Accepted: 26 March 2020; Published: 1 April 2020

Abstract: This paper reports on a significant further improvement of the high temperature stability of RuAl thin films (110 nm) on the piezoelectric $Ca_3TaGa_3Si_2O_{14}$ (CTGS) and $La_3Ga_5SiO_{14}$ (LGS) substrates. RuAl thin films with AlN or SiO_2 cover layers and barriers to the substrate (each 20 nm), as well as a combination of both were prepared on thermally oxidized Si substrates, which serve as a reference for fundamental studies, and the piezoelectric CTGS, as well as LGS substrates. In some films, additional Al layers were added. To study their high temperature stability, the samples were annealed in air and in high vacuum up to 900 °C, and subsequently their cross-sections, phase formation, film chemistry, and electrical resistivity were analyzed. It was shown that on thermally oxidized Si substrates, all films were stable after annealing in air up to 800 °C and in high vacuum up to 900 °C. The high temperature stability of RuAl thin films on CTGS substrates was improved up to 900 °C in high vacuum by the application of a combined AlN/SiO_2 barrier layer and up to 800 °C in air using a SiO_2 barrier. On LGS, the films were only stable up to 600 °C in air; however, a single SiO_2 barrier layer was sufficient to prevent oxidation during annealing at 900 °C in high vacuum.

Keywords: SAW sensors; interdigital transducer material; high temperature stability; RuAl; thin films; CTGS; LGS

---

1. Introduction

The development of sensors working at high temperatures is an important research field since the knowledge of process parameters at high temperatures is required to control and optimize high temperature processes. One route to realize such sensors is to use the principle of the surface acoustic waves (SAW) technology, which applies the piezoelectric effect to transfer a high frequency voltage into a mechanical wave and vice versa with electrodes that are structured with a certain geometry on a piezoelectric substrate to serve as interdigital transducers (IDTs). For the application of such sensors at high temperatures, a suitable metallization with a high thermo-mechanical stability, a low electrical resistivity, and a high oxidation and corrosion resistance, as well as a high temperature stable piezoelectric substrate are required. Several metallization systems have been investigated during the last few years concerning their high temperature stability, e.g., Pt- or Ir-based materials [1–4], oxide dispersion hardened materials [5,6], or refractory metals [7].

Another alternative material for high temperature stable electrodes is the RuAl alloy with its high melting temperature of 2050 °C [8] and strong oxidation and corrosion resistance [9]. During the last few years, we investigated the high temperature stability of RuAl thin films on the high temperature stable piezoelectric $Ca_3TaGa_3Si_2O_{14}$ (CTGS), as well as $La_3Ga_5SiO_{14}$ (LGS) substrates. The experiments revealed that an oxidation barrier between the substrates and the RuAl film was required to prevent a chemical reaction between the Al and the CTGS or LGS if the samples were annealed at 800 °C in high vacuum (HV) [10,11]. It was shown that this reaction was suppressed if a 10 nm thick sputtered SiO_2 layer was added on top of the substrate [12].

Reference RuAl films on thermally oxidized Si substrates (SiO_2/Si) were used to study the oxidation behavior of the thin films in air, since in combination with these substrates, there is hardly any chemical reaction between film and substrate, and therefore, only a chemical reaction of the RuAl with the oxygen of the surrounding atmosphere can take place (at 800 °C, a slight reduction of the SiO_2 to Si and the formation of Al_2O_3 are observed at the interface between the substrate and the RuAl film; however, this is in such a small amount that it has hardly any influence on the performance of the film). While a 20 nm thick sputtered Al_2O_3 cover layer fails during annealing at 800 °C in air and leads to a strong oxidation of the RuAl film, a 20 nm thick SiO_2 cover layer successfully prevents the reaction between the film and the surrounding atmosphere at this temperature for at least 10 h [13].

However, further improvements are necessary to realize RuAl-based SAW sensors on CTGS and LGS substrates that can operate at high temperatures in ambient condition. In this paper, extended RuAl thin films in combination with barrier layer systems on CTGS and LGS substrates are investigated as a necessary step towards the production of structured IDTs. AlN thin films are tested as an alternative oxidation barrier as individual layers, as well as in combination with a SiO_2 barrier. Former experiments showed that an increase in the Al content of the RuAl layer led to an improved high temperature stability [14]. As an alternative approach to the homogeneous increase in the Al content, 10 nm thick Al layers are added on top and below the RuAl film, and their influence on the film behavior after the annealing is analyzed.

2. Materials and Methods

RuAl thin films with a thickness of 110 nm were prepared by co-sputtering from elemental targets on $Ca_3TaGa_3Si_2O_{14}$ (CTGS) and $La_3Ga_5SiO_{14}$ (LGS) substrates with either a 20 nm AlN or SiO_2 cover and barrier layer, as well as a combination of both (total thickness 40 mm). Al layers with a thickness of 10 nm were included below and above the RuAl layer. Another layer stack consisting of 7 nm Al, 55 nm RuAl, 7 nm Al, and 55 nm RuAl covered by 7 nm Al, which was in sum again about 20 nm Al and 110 nm RuAl, was prepared as a sample with an alternative distribution of the Al. A RuAl film with a SiO_2 cover and barrier layer without additional Al layers served as a reference sample. To reveal if oxidation processes arose from the CTGS or LGS substrate, which were known to oxidize, e.g., Al films deposited directly on these substrates, or were due to the surrounding atmosphere, the same sample systems were simultaneously deposited on thermally oxidized Si substrates (SiO_2/Si). Details of the sputter deposition of the RuAl films were described in [10]. The AlN layers were prepared by sputter deposition from an AlN target at room temperature (RT) with a mixture of N_2 and Ar of 1 : 11. SiO_2 was deposited from a SiO_2 target at a temperature of 180 °C with a mixture of O_2 and Ar with a ratio of 1 : 6 as the sputtering gas. Figure 1 presents an overview of the analyzed samples and their notation.

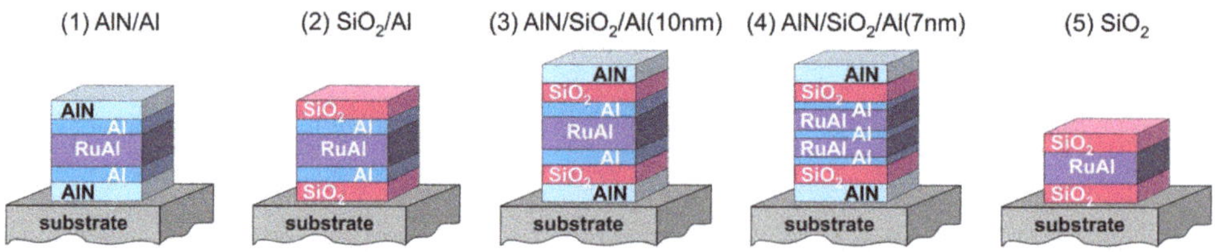

Figure 1. Architecture and notation of the different samples. As substrate materials thermally oxidized Si (SiO_2/Si), CTGS and LGS are used.

The samples were annealed for 10 h at up to 900 °C in air and for comparison at 900 °C in HV. The phase formation was analyzed with X-ray diffraction in Bragg–Brentano geometry (XRD, Philips X'Pert, Co-Kα). To obtain the full texture information, pole figure measurements were performed

(Philips X'Pert, Cu-Kα). Cross-sections of the samples were fabricated in a focused ion beam device (FIB, Zeiss 1540 XB CrossBeam, Carl Zeiss Microscopy GmbH, Oberkochen, Germany) and imaged by scanning electron microscopy (SEM) in the same instrument. SEM images of the surfaces of the samples were obtained with a Zeiss Ultra Plus (Carl Zeiss Microscopy GmbH, Oberkochen, Germany).

The distribution of the elements across the sample thickness was determined by Auger electron spectroscopy (AES, JEOL JAMP-9500F Field Emission Auger Microprobe). AES depth profiles were realized by alternately sputtering with Ar ions with an energy of 1 kV for 120 s and measuring of the AES spectra. High resolution images of the sample cross-sections were recorded by annular dark field scanning transmission electron microscopy (ADF-STEM, Tecnai F30, FEI company, Hillsboro, OR, USA) in combination with energy dispersive X-ray spectroscopy (EDX, Octane T Optima, EDAX Company, Mahwah, NJ, USA) to reveal the local chemical composition.

The electrical resistivity of the films was determined by the van der Pauw technique (vdPauw, W tips). An electrical current of 5 mA was injected into the sample, and its polarity was changed after each measurement. A nanovoltmeter (2182A-Nanovoltmeter, KEITHLEY-TEKTRONIX Inc., Beaverton, OR, USA) was used to measure the voltage. The electrical resistivity was calculated from the measured resistance and the sample thickness.

3. Results and Discussion

3.1. Film Morphology

To analyze the high temperature stability and possible reactions of the sample with the substrate and the surrounding atmosphere, cross-section images of the substrate-sample systems were evaluated. The SEM images of the cross-sections prepared with FIB for the films on SiO_2/Si, CTGS, and LGS, respectively, together with an exemplary SEM image of the surface of a sample on CTGS are summarized in Figures 2–5.

After heat treatment at 600 °C in air, the samples on SiO_2/Si were not degraded (Figure 2a). Only the sample with SiO_2 barriers without additional Al layers ((5)-SiO_2) showed some oxide formation, which is visible from the small bright grains at the surface, which consist of Ru, as shown in former work. Such Ru-rich grains are formed if Al is oxidized and lacks for the formation of the RuAl phase. The sample morphology strongly changed after heat treatment at 800 °C in air (Figure 2b). The (1)-AlN/Al sample contained many pores, which indicates strong oxidation effects. The (2)-SiO_2/Al, (3)-AlN/SiO_2/Al(10 nm), and (4)-AlN/SiO_2/Al(7 nm) samples were still intact and only contained small pores. The (5)-SiO_2 sample was also still intact; however, showing a rough surface.

The heat treatment at 900 °C in air led to a strong degradation of all samples (Figure 2c). In the films with a single barrier, it can be seen that there is no clear interface to the substrate any more. This indicates a strong interaction between the film and the substrate material. The films with the combined barrier layers were slightly less oxidized as compared to the other samples.

After annealing at 900 °C in HV, some small brighter grains were only visible for the (1)-AlN/Al sample at the film surface, which again indicates a partial oxidation. Besides this, none of the samples were degraded (Figure 2d).

The air annealed samples on CTGS (as-prepared state shown in Figure 3a) behaved like the corresponding samples on the thermally oxidized Si substrates. All samples were intact after annealing at 600 °C in air (Figure 3b). The 800 °C annealing in air led to a degradation of the film with AlN barrier, while the other samples were not damaged, and only small pores were visible (Figure 3c). After annealing at 900 °C in air, all samples were strongly degraded (Figure 3d). However, damages in the CTGS substrate were only present for both samples with the SiO_2 barrier with and without additional Al layers. SEM images of the surface of the samples revealed the formation of bulges for all systems. Such bulges were found for the (1)-AlN/Al system already for the annealings at a lower temperature in air. On the sample surface of the films annealed at 900 °C in air, locally, structures with sharp needles formed, as shown in Figure 4. An EDX analysis revealed that these structures mainly consisted of Ga oxides, Ga-Al oxides, and Si oxides.

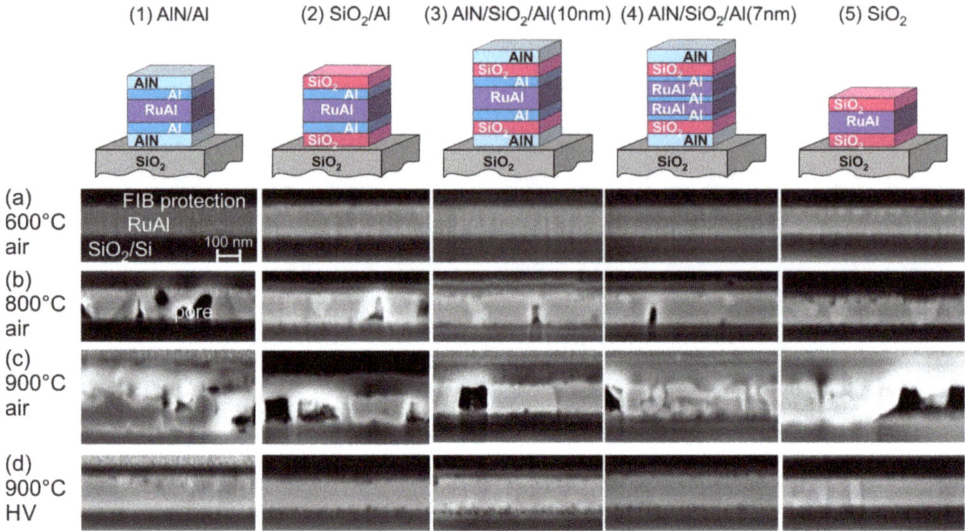

Figure 2. SEM images (inLens, 3 kV) of cross-sections of the RuAl thin films with various protection layers on SiO_2/Si after annealing in air for 10 h at (**a**) 600 °C, (**b**) 800 °C, (**c**) 900 °C, and (**d**) in high vacuum at 900 °C.

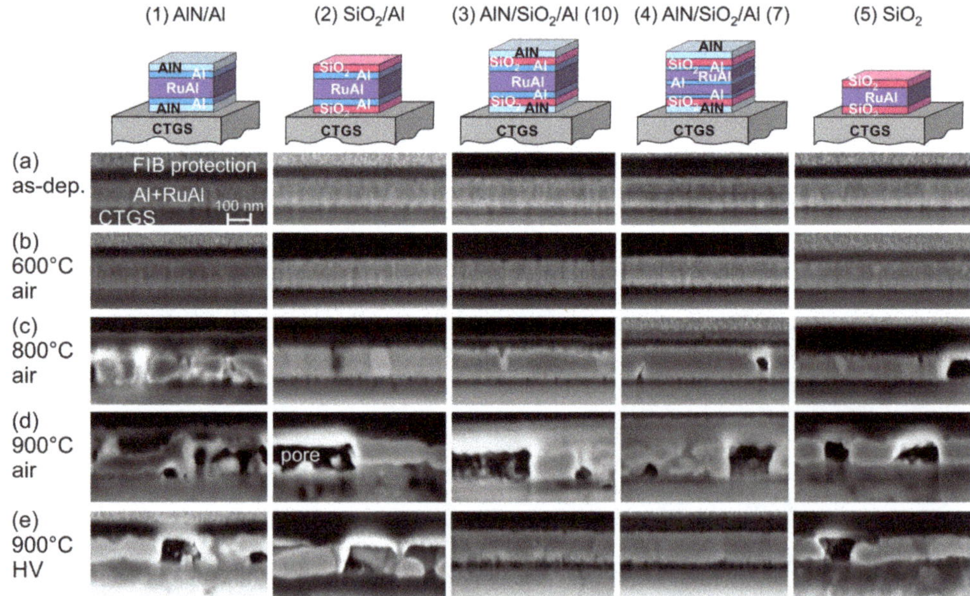

Figure 3. SEM images (inLens, 3 kV) of cross-sections of the RuAl thin films with various protection layers on CTGS in (**a**) the as-deposited state and after annealing in air for 10 h at (**b**) 600 °C, (**c**) 800 °C, (**d**) 900 °C, and (**e**) in high vacuum at 900 °C.

The influence of the CTGS substrate became visible in the samples annealed at 900 °C in high vacuum (Figure 3e). While all samples on SiO_2/Si were stable after this annealing, on CTGS, the (1)-AlN/Al, (2)-SiO_2/Al, and (5)-SiO_2 samples were oxidized. The film with the SiO_2 barrier without additional Al layers appeared less oxidized as compared to the film with the additional Al layers. The images also show a degradation of the surface of the CTGS substrate. Only the two samples with the combined barrier layers (3)-AlN/SiO_2/Al(10 nm) and (4)-AlN/SiO_2/Al(7 nm) were not oxidized. Although a 10 nm thick SiO_2 barrier layer on top of the CTGS substrate was sufficient to prevent the diffusion of oxygen out of the substrate into the RuAl film if the sample was annealed at 800 °C in HV [12], at 900 °C, a SiO_2 film with even twice the thickness was not sufficient.

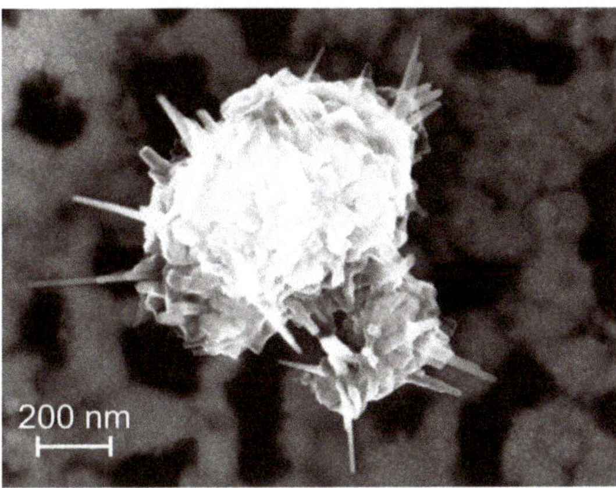

Figure 4. SEM image (inLens, 3 kV) of the surface of the sample with the (5)-SiO_2 barrier layer on CTGS at a position with a defect.

Figure 5 summarizes the cross-section images of the films on the LGS substrates. For this kind of substrate also, the annealing at 600 °C in air did not degrade the films (Figure 5a). However, in contrast to the films on the other substrates after annealing at 800 °C in air, all films were strongly oxidized and peeled off the substrate to a large extent. The cross-section images show regions of the films that were still in contact with the LGS (Figure 5b). Only the films with the SiO_2 barrier were still attached to a large extent to the substrate. The (4)-AlN/SiO_2/Al(7 nm) sample was the most destroyed film with hardly any film in contact with the substrate. However, in contrast to the films with CTGS, no inhomogeneities were observed in the surface region of the substrate.

The annealing at 900 °C in air likewise led to a strong degradation of the films (Figure 5c) and to a large-scale peeling of the film off the substrate. Again, the cross-section images represent regions of the film that were still attached to the LGS.

The annealing at 900 °C in HV led to a strong degradation of the (1)-AlN/Al sample (Figure 5d), and locally, defect formation in the LGS substrate up to a depth of a few μm was observed. For all other samples, the RuAl layer was still visible. The presence of the additional Al layers in the sample with SiO_2 barriers resulted in a stronger chemical reaction between the film and the LGS substrate, which was visible from the irregular rough interface between the film and the substrate, while in the sample without additional Al, still, a smooth interface was present. There was also a local degradation of the LGS for the (2)-SiO_2/Al sample.

The formation of bulges in the film was observed for almost all annealed samples on LGS.

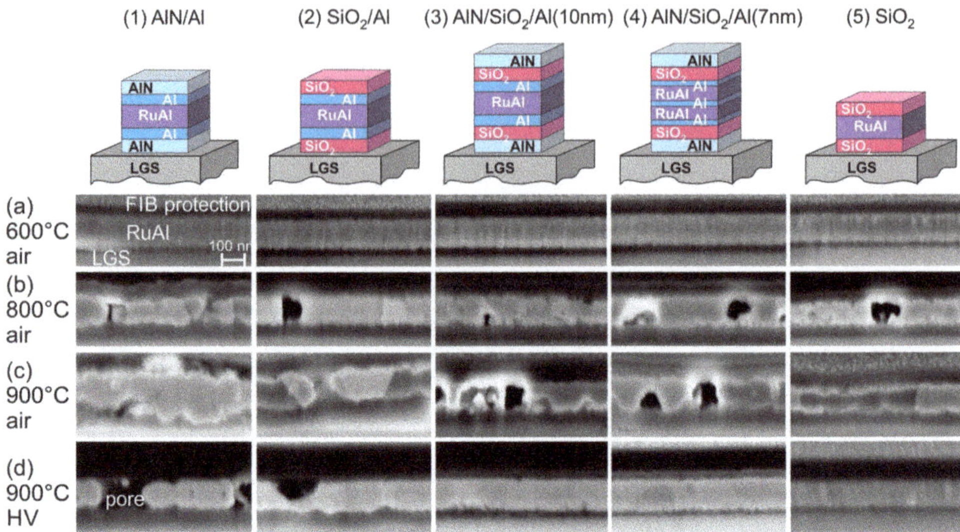

Figure 5. SEM images (inLens, 3 kV) of cross-sections of the RuAl thin films with various protection layers on LGS after annealing in air for 10 h at (**a**) 600 °C, (**b**) 800 °C, (**c**) 900 °C, and (**d**) in high vacuum at 900 °C.

3.2. Phase Formation

The XRD measurements of the different film systems on SiO_2/Si, CTGS, and LGS are presented in Figures 6–8.

All samples deposited on SiO_2/Si showed a (100) RuAl peak in the as-deposited state, which was located at a slightly lower angle (shift up to $-0.23°$) as compared to the bulk RuAl value ($2\theta = 34.8°$). Annealing at 600 °C in air led to an increase in the intensity of the RuAl (100) peak for all samples and to a shift to higher 2θ values than the theoretical position of the RuAl (100) (up to 0.36°) except for the (4)-AlN/SiO_2/Al(7 nm) film, which showed the peak at 34.74°. In the case of the (1)-AlN/Al (Figure 6a) and (3)-AlN/SiO_2/Al(10 nm) (Figure 6c) samples, a $RuAl_2$ peak became visible at a 2θ position of 48.4°, which is slightly shifted with respect to the bulk value of the $RuAl_2$ (004) reflex ($2\theta = 48.2°$). A very small $RuAl_2$ peak also appeared for the (2)-SiO_2/Al sample (Figure 6b).

The heat treatment at 800 °C in air resulted in a further shift of the RuAl (100) peak to higher 2θ values (up to 0.46°) and the disappearance of the $RuAl_2$ peak. In the case of the (1)-AlN/Al sample, the RuAl (100) peak intensity strongly decreased, which is in accordance with the strong degradation, which was visible in the cross section image (Figure 2b). The oxidation of Al also explains the appearance of Ru peaks. The RuAl peak intensity was slightly reduced for the (3)-AlN/SiO_2/Al(10 nm) film and was increased for both films with the SiO_2 barrier. Surprisingly, in the case of the (4)-AlN/SiO_2/Al(7 nm) film, no RuAl peak was visible any more, also for higher annealing temperatures, although the cross-section images were comparable to the (3)-AlN/SiO_2/Al(10 nm) sample and did not show a stronger degradation.

After annealing at 900 °C in air, the RuAl peak disappeared for the (1)-AlN/Al and (5)-SiO_2 sample. For the (1)-AlN/Al sample, a RuO_2 (110) peak at 2θ of 32.7° appeared. The (2)-SiO_2/Al and (3)-AlN/SiO_2/Al(10 nm) sample still showed a small RuAl peak, and a very small Ru peak became visible. The AlN/SiO_2/Al(7 nm) showed a stronger Ru peak.

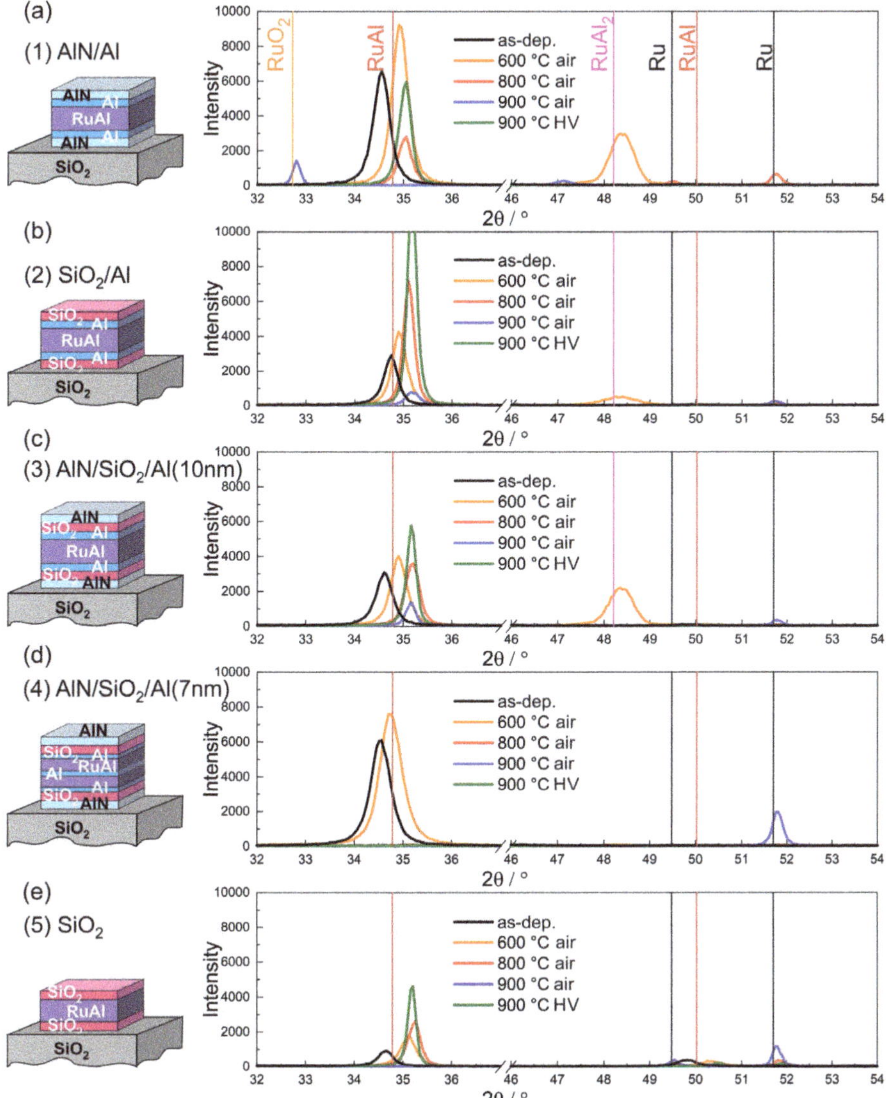

Figure 6. Results of the XRD measurements (Bragg–Brentano, Co-Kα) for the samples with the different cover and barrier layers on SiO_2/Si substrates: (**a**) (1)-AlN/Al, (**b**) (2)-SiO_2/Al, (**c**) (3)-AlN/SiO_2/Al(10 nm), (**d**) (4)-AlN/SiO_2/Al(7 nm), and (**e**) (5)-SiO_2.

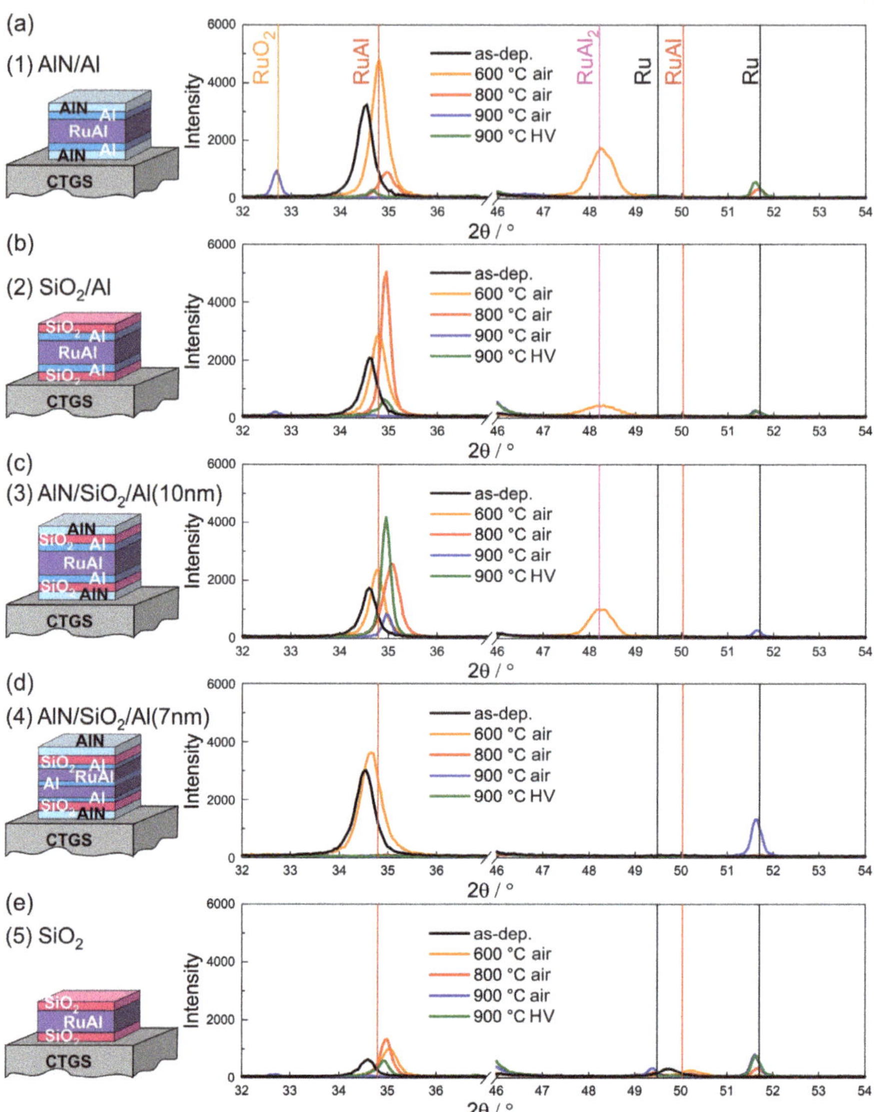

Figure 7. Results of the XRD measurements (Bragg–Brentano, Co-Kα) for the samples with the different cover and barrier layers on CTGS substrates: (**a**) (1)-AlN/Al, (**b**) (2)-SiO_2/Al, (**c**) (3)-AlN/SiO_2/Al(10 nm), (**d**) (4)-AlN/SiO_2/Al(7 nm), and (**e**) (5)-SiO_2.

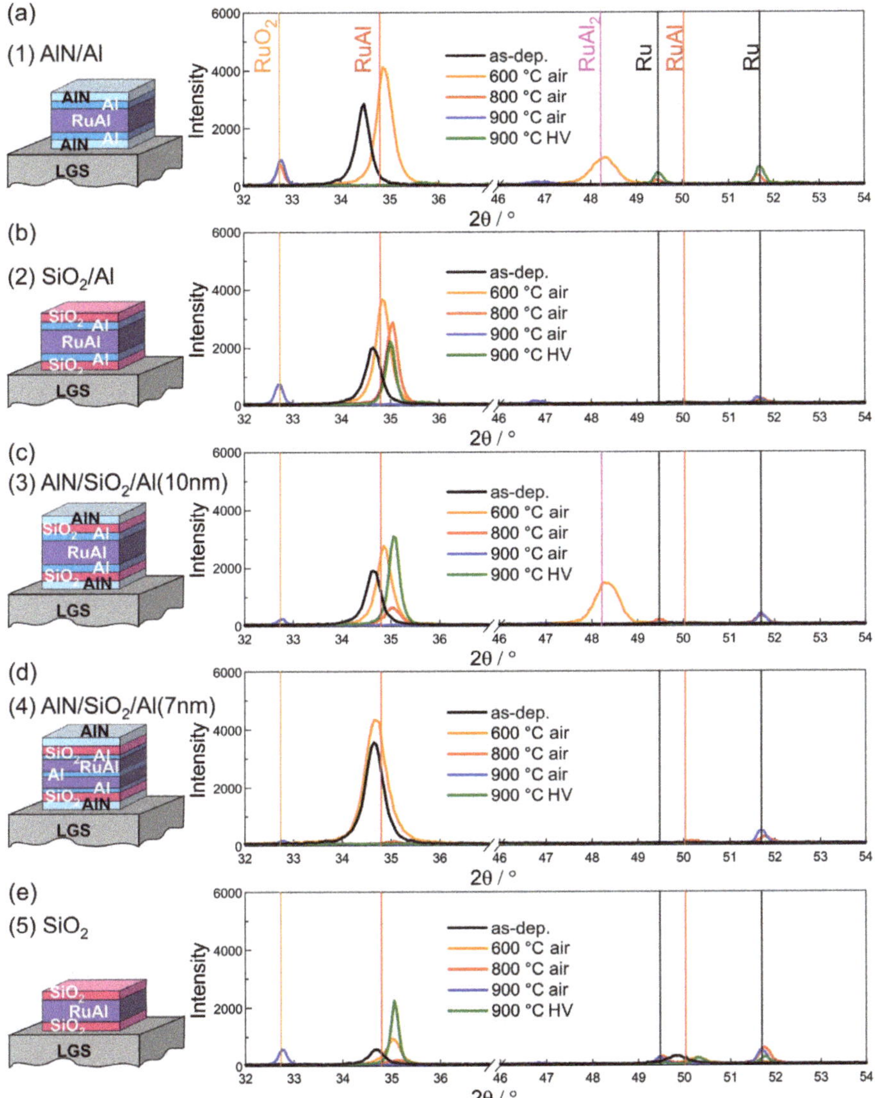

Figure 8. Results of the XRD measurements (Bragg–Brentano, Co-Kα) for the samples with the different cover and barrier layers on LGS substrates: (**a**) (1)-AlN/Al, (**b**) (2)-SiO_2/Al, (**c**) (3)-AlN/SiO_2/Al(10 nm), (**d**) (4)-AlN/SiO_2/Al(7 nm), and (**e**) (5)-SiO_2.

Annealing in high vacuum at 900 °C resulted in a RuAl peak that was higher than for the other annealing procedures in the case of the (2)-SiO_2/Al, (3)-AlN/SiO_2/Al(10 nm), and (5)-SiO_2 samples. The (1)-AlN/Al sample showed a RuAl peak with a reduced intensity as compared to the as-prepared state.

In summary, the XRD measurements confirmed that AlN is not a suitable barrier layer for RuAl thin films. Comparing the samples (2)-SiO_2/Al and (5)-SiO_2 revealed that the additional Al layers are beneficial for the RuAl phase formation. A comparison between the samples (3)-AlN/SiO_2/Al(10 nm) and (4)-AlN/SiO_2/Al(7 nm) showed that despite a similar film morphology visible from the cross-section images, there is a strong difference in phase formation. This issue will be discussed below.

The formation of the $RuAl_2$ phase during the annealing at 600 °C was the result of the interdiffusion between the Al and RuAl layers. At the interface, a region with a higher Al content developed, which led to the formation of the Al rich $RuAl_2$ phase. A comparison between the (3)-AlN/SiO_2/Al(10 nm) and (4)-AlN/SiO_2/Al(7 nm) films indicated that the reduced thickness of the Al layer of 7 nm was not sufficient to realize the formation of $RuAl_2$. In both films, the same total amount of Al was added so that the absence of the $RuAl_2$ in the (4)-AlN/SiO_2/Al(7 nm) film can only be explained by the different local distribution. During the annealing at higher temperatures, two processes took place: The further interdiffusion of the Al within the RuAl film led to a lower local concentration of the Al. On the other hand, first oxidation processes took place, which also reduced the Al concentration, so that the Al content was not sufficient to form the $RuAl_2$ phase any more.

The results of the XRD measurements of the films on the CTGS substrates are shown in Figure 7. As for the films on SiO_2/Si, all films showed the RuAl (100) peak at a slightly lower 2θ value in the as-prepared state. Annealing at 600 °C in air led to an increase in peak intensity and to a shift to higher 2θ values for all samples. In this case also, the $RuAl_2$ phase was formed for the (1)-AlN/Al and (3)-AlN/SiO_2/Al(10 nm) films and with a very low amount also for the (2)-SiO_2/Al film. Furthermore, the behavior after annealing at 800 °C in air was similar to that of the films on the thermally oxidized Si substrates: the peak intensity was increased for the films with the SiO_2 barrier, slightly increased for the (3)-AlN/SiO_2/Al(10 nm) sample, and strongly reduced for the (1)-AlN/Al film. Again, no RuAl peaks were measured for the (4)-AlN/SiO_2/Al(7 nm) samples for 800 °C and higher temperatures.

Differences of the films on CTGS to the samples on the thermally oxidized Si substrates became obvious for the samples annealed at 900 °C in air and in HV. For the samples annealed in air, only for the (3)-AlN/SiO_2/Al(10 nm) film, a small RuAl reflex remained. For the samples on SiO_2/Si, RuAl was additionally still visible for the (2)-SiO_2/Al film. While for all samples on the SiO_2/Si substrate annealed at 900 °C in HV, a strong RuAl peak was measured, in the case of the CTGS substrate, a strong RuAl peak was only visible for the (3)-AlN/SiO_2/Al(10 nm) sample system, and only very small peaks appeared for both films with the SiO_2 barrier.

Figure 8 summarizes the results of the XRD measurements of the different films on the LGS substrates. As for the other two systems for all films, a RuAl peak was visible in the as-prepared state. The intensity increased after annealing at 600 °C in air, and a $RuAl_2$ peak appeared for the (1)-AlN/Al and (3)-AlN/SiO_2/Al(10 nm) sample system. However, after annealing at 800 °C in air, where for the other two substrates, RuAl was visible for all films, with decreased intensity for the (1)-AlN/Al sample, in the case of LGS, the RuAl peak was only visible for the (2)-SiO_2/Si and with a strongly reduced intensity for the (3)-AlN/SiO_2/Al(10 nm) layer stack. The film with the SiO_2 barrier without the additional Al layer did not show RuAl anymore. For the film with the AlN barrier, already at this temperature, RuO_2 was formed, which was not detected in any of the films on the other substrates.

After annealing at 900 °C in air, for none of the samples RuAl was measured. Instead, besides the (4)-AlN/SiO_2/Al(7 nm) sample, all films showed the formation of RuO_2.

These results revealed that there is an improved high temperature stability of CTGS as compared to LGS concerning heat treatments in air. On the other hand, the films on LGS showed a stronger phase formation and an improved film stability after annealing at 900 °C in high vacuum. In the case of LGS, a strong RuAl peak was not only measured for the (3)-AlN/SiO_2/Al(10 nm) layer stack, as was the case for the CTGS substrate, but also for both samples with the SiO_2 barrier without and with the additional Al layers. The better stability under high vacuum conditions was already visible in the cross-section images in Figure 5, which showed a more homogeneous film structure and less reactions with the substrate for these samples.

Common for all substrates is the different behavior of both systems with the combined barrier layers with the varying distribution of the Al. While the layer stack (3)-AlN/SiO_2/Al(10 nm) showed a RuAl peak for all annealing conditions (except for LGS, where no peak was visible for the 900 °C air annealed sample), in the case of the (4)-AlN/SiO_2/Al(7 nm) sample, the RuAl peak was measured only for the as-prepared and 600 °C annealed sample. For the evaluation of these data, however, one has to be aware of the fact that the measurements in Bragg–Brentano geometry are only sensitive to lattice planes, which are oriented parallel to the film surface. To reveal the whole texture information, pole figure measurements were conducted. The pole figures of the (100) RuAl peak for the as-prepared state and after annealing at 600 and 800 °C in air and 900 °C in HV are presented in Figure 9a and b for the (3)-AlN/SiO_2/Al(10 nm) and (4)-AlN/SiO_2/Al(7 nm) systems, using the example of the films on CTGS. In addition, the RuAl (110) pole was measured.

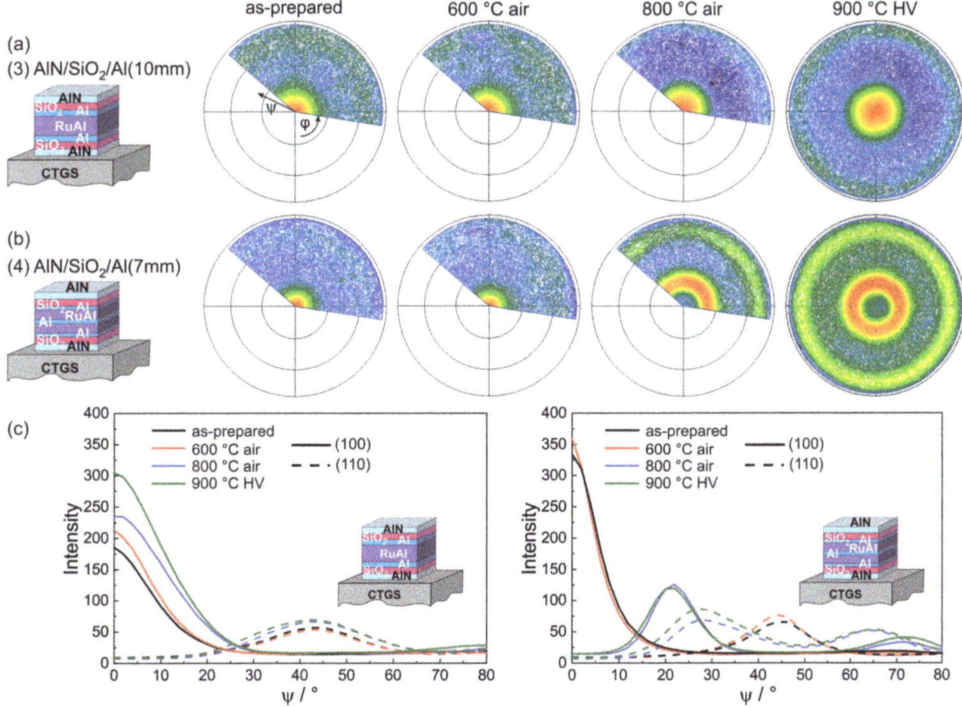

Figure 9. Pole figure measurements of the RuAl (100) pole for (**a**) the system (3)-AlN/SiO_2/Al(10 nm) and (**b**) the system (4)-AlN/SiO_2/Al(7 nm) on CTGS in the as-prepared state and after annealing at 600 and 800 °C in air and 900 °C in high vacuum. (**c**) shows the averaged intensity of the (100) and (110) pole figures dependent on the tilt angle ψ for both systems.

For both layer systems, the pole figures revealed a (100) fiber texture in the as-prepared state and after annealing at 600 °C in air. While this (100) fiber texture was maintained in the case of the (3)-AlN/SiO_2/Al(10 nm) sample for the heat treatments at higher temperatures (Figure 9a), a change in the orientation of the texture appeared in the (4)-AlN/SiO_2/Al(7 nm) system (Figure 9b). The pole figure showed a fiber texture with a tilt of the (100) planes of 22° with respect to the film normal, and the corresponding second ring of the fiber texture appeared at a ψ of 74°. These results explain why no intensity was measured in the Bragg–Brentano XRD: the measured intensity in this case corresponded to the intensity at $\psi = 0$ (no tilt of the sample) in the pole figures.

The images of the pole figures are scaled with respect to their maximum intensity. To allow a comparison between the textures of the samples, in Figure 9c the azimuthal averaged intensity versus the tilt angle ψ is plotted for the (100) and the (110) pole. In case of the (3)-AlN/SiO_2/Al(10 nm) sample, the highest intensity was achieved for the sample annealed at 900 °C in HV, which is in agreement with the XRD results obtained in Bragg–Brentano geometry. The half width at half maximum increased with increasing annealing temperature, which means that the quality of the texture decreased. The plot for the (4)-AlN/SiO_2/Al(7 nm) system illustrates the similar texture for the as-prepared and 600 °C sample. There is a better quality of the texture as compared to the other layer stack. The plots of the samples annealed at the higher temperatures clearly demonstrated the tilt of the lattice planes with almost the same quality of texture.

To conclude, the XRD measurements in Bragg–Brentano geometry and the pole figure measurements revealed the following findings:

- The presence of additional Al layers improves the RuAl phase formation. This became obvious comparing the sample systems (2)-SiO_2/Al and (5)-SiO_2.
- Changing the distribution of Al within the layer system can change the texture of the RuAl phase after annealing at higher temperatures, leading to a tilted orientation of the (100) lattice planes.

3.3. *Film Chemistry*

Figure 10 summarizes the results of the AES depth profile measurements of the samples after the heat treatment at 600 and 800 °C in air and at 900 °C in HV. Since the CTGS substrates are most promising for future applications in high temperature sensors, the AES measurements were conducted for these substrates. The evaluation of the AES was done using standard relative sensitivity factors. Together with the effect of a preferential sputtering, the calculations of the atomic concentration of the RuAl phase pretended a much higher Ru content. Therefore, the AES measurements are only suited to compare the samples and to follow the oxidation processes. Due to charging effects, it was not possible to measure the barrier layers at the interface to the substrate and the insulating CTGS itself.

After annealing at 600 °C in air, all samples showed an intact RuAl layer. The measurement of the (1)-AlN/Al layer (Figure 10a) revealed that the cover layer consisted of Al, N, and O. Already in the as-prepared state, the AlN layer contained some O (about 20 at% according to the AES evaluation). Despite the presence of O, we denote this layer material as AlN. In both films with the SiO_2 cover layer with (Figure 10b) and without (Figure 10e) the additional Al layer, an Al_2O_3 layer formed below the SiO_2. Its thickness was independent of the presence of the additional Al. However, in the case of the film with additional Al, a constant concentration of Ru and Al across the RuAl film thickness was observed, while in case of the film without the additional Al, the Ru content was slightly higher at the surface and was slowly reduced approaching the substrate; correspondingly, the Al content increased.

The films with the combined cover/barrier showed only a small oxidation of the Al at the film surface for the (3)-AlN/SiO_2/Al(10 nm) system and hardly any oxidation for the (4)-AlN/SiO_2/Al(7 nm) sample (Figure 10c,d). For the sample with the intermediate Al layer, an Al maximum was visible in the center of the RuAl film, which shows that the interdiffusion of the Al was not completed at this temperature. Regarding the distribution of the N in the cover layers, a difference was visible between both systems: while for the (4)-AlN/SiO_2/Al(7 nm) sample, a clear N signal was measured in the uppermost layer, the N signal was very low in the (3)-AlN/SiO_2/Al(10 nm) system, and a low N concentration was measured in the SiO_2 layer.

The time needed to sputter the RuAl film was comparable for all films with the additional Al layers. In contrast to this, the sputter time was about 25% longer for the film without additional Al (Figure 10e).

The heat treatment at 800 °C in air led to a strong oxidation of the sample with the AlN cover/barrier layer (Figure 10a), which was already visible in the cross-section image (Figure 3b). Al_2O_3 was formed at the sample surface, followed by a layer consisting of Al, O, and a small content of N. In the lower region of the film Ru, Al, O, and N were measured. According to the XRD measurements, the sample contained RuAl, as well as Ru grains. Therefore, the measured AES signal can be explained by a superposition of RuAl, Ru, and Al_2O_3 grains (Al_2O_3 grew with an amorphous structure and was not visible in the XRD) and a partial diffusion of N from the lower barrier into the overlying film.

In both samples with SiO_2 barrier layers, the thickness of the Al_2O_3 layer below the SiO_2 increased and was the same for both systems. However, the Ru concentration in the RuAl layer was slightly higher in the film without the additional Al layers (Figure 10e) and slightly decreased approaching the substrate.

The AES measurements of both samples with the combined barriers were similar despite the different initial distribution of the additional Al. An Al_2O_3 layer was now present at the sample surface, followed by a SiO_2 layer and another Al_2O_3 layer, which contained some N. In contrast to the film annealed at 600 °C in air, the annealing at 800 °C was sufficient to realize a homogeneous interdiffusion of the Al into the RuAl film.

These results demonstrated that even for the combined barrier layers, a partial diffusion of O into the RuAl layer took place during the annealing at 800 °C in air. However, the time that was necessary to sputter the RuAl layer was only slightly reduced, indicating almost the same thickness of the RuAl layer (small deviations in sputtering time can also arise from the sputtering process itself).

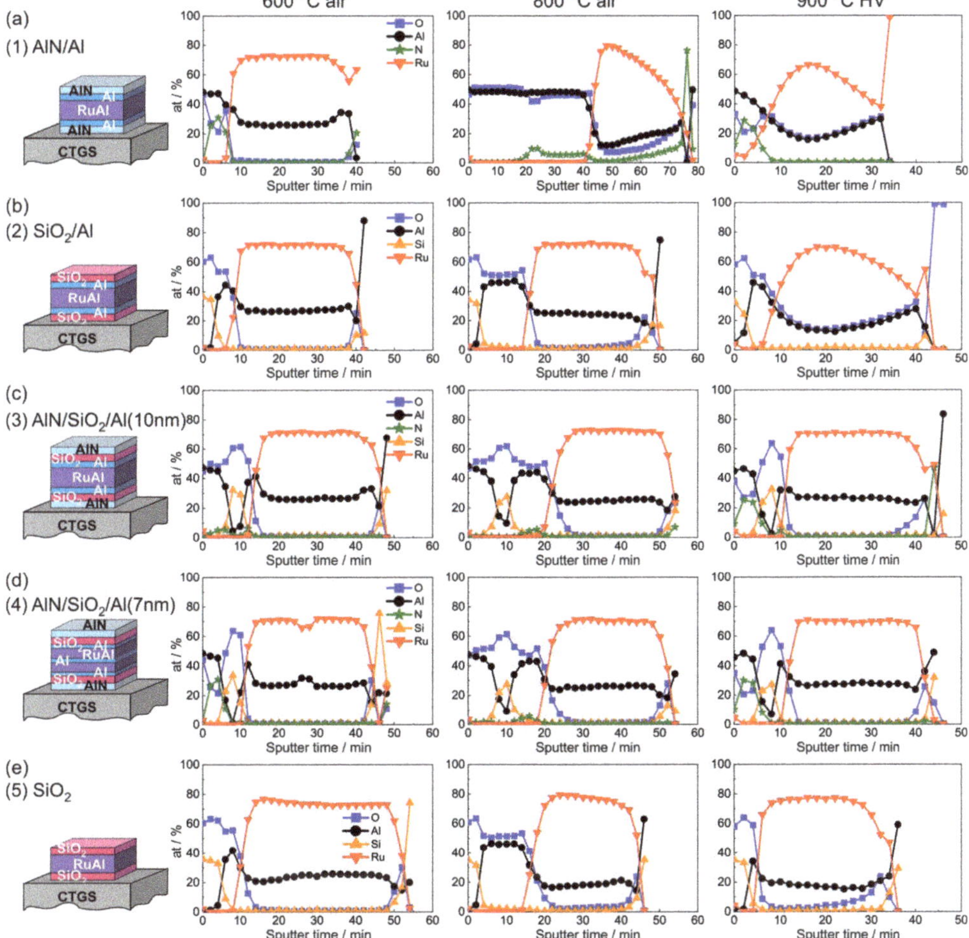

Figure 10. Results of the AES measurements for the samples with the different cover and barrier layers on CTGS substrates after annealing at 600 and 800 °C in air and 900 °C in HV: (**a**) (1)-AlN/Al, (**b**) (2)-SiO_2/Al, (**c**) (3)-AlN/SiO_2/Al(10 nm), (**d**) (4)-AlN/SiO_2/Al(7 nm), and (**e**) (5)-SiO_2.

After annealing at 900 °C in high vacuum, the film with the AlN or SiO_2 cover/barrier with the additional Al layers showed the presence of oxygen across the whole film. Already in the cross-section images, these samples showed a clear morphology change and the formation of pores. The sample (5)-SiO_2 without the additional Al was significantly less oxidized as the corresponding sample with Al. This film showed only a thin Al_2O_3 layer below the SiO_2 cover layer and just a slight increase in O approaching the interface to the substrate (Figure 10e). This increase in O may arise from a simultaneous sputtering of the RuAl film and the O rich substrate because of the presence of pores in the film.

Both films with the combined cover/barrier layer did not contain O in the RuAl layer, and only a very thin Al_2O_3 layer formed below the cover layers. In addition, it was obvious that the barrier layers at the surface of the film were still visible as AlN followed by SiO_2, which was in contrast to the films annealed at 800 °C in air where hardly any N was detected and a three layer system Al_2O_3–SiO_2–Al_2O_3 was formed. The systems with the combined barrier layers showed a difference at the interface to the lower barrier layers. In the case of the (3)-AlN/SiO_2/Al(10 nm) sample, there was a stronger increase in O in the RuAl film, indicating a partial oxidation of Al to Al_2O_3, which was more pronounced as compared to the (4)-AlN/SiO_2/Al(7 nm) layer stack.

3.4. TEM Studies on Layer Stacks (3) and (4)

In the cross-section images and AES measurements, hardly any differences between the layer stacks (3)-AlN/SiO_2/Al(10 nm) and (4)-AlN/SiO_2/Al(7 nm) were visible. To analyze more in detail the local morphology and chemistry, STEM measurements in combination with EDX were performed. These measurements also allowed analyzing the barrier layers between the film and the CTGS substrate, which cannot not be measured in AES due to the charging effects.

Figure 11 shows the images for the samples after annealing at 800 °C in air. These results revealed that there were differences between both sample systems. The images with predominant chemical contrast (Figure 11a,b) showed that the (4)-AlN/SiO_2/Al(7 nm) film had a much more homogeneous structure with significantly less Ru rich grains.

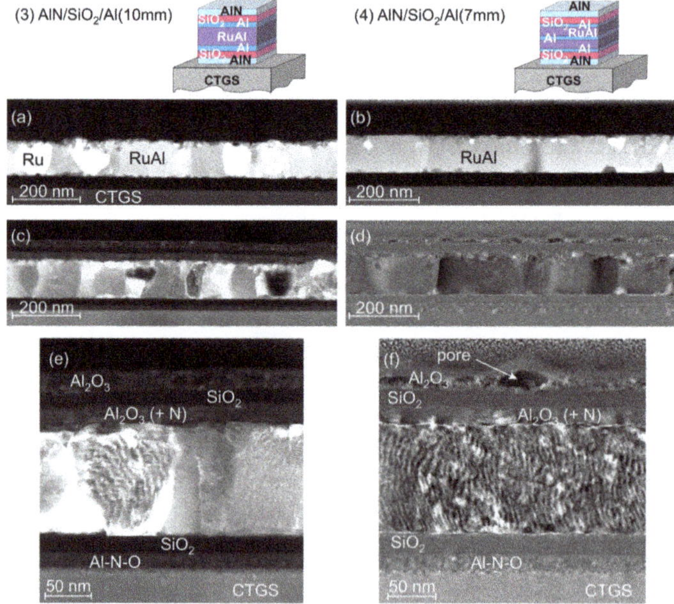

Figure 11. STEM images of layer stacks (3)-AlN/SiO_2/Al(10 nm) and (4)-AlN/SiO_2/Al(7 nm) on CTGS after annealing at 800 °C in air: (**a**,**b**) with predominant chemical contrast and (**c**–**f**) with predominant orientation contrast.

In the images with predominant orientation contrast, it could be seen that in the (3)-AlN/SiO_2/Al(10 nm) sample, the grains were mostly extended across the thickness of the layer and had an in-plane grain size between a few tens and about 100 nm (Figure 11c). In the (4)-AlN/SiO_2/Al(7 nm) sample, all grains were extended across the whole film thickness and had an in-plane grain size up to several 100 nm (Figure 11d). The images with higher magnification (Figure 11e,f) showed that the structure of the cover layers was also slightly different. In the (3)-AlN/SiO_2/Al(10 nm) sample, a three layer structure consisting of a Al_2O_3 layer on top, followed by a SiO_2 layer, and finally, another Al_2O_3 layer, which contained a small amount of N, was visible, which was already derived from the AES measurements. Although this three layer structure was likewise measured in AES for the (4)-AlN/SiO_2/Al(7 nm) layer stack, the STEM measurements revealed that the top Al_2O_3 layer had a higher roughness and contained pores.

At the interface to the substrate, the initial order of the barrier layers (the AlN (AlNO) layer on top of the CTGS followed by SiO_2) was kept for both samples, and there were hardly any inhomogeneities in the surface region of the CTGS.

The STEM images of the samples after annealing at 900 °C in HV are summarized in Figure 12. For both kinds of samples, the images with predominant chemical contrast showed a more homogeneous structure as compared to the samples annealed at 800 °C in air. However, for the (3)-AlN/SiO_2/Al(10 nm) layer stack, inhomogeneities were visible in the RuAl film at the lower interface (Figure 12a). These darker regions represent Al_2O_3 grains. For this kind of sample, inhomogeneities were present in the upper region (about 50 nm) of the CTGS (Figure 12a,e).

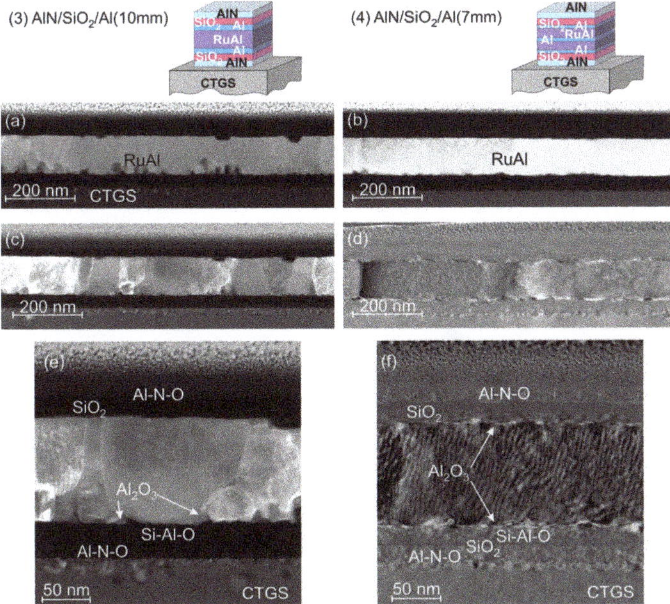

Figure 12. STEM images of layer stacks (3)-AlN/SiO_2/Al(10 nm) and (4)-AlN/SiO_2/Al(7 nm) on CTGS after annealing at 900 °C in HV: (**a**,**b**) with predominant chemical contrast; (**c**–**f**) with predominant orientation contrast.

In the case of the (4)-AlN/SiO_2/Al(7 nm) sample, a very thin layer (a few nm) of Al_2O_3 was formed at the interface between the RuAl and upper and lower SiO_2 layer (Figure 12f). There were significantly less inhomogeneities in the upper region of the CTGS as compared to the other system. The stronger oxidation at the interface to the substrate of the (3)-AlN/SiO_2/Al(10 nm) film as compared to the (4)-AlN/SiO_2/Al(7 nm) layer stack was already derived from the AES measurements. This difference in oxidation is in agreement with the different degrees of degradation of the CTGS at the interface to the film.

Regarding the film morphology, it was again obvious that the grain structure was more inhomogeneous in the (3)-AlN/SiO_2/Al(10 nm) system, and there were grains that were not extended across the film thickness. The grain size distribution ranged from a few 10 nm up to 200 nm. Again, the grains in the (4)-AlN/SiO_2/Al(7 nm) sample had a much larger lateral expansion ($\approx$ 100–500 nm).

The EDX analysis of the barrier and cover layers revealed that during the HV annealing, the upper barriers remained unchanged, so that there was still the AlN-SiO_2 bilayer. However, in the lower barrier, there was an interdiffusion of Al into the SiO_2. In the case of the (3)-AlN/SiO_2/Al(10 nm) system, Al was distributed homogeneously in the SiO_2 layer, while in the (4)-AlN/SiO_2/Al(7 nm) sample, Al was only found in the upper region of the SiO_2 layer, and the lower region was free of Al.

In summary, the STEM investigations revealed differences in the film morphology between the (3)-AlN/SiO_2/Al(10 nm) and (4)-AlN/SiO_2/Al(7 nm) sample systems. The latter was more homogeneous with larger grains and less oxidation effects at the boundaries of the RuAl film.

3.5. Electrical Properties

For the application as SAW electrodes, the electrical resistivity R_{el} of the metallization has to be known to optimize the device layout. The sheet resistance of the films on the different substrates was determined by the van der Pauw method, and then, the electrical resistivity was calculated using the film thickness. The thickness of the RuAl and Al layers, however, was only known for the as-deposited state, and the effective thickness of the conductive layer after the heat treatments was reduced due to the oxidation effects. Despite this, the initial thickness (130 nm for the film systems (1)–(4) and 110 nm for the film system (5)) was used to calculate the resistivity, which meant that the actual resistivity of the annealed films was slightly lower. Figure 13 summarizes the values of the electrical resistivity for the films on the different substrates.

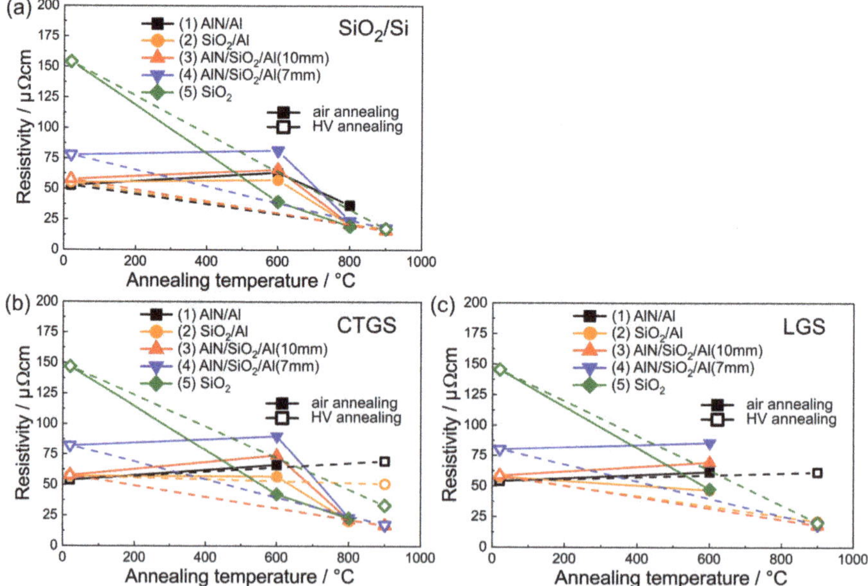

Figure 13. Electrical resistivity of the different film systems on (**a**) SiO_2/Si, (**b**) CTGS, and (**c**) LGS substrates in the as-prepared state and after annealing. For SiO_2/Si and CTGS, the values for the air annealed samples are shown up to 800 °C and for LGS up to 600 °C. The full symbols represent the annealing in air, the open symbols in HV. The straight lines visualize samples of the same set.

In the as-prepared state, R_{el} was the lowest for the films with the two times 10 nm additional Al layers and hardly depended on the applied substrate and barrier/cover layers (53–59 μΩcm range for the different substrates and barrier layers, deviation less than 15 %). The (4)-AlN/SiO_2/Al(7 nm) films had a slightly higher R_{el} (78–82 μΩcm), and the film without additional Al showed the highest R_{el} of 145–154 μΩcm. These results proved that in the as-prepared state, the electrical resistivity was dominated by the Al layers with their excellent electrical properties. Two thicker Al films led to a lower resistivity than three Al layers with the same total thickness due to an increased scattering of the electrons because of the reduced film thickness and at the additional two interfaces between the Al and the RuAl.

After annealing at 900 °C in HV, all films on the SiO_2/Si substrates had almost the same R_{el} (16–18 μΩcm), which was much lower as compared to the as-prepared state (13a). The decrease of R_{el} is explained by the grain growth and reduction of defects in the films during the heat treatment. It is concluded that all barrier layers were equally able to prevent the oxidation of the films in HV. After annealing, R_{el} of the films with additional Al was not lower than that of the film without additional Al.

For the films on the CTGS substrate after annealing at 900 °C in HV, R_{el} strongly depended on the applied barrier/cover layers (Figure 13b). As already observed in the cross-section images, only the two films with the combined cover/barrier layer were not oxidized. These two films reached the same low R_{el} (17 μΩcm) as compared to the films on SiO_2/Si. Both films with the SiO_2 barrier behaved differently. The addition of Al layers led to a higher R_{el} (33 μΩcm without, 50 μΩcm with Al layers). A stronger oxidation was also observed in the cross-section images. The highest R_{el} was measured for the film with AlN barrier layers (70 μΩcm).

In the case of the LGS substrate after annealing at 900 °C in HV, for all films except those covered with AlN, which had a higher value of 61 μΩcm, a low resistivity comparable to the films on SiO_2/Si was measured (18–21 μΩcm) (Figure 13c).

For all samples with additional Al layers, the annealing in air at 600 °C led to an increase of R_{el}. There were only slight differences between the different substrates. The highest value was reached for the (4)-AlN/SiO_2/Al(7 nm) sample system (69–76 μΩcm), followed by (3)-AlN/SiO_2/Al(10 nm) (56–63 μΩcm), (1)-AlN (52-56 μΩcm), and (2)-SiO_2 (40-48 μΩcm). The lowest value was measured for the SiO_2 covered film without additional Al (40-47 μΩcm). The lowest value for this sample can be explained by the formation of $RuAl_2$ in the other samples at the interfaces between the Al and the RuAl, which is a semiconductor with a low electrical resistivity at RT [15].

Since after annealing at 800 °C in air, the films on LGS were mostly destroyed, the electrical resistivity after this heat treatment was only measured for the films on SiO_2/Si and CTGS. Except for the (1)-AlN/Al sample system, all other films had a low electrical resistivity (19-24 μΩcm), which was only slightly higher than that measured for the HV annealed films. It was not possible to measure an electrical resistivity for the (1)-AlN/Al sample system on CTGS, and on SiO_2, a higher value (36 μΩcm) than for the other samples was determined. For the (1)-AlN/Al sample, the cross-section images, AES, and XRD measurements indicated that the film had a thick Al_2O_3 layer on top and the lower region of the film consisted of Ru and RuAl grains, which were isolated by Al_2O_3 so that the films were not conductive despite the presence of metallic grains.

After annealing at 900 °C, all films were more or less destroyed, so that no electrical measurements were performed.

4. Conclusions

This paper reports on the further significant improvement of high-temperature stable RuAl thin films on the piezoelectric CTGS and LGS in comparison to reference thermally oxidized Si substrates. In former work for RuAl films on SiO_2/Si, a high temperature stability up to 800 °C in air for 10 h was reached if a 20 nm thick SiO_2 cover layer was applied. Now, the focus was on stabilizing the films against oxidation originating from the CTGS or LGS substrate.

In the case of the air annealed samples, the results showed that the films on CTGS were more stable than the films on LGS. On the other hand, in the case of annealing in HV, the films on LGS were less oxidized than the films on CTGS. In contrast to this, the results of our former work on RuAl thin films directly deposited on the substrates after annealing at 800 °C in HV showed a stronger oxidation for the films on LGS [10]. The opposite behavior observed in this work for the annealings at 900 °C in HV might originate from a different chemical reactivity between the substrates and the Al of the RuAl (set up in the former work) as compared to that between the substrate and the respective barrier material presented in this paper. Regarding the activation energy of the oxygen vacancy diffusion, it is known that in LGS, it is lower (0.8–1 eV) as compared to CTGS (1.3 eV), as, e.g., described by Zhang et al. [16]. However, a detailed explanation of the observed behavior cannot be given here, and further research by experts in the field of the high temperature stable crystals is needed.

A comparison of the cover layers revealed that a single AlN layer was least suited to protect RuAl thin films against oxidation. On SiO_2/Si substrates, all barrier layers led to stable RuAl films in the case of HV annealing up to 900 °C, and except the film with AlN barrier layers, all films were stable during annealing in air up to 800 °C for at least 10 h. On CTGS, also except the film with AlN barrier layers, all other films were stable during annealing in air up to 800 °C. However, during annealing in HV, only the films with the combined barrier layers were not oxidized. In this case, the combination of both barrier layers was required to protect the RuAl film from oxygen, which diffused out of the substrate. The films on LGS were strongly oxidized already during annealing in air at 800 °C. During annealing in HV at 900 °C, all films except the AlN covered samples were stable.

In summary, we conclude that a combined barrier layer consisting of 20 nm AlN and 20 nm SiO_2 is able to prevent the oxidation of the RuAl film on CTGS up to 800 °C in air and 900 °C in HV. On LGS, stable films are realized up to 600 °C in air and 900 °C in HV.

The addition of 10 nm Al layers on the top and bottom of the RuAl film results in an improved phase formation as was seen in the XRD measurements of the samples with the SiO_2 cover with and without the addition of Al layers.

Comparing the two film architectures (10 nm Al)–(110 nm RuAl)–(10 nm Al) and (7 nm Al)–(55 nm RuAl)–(7 nm Al)–(55 nm RuAl)–(7 nm Al) revealed that after annealing in the latter films, the grains were larger and a texture with a tilt of the (100) axis of about 22° with respect to the film normal developed. In addition, it was found that at the interface to the substrate, a stronger chemical reaction with the substrate material took place for the (3)-AlN/SiO_2/Al(10 nm) system, which was seen in the STEM images.

These results represent an important progress in the realization of a material system for high temperature SAW applications. The next step is the investigation of structured RuAl electrodes and the development of SAW devices based on the RuAl films with the combined barrier layer system on CTGS, which is the subject of our ongoing work.

Funding: The work was supported by the German Federal Ministry of Education and Research (BMBF) under grant InnoProfile-Transfer 03IPT610Y and Federal Ministry of Economic Affairs and Energy (BMWI) under grant 03ET1589 A.

Acknowledgments: The author gratefully acknowledges Erik Brachmann for the sample preparation, Thomas Wiek, as well as Dina Bieberstein for FIB cuts and TEM lamella preparation and Steffi Kaschube for AES measurements. Thomas Wiek is also acknowledged for the measurements of the electrical resistivity.

Conflicts of Interest: The author declares no conflict of interest. The funders had no role in the design of the study; in the collection, analyses, or interpretation of data; in the writing of the manuscript; nor in the decision to publish the results.

Abbreviations

The following abbreviations are used in this manuscript:

AES	Auger electron spectroscopy
as-dep.	as-deposited
EDX	energy dispersive X-ray spectroscopy
FIB	focused ion beam
HV	high vacuum
IDT	interdigital transducer
RT	room temperature
SAW	surface acoustic wave
SEM	scanning electron microscopy
TEM	transmission electron microscopy
STEM	scanning electron transmission microscopy
vdP	van der Pauw
XRD	X-ray diffraction

References

1. Thiele, J.A.; Pereira da Cunha, M. Platinum and palladium high-temperature transducers on langasite. *IEEE Trans. Ultrason. Ferroelect. Freq. Contr.* **2005**, *52*, 545–549. [CrossRef] [PubMed]
2. Pereira da Cunha, M.; Moonlight, T.; Lad, R.; Bernhardt, G.; Frankel, D.J. Enabling Very High Temperature Acoustic Wave Devices for Sensor & Frequency Control Applications. *2007 IEEE Ultrasonics Symp. Proc.* **2007**, *P4L-1*, 2107–2110.
3. Aubert, T.; Elmazira, O.; Assouar, B.; Bouvot, L.; Hehn, M.; Weber, S.; Oudich, M.; Geneve, D. Behavior of platinum/tantalum as interdigital transducers for SAW devices in high-temperature environments. *IEEE Trans. Ultrason. Ferroelect. Freq. Contr.* **2011**, *58*, 603–610. [CrossRef] [PubMed]
4. Taguett, A.; Aubert, T.; Elmazria, O.; Bartoli, F.; Lomello, M.; Hehn, M.; Ghanbaja, J.; Boulet, P.; Mangin, S.; Xu, Y. Comparison between Ir, $Ir_{0.85}Rh_{0.15}$ and $Ir_{0.7}Rh_{0.3}$ thin films as electrodes for surface acoustic waves applications above 800 °C in air atmosphere. *Sensor. Actuat. A-Phys.* **2017**, *266*, 211–218. [CrossRef]
5. Moulzolf, S.C.; Frankel, D.J.; Pereira da Cunha, M.; Lad, R. High temperature stability of electrically conductive Pt-Rh/ZrO_2 and Pt-Rh/HfO_2 nanocomposite thin film electrodes. *Microsyst. Technol.* **2014**, *20*, 523–531. [CrossRef]
6. Menzel, S.B.; Seifert, M.; Priyadarshi, A.; Rane, G.K.; Park, E.; Oswald, S.; Gemming, T. Mo-La_2O_3 Multilayer Metallization Systems for High Temperature Surface Acoustic Wave Sensor Devices. *Materials* **2019**, *12*, 2651. [CrossRef] [PubMed]
7. Rane, G.K.; Seifert, M.; Menzel, S.; Gemming, T.; Eckert, J. Tungsten as a Chemically-Stable Electrode Material on Ga-Containing Piezoelectric Substrates Langasite and Catangasite for High-Temperature SAW Devices. *Materials* **2016**, *9*, 101. [CrossRef] [PubMed]
8. Okamoto, H. Al-Ru (aluminum-ruthenium). *J. Phase Equilib.* **1997**, *18*, 105. [CrossRef]
9. Mücklich, F.; Ilić, N.; Woll, K. RuAl and its alloys, Part II: Mechanical properties, environmental resistance and applications. *Intermetallics* **2008**, *16*, 593–608. [CrossRef]
10. Seifert, M.; Menzel, S.; Rane, G.K.; Hoffmann, M.; Gemming, T. RuAl thin films on high-temperature piezoelectric substrates. *Mater. Res. Express* **2015**, *2*, 085001. [CrossRef]
11. Seifert, M.; Menzel, S.B.; Rane, G.K.; Gemming, T. TEM studies on the changes of the composition in LGS and CTGS substrates covered with a RuAl metallization and on the phase formation within the RuAl film after heat treatment at 600 and 800 °C. *J. Alloys Compd.* **2016**, *664*, 510–517. [CrossRef]
12. Seifert, M.; Menzel, S.B.; Rane, G.K.; Gemming, T. The influence of barrier layers (SiO_2, Al_2O_3, W) on the phase formation and stability of RuAl thin films on LGS and CTGS substrates for surface acoustic wave technology. *J. Alloys Compd.* **2016**, *688*, 228–240. [CrossRef]
13. Seifert, M.; Rane, G.K.; Menzel, S.; Gemming, T. Improving the oxidation resistance of RuAl thin films with Al_2O_3 or SiO_2 cover layers. *J. Alloys Compd.* **2019**, *776*, 819–825. [CrossRef]
14. Seifert, M.; Rane, G.K.; Oswald, S.; Menzel, S.; Gemming, T. The Influence of the Composition of $Ru_{100-x}Al_x$ ($x = 50, 55, 60, 67$) Thin Films on Their Thermal Stability. *Materials* **2017**, *10*, 277. [CrossRef] [PubMed]

15. Mandrus, D.; Keppens, V.; Sales, B.C.; Sarrao, J.L. Unusual transport and large diamagnetism in the intermetallic semiconductor $RuAl_2$. *Phys. Rev. B.* **1998**, *58*, 3712–3716. [CrossRef]
16. Zhang, S.; Zheng, Y.; Kong, H.; Xin, J.; Frantz, E.; Shrout, T.R. Characterization of high temperature piezoelectric crystals with an ordered langasite structure. *J. Appl. Phys.* **2009**, *105*, 114107. [CrossRef]

Article

Phase Formation and High-Temperature Stability of Very Thin Co-Sputtered Ti-Al and Multilayered Ti/Al Films on Thermally Oxidized Si Substrates

Marietta Seifert *, Eric Lattner, Siegfried B. Menzel, Steffen Oswald and Thomas Gemming

Leibniz IFW Dresden, Helmholtzstraße 20, 01069 Dresden, Germany; e.lattner@ifw-dresden.de (E.L.); s.menzel@ifw-dresden.de (S.B.M.); s.oswald@ifw-dresden.de (S.O.); t.gemming@ifw-dresden.de (T.G.)

* Correspondence: marietta.seifert@ifw-dresden.de; Tel.: +49-351-4659-639

Received: 24 March 2020; Accepted: 23 April 2020; Published: 27 April 2020

Abstract: Ti-Al thin films with a thickness of 200 nm were prepared either by co-sputtering from elemental Ti and Al targets or as Ti/Al multilayers with 10 and 20 nm individual layer thickness on thermally oxidized Si substrates. Some of the films were covered with a 20-nm-thick SiO_2 layer, which was used as an oxidation protection against the ambient atmosphere. The films were annealed at up to 800 °C in high vacuum for 10 h, and the phase formation as well as the film architecture was analyzed by X-ray diffraction, cross section, and transmission electron microscopy, as well as Auger electron and X-ray photoelectron spectroscopy. The results reveal that the co-sputtered films remained amorphous after annealing at 600 °C independent on the presence of the SiO_2 cover layer. In contrast to this, the γ-TiAl phase was formed in the multilayer films at this temperature. After annealing at 800 °C, all films were degraded completely despite the presence of the cover layer. In addition, a strong chemical reaction between the Ti and SiO_2 of the cover layer and the substrate took place, resulting in the formation of Ti silicide. In the multilayer samples, this reaction already started at 600 °C.

Keywords: TiAl; thin films; surface acoustic waves; high-temperature stability; phase formation

---

1. Introduction

Ti-Al-based materials are widely used in industry, especially for aerospace applications in aircraft engines [1] or gas turbines [2] due to their high thermal stability in combination with a low density. Therefore, much research has been performed on Ti-Al based bulk material and on Ti-Al coatings. There is also literature on Ti-Al thin films; however, in general, this refers to films with a thickness of a few or tens of µm [3–5]. Besides the application as bulk material or µm-thick coating, Ti-Al films with a thickness of a few 100 nm are interesting as a metallization in surface acoustic wave (SAW) devices operating at elevated temperatures (above 400 °C).

Current research on high temperature stable metallizations for SAW devices mainly focuses on Pt based materials due to their noble character [6,7], or among others, Ir or Ir-based alloys [8,9]. Since Pt has a strong tendency to agglomerate, oxide dispersion hardening (ODS) was applied to stabilize the films [10]. Another alternative are ODS stabilized Mo films [11], because of the very high melting temperature of Mo, and therefore expected low creep at the operation temperature of the devices (600–800 °C). In our group, during the last years, many efforts were made to improve the high temperature stability of RuAl thin films. RuAl is a promising material due to its high melting temperature of 2050 °C [12] and its strong oxidation and corrosion resistance [13]. Recently, we realized RuAl thin films that were stable up to 800 °C in air and 900 °C in HV for at least 10 h [14].

The mass of the metal used for electrodes determines the velocity of the propagating SAW, and, thus, the coupling coefficient. Furthermore, heavy metals are especially used for reflective electrodes as, for instance, in resonator structures in the filter segment of SAW technology. In contrast, in other applications of SAW technology, e.g., in sensors, SAW-driven microfluidic actuators or SAW-based tags, where delay lines or reflective delay lines are often used, a more lightweight metal is often preferred to reduce the reflection of the acoustic energy at each of the finger electrodes.

Therefore, a wide range of mass loading provided by the metals used for the finger electrodes is essential to enhance the range of application in SAW technology. The above-listed materials, on which current research for high temperature metallizations focuses, all possess a high density (e.g., Pt: 21.45 g/cm^3; Mo: 10.28 g/cm^3; and RuAl: $\approx 8\ g/cm^3$). In contrast to this, Ti-Al has a much lower density (3.9–4.3 g/cm^3 [15]), which allows different design variations and with this additional applications of the devices.

For the very thin Ti-Al films (few 100 nm), the realization of the high temperature stability is more challenging than for the thick coatings or bulk materials. In the latter, during heat treatment at elevated temperatures, Al is oxidized and a several µm-thick protective oxide scale is formed on top of the sample. As a consequence, a thin Ti-Al film is destroyed completely under these conditions. Therefore, an additional barrier layer on top of the Ti-Al film is required to prevent oxidation. In this work, we applied SiO_2 as a cover layer material, since it successfully protected RuAl thin films from oxidation up to 800 °C in air for at least 10 h [14,16]. A cover layer of Al_2O_3 was tested in the former work for RuAl thin films as well; however, it led to a stronger oxidation of the RuAl film as compared to SiO_2 [16].

This paper presents a study on very thin (200 nm) Ti-Al films with focus on the high-temperature stability as well as on the γ-TiAl phase formation. Co-sputtered Ti-Al films with and without SiO_2 protection layer were compared with multilayered Ti/Al films with SiO_2 cover.

2. Materials and Methods

Ti-Al thin films with a thickness of 200 nm were prepared on Si substrates with 1 µm of thermally grown SiO_2 on top by two different routines using DC sputtering. On the one hand, co-sputtering from elemental Ti and Al targets (100 mm target, 3.5×10^{-3} mbar, 500 W for Ti and 280 W for Al) was applied to deposit Ti-Al alloy films with a composition of 50:50. On the other hand, multilayer systems were prepared by subsequent sputtering of Ti and Al layers (3.2×10^{-3} mbar, 500 W). The multilayer (ML) stacks were started with a Ti layer so that the uppermost layer consisted of Al. Two ML systems with a thickness of the individual layers of 10 or 20 nm were deposited. Since the total thickness of the film was kept constant at 200 nm, the $Ti_{10nm}Al_{10nm}$ bilayer was repeated 10 times and the $Ti_{20nm}Al_{20nm}$ 5 times. A 20-nm-thick SiO_2 cover layer (RF sputtering, 200 mm target, 3.5×10^{-3} mbar, 2000 W) was added on top of some of the samples as an oxidation barrier. The SiO_2 films were deposited by sputtering from a SiO_2 target at a deposition temperature of 180 °C with a sputtering gas of Ar and O_2 (ratio of 6:1). The depositions of the metallization and the cover layers were carried out in different chambers of the same cluster tool (CREAMET 350-CL 6, CREAVAC-Creative Vakuumbeschichtung GmbH, Dresden, Germany) without interruption of the high vacuum.

The films were annealed in high vacuum (HV, pressure lower than 10^{-5} mbar) for 10 h at 400, 600, and 800 °C. X-ray diffraction in Bragg Brentano geometry (XRD, Philips X'pert, Co Kα) was applied to analyze the phase formation. To reveal the full texture information and to confirm the formation of the γ-TiAl phase, pole figure measurements of the (101) and (110) lattice planes were performed (Philips X'Pert MRD, Cu Kα). For the γ-TiAl phase, the diffraction angle 2θ is 38.7° for the (101) and 45.3° for the (110) pole. To realize an optimum quality of these analyses, prior to each measurement, overview scans were performed with this theoretical value to identify the respective pole. Then, for each sample, the 2θ value was measured at this position, since it might slightly deviate from the theoretical value due to a small change of the composition or due to stresses in the thin film during growth and subsequent annealing. This optimized 2θ value was used for the texture measurement. The theoretical peak position was used if it was not possible to identify a pole with the overview scan.

Cross sections of the samples were prepared by the focussed ion beam technique (FIB) and imaged by scanning electron microscopy (SEM) in the same device (Zeiss 1540 XB Cross Beam, Carl Zeiss Microscopy GmbH, Oberkochen, Germany).

The distribution of the elements across the film thickness was analyzed by Auger electron spectroscopy (AES, JEOL JAMP-9500 Field Emission Auger Microprobe). Ar sputtering (energy of Ar ions: 1 keV, current of 0.7×10^{-6} A) was done for a time span between 30 and 120 s, and, subsequently, the AES spectra were recorded. These processes were repeated until the substrate was reached. The evaluation of the spectra was carried out using standard single element sensitivity factors of the PHI-Multipak software [17]. The analysis of the AES peak shape and position additionally allows deriving conclusions on the oxidation state of the metal elements.

For selected samples, X-ray photoelectron spectroscopy (XPS, PHI 5600 CI-System, Physical Electronics) was performed. Non-monochromatic MgKα X-rays (400 W) were used to excite the sample and a hemispherical electron analyzer working at a pass energy of 29 eV was applied to record the spectra. XPS allows a more detailed analysis of the oxidation state of the elements. In this case, depth profiles were also obtained by alternately sputtering with Ar ions (3.5 keV) and measuring the spectra.

Annular dark field scanning transmission electron microscopy (ADF-STEM, Technai F30, FEI company, Hillsboro, OR, USA) was used to image the local morphology. The local composition was determined by energy dispersive X-ray spectroscopy in the TEM (Octane T Optima, EDAX Company, Mahwah, NJ, USA). The TEM lamella were prepared using the FIB technique.

3. Results and Discussion

3.1. Phase Formation

Figure 1 presents the results of the XRD measurements of the different samples in the as-prepared state and after annealing at 400, 600, and 800 °C in HV. It can be seen that for both co-sputtered films no peaks were visible for all sample states (Figure 1a,b). In contrast to this, clear XRD peaks were present for the ML films. In the as-prepared state, a superposition of the Ti (002, theoretical peak position: $2\theta = 44.96°$) and Al (111, theoretical peak position: $2\theta = 45°$) appeared (Figure 1c,d). The measured peaks were slightly shifted with respect to the theoretical position, which might be due to stresses in the thin films. Annealing at 400 °C led to a decrease of the peak intensities. However, there was still a superposition of two XRD peaks.

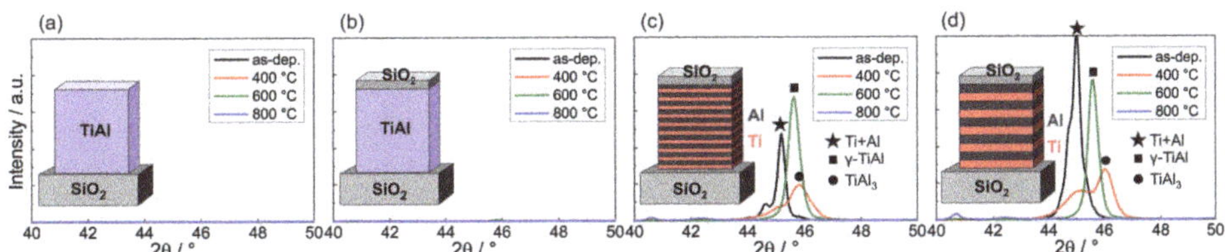

Figure 1. XRD measurements on: (**a**) a co-sputtered Ti-Al film; (**b**) a co-sputtered Ti-Al film with SiO_2 cover layer; (**c**) a Ti/Al ML film with 10 nm individual layer thickness; and (**d**) a Ti/Al ML film with 20 nm individual layer thickness, both with SiO_2 cover layer.

The phase formation in µm thick Ti/Al ML films was analyzed by various authors. For very thin individual layers (2 nm), a transition from (Ti) + (Al) $\rightarrow$ disordered TiAl + (Ti) $\rightarrow$ γ-TiAl + α2-Ti_2Al was described by Ramos et al. [18]. In contrast to this, for thicker individual layers (100 nm), the authors described the transition from Al and Ti to γ-TiAl via a $TiAl_3$ intermediate phase. The occurrence of $TiAl_3$ in the phase sequence during annealing of Ti/Al ML samples was also described by Illekova et al. for an individual layer thickness of 20 nm or more [5]. For Al rich Ti-Al ML samples (substrate/40 nm Ti/180 nm Al/40 nm Ti/180 nm Al), the formation of $TiAl_3$ was also observed after a heat treatment at 450 °C [19,20]. The two peaks measured in our work after annealing at 400 °C are therefore ascribed to Ti (002, now measured at a slightly higher 2θ of 45.1°) and $TiAl_3$ (103 or 112, theoretical 2θ = 45.82°, measured at 46.0°).

After annealing at 600 °C, one single peak was measured for both ML samples, which can be attributed to the γ-TiAl (101) peak (theoretical position: 2θ = 45.27°, measured at 45.6°). After annealing at 800 °C, the TiAl peak disappeared.

Pole figure measurements were performed to confirm the formation of the γ-TiAl phase in the ML films. In addition, these measurements allowed to exclude that in the co-sputtered films the γ-TiAl phase has formed with a tilted texture, which cannot be detected by XRD in Bragg Brentano geometry. Figure 2 summarizes the pole figures of the (110) and (101) γ-TiAl poles of the four different samples annealed at 600 °C for 10 h. Since the pole figures are scaled to their respective maximum, in addition, the azimuthally averaged intensity versus the tilt angle Ψ is presented to allow a comparison between the different samples. It can be seen that both co-sputtered samples with or without SiO_2 cover layer did not show any phase formation (see Figure 2a,b). The pole, which was visible in the TiAl (110) pole figure at Ψ = 54° was caused by the Si (220) lattice planes. Their 2θ value of 47.3° (Cu Kα) is close to that of the TiAl (110), so that it is also measured. Besides this, no significant intensity was detected.

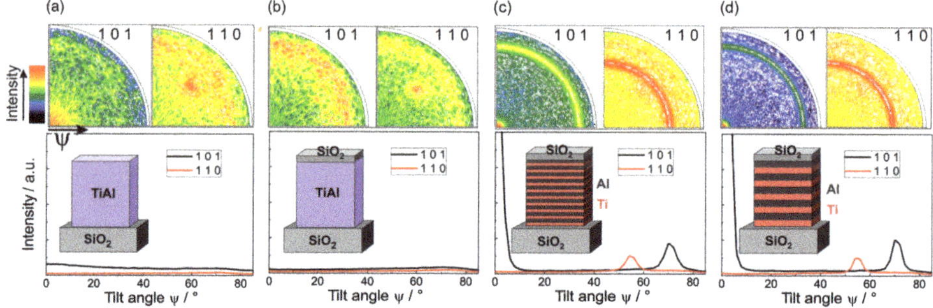

Figure 2. Results of the texture measurements of: (**a**) a co-sputtered Ti-Al film; (**b**) a co-sputtered Ti-Al film with SiO_2 cover layer; (**c**) a ML Ti/Al film with 10 nm individual layer thickness; and (**d**) a ML Ti/Al film with 20 nm individual layer thickness, both with SiO_2 cover layer, after annealing at 600 °C. The graphs represent the azimuthally averaged intensity versus the tilt angle Ψ.

In contrast to this, both ML samples showed a strong intensity in the center of the (101) pole figure and a clear ring at Ψ = 72° (see Figure 2c,d). Corresponding to this, the TiAl (110) pole figure contained a ring at Ψ = 54°. These two measurements confirmed the formation of the γ-TiAl phase.

It is known from the literature that Ti-Al films co-sputtered at room temperature are in general an amorphous mixture of Ti and Al [4,21,22]. The formation of crystalline TiAl phases in amorphous Ti-Al or multilayer Ti/Al thick films in the range of 1.6 µm to 150 µm was investigated by several authors. Illekova et al. studied the activation energy of the phase formation in Ti/Al multilayers with various thicknesses of the individual layer (from 4 to 1000 nm). For a thickness of the individual layer of 20 nm, they derived a value of 169 ± 3 kJ/mol [5]. They also described that for very thin individual layers (4 nm) at first the material becomes amorphous, so that the start of the crystallization is delayed. Senkov et al. analyzed amorphous Ti-Al films and determined an activation energy for the transition from the amorphous state to the first crystalline phase of 315 ± 5 kJ/mol [23]. Although these values were extracted for films with a much larger thickness, the higher activation energy for the formation of the crystalline TiAl phase in amorphous films as compared to ML films can explain the observed difference in the behaviour of the thin films studied in this work.

3.2. Film Morphology

Figure 3 shows the results of the evaluation of the measured AES depth profiles for the amorphous and ML Ti-Al samples in the as-prepared state and after annealing at 400, 600, and 800 °C in HV. The composition was calculated from the measured AES spectra using the standard single element relative sensitivity factors (RSF). As can be seen for the co-sputtered films in the as-prepared state (Figure 3a,b), the calculation led to a higher atomic concentration of Ti. In contrast to this, an EDX analysis of a thin lamella of such a sample in the TEM resulted in an atomic composition with slightly more Al than Ti. This deviation of the results can be explained by different actual values of the RSF for the Ti-Al alloy and by a preferential sputtering. The AES results therefore mainly allow a relative comparison of the different samples and annealing states. Besides this, from the analysis of the peak shape and position, the oxidation state of Ti, Al, and Si was derived.

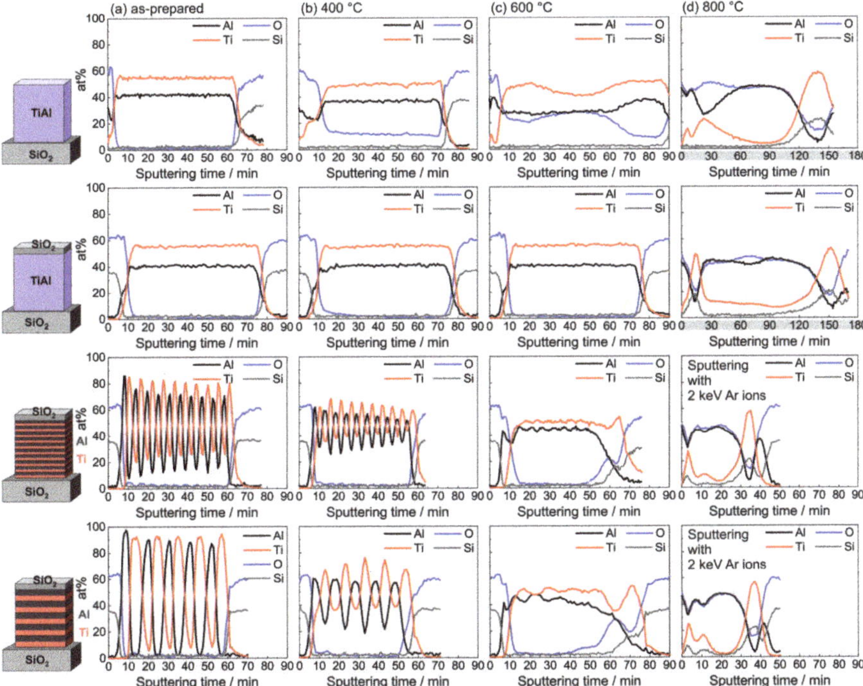

Figure 3. Results of the AES measurements of the co-sputtered Ti-Al film, co-sputtered Ti-Al film with SiO_2 cover layer, ML Ti/Al film with 10 nm and 20 nm individual layer thickness, both with SiO_2 cover layer in: (**a**) the as-prepared state; (**b**) after annealing at 400 °C; (**c**) after annealing at 600 °C; and (**d**) after annealing at 800 °C. In (**d**), the gray contrast of the abscissa in the figures of the alloy samples points out the longer sputtering time as compared to the other graphs. In contrast to this, the sputtering of the ML samples was performed with a higher sputtering energy (2 keV instead of 1 keV), which resulted in a shorter total sputtering time.

In Figure 3a, the results of the measurements of the samples in the as-prepared state are presented. In the uncovered co-sputtered film, a thin Al_2O_3 layer formed at the sample surface, which can be explained by the lower free energy of the formation of Al oxide as compared to that of TiO_2 for pure elements [24]. (It is reported in the literature that in Ti-Al materials the oxide formation depends on the composition and the phases which are present in the material. Rahmel and Spencer determined that Al oxide is more stable than TiO_2 in Al rich Ti-Al materials [25].) Across the whole sample thickness, the concentration of Al and Ti was constant.

In the SiO_2 covered alloy film as well, a thin Al_2O_3 layer developed at the interface between the cover layer and the Ti-Al film. For both ML samples, the sequence of the Ti and Al layers was clearly visible. The concentration of both elements did not reach 100 at% in the separate layers and, especially for the film with 10 nm individual layer thickness, a decrease of the amplitudes of the element concentration with increasing sputtering time was seen. These findings can be explained by a partial intermixing of the elements at the interfaces during the sputtering process and by a roughening of the sample surface due to the Ar sputtering, which increases with increasing sputtering time. This roughening led to a simultaneous measurement of different depths, which resulted in an apparent mixture of Ti and Al. In addition, some intermixing can result from the high ion energy during the depositions of the layers. In these ML samples as well, Al_2O_3 was present below the SiO_2 cover layer. The SiO_2 layer was deposited at 180 °C. Obviously, this temperature is sufficient to initiate a reduction reaction of the SiO_2 to Si by Al which forms Al_2O_3. This effect, which has been widely studied (see, e.g., [26]), was proven for these samples by XPS measurements.

The AES profiles of the samples after annealing at 400 °C in HV are shown in Figure 3b. For the uncovered co-sputtered sample, a strong oxide peak was visible at the sample surface, and a constant O signal was measured across the whole sample thickness. However, except for the sample surface, the AES showed Al and Ti only in the elemental and not in the oxidized state, which was derived from the AES peak position. This indicated that the O was solved within the amorphous Ti-Al layer. In contrast to this, in the co-sputtered film with SiO_2 cover layer as well as in the ML films no O was measured within the sample. In the ML films, a partly interdiffusion of Ti and Al occurred. In addition, at the Ti-SiO_2 interface at the bottom of the film some O was detected in the Ti layer.

After annealing at 600 °C, in the co-sputtered film without cover layer the O signal strongly increased. Within the first 5 min of sputtering, Ti was present in the oxidized and for the rest of the measurement in the elemental state. Oxidized Al was found at the sample surface. In the regime between 10 and 20 min of sputtering, Al was present in the elemental state, and the O which was detected there was again solved in the TiAl without forming an oxide. With increasing sputtering time, the amount of oxidized Al increased again, and all Al was present in the oxidized state for the sputtering time between about 35 and 50 min. Afterwards, the part of elemental Al increased again, and in the time regime between about 70 and 85 min only elemental Al was found. Again, O was just solved without forming an oxide. At this temperature, the SiO_2 cover layer still successfully acted as a diffusion barrier in the case of the co-sputtered Ti-Al film, which was seen from the constant thickness of the Al_2O_3 layer below it. However, in the case of the ML samples, the thickness of the Al_2O_3 interlayer between the SiO_2 cover and the TiAl layer increased. At this interfacial region as well, XPS proved the presence of elemental Si. The interdiffusion of the Al and Ti layers in the ML samples was almost completed and only a small variation of the composition remained visible. The annealing at 600 °C, however, led to a strong reaction in the Ti/Al ML samples with the SiO_2 at the substrate surface. The measurements revealed a layer consisting of elemental Ti and Si on top of the substrate, followed by an Al_2O_3 layer.

The annealing at 800 °C (Figure 3d) led to a degradation of all films and all Al was oxidized. The formed Al_2O_3 was mainly present in the center of the film. For all samples, at the interface to the substrate a strong Ti and Si signal was measured. In the samples with SiO_2 cover layer, Al_2O_3 was found at the surface of the film. In the case of the co-sputtered film, below this Al_2O_3, there was a defined layer consisting of Ti and Si. In the ML samples, Ti and Si were distributed across a broader thickness in the upper region of the film, and no oxidized Ti was found at all.

The time necessary to sputter the film increased significantly as compared to the previous measurements. One reason was the low sputter rate of Al_2O_3. In addition, due to the insulating material, electric charges were accumulated, leading to electric fields which further decreased the sputter rate. Therefore, for the ML samples an increased energy of the Ar ions (2 keV instead of 1 keV) was applied to achieve higher sputter rates.

The AES results showed that the chemical reaction between the Ti-Al film and the SiO_2 cover layer and substrate strongly depended on the initial distribution of the elements Ti and Al in the film. In the ML samples, stronger reactions took place at the respective interfaces between the pure Al layer on top and the pure Ti layer at the bottom and the SiO_2: a pure Ti layer at the bottom of the sample resulted in a marginal reaction between Ti and Si already at 400 °C (see the O signal in the bottom Ti layer) and a strong reaction at 600 °C. In contrast to this, in the case of the Ti-Al alloy, such a reaction did not take place at these temperatures. There was also a stronger oxidation of the Al at the interface with the SiO_2 cover layer in case of the pure Al layer, as can be seen from the thicker Al_2O_3 interlayer.

XPS was used to identify the chemical state of the elements in the degraded films. The result of an XPS depth profile analysis of the ML Ti/Al sample with 20 nm individual layer thickness and SiO_2 cover layer after annealing at 800 °C is shown in Figure 4a. The measurements revealed that Al was only present in the oxidized state (marked by Al_{ox}), and the intensity curve follows that of O. Ti was only present in the elemental state (marked by Ti_{el}) and was not oxidized at all. Unoxidized Si was found in the upper and lower region of the film, where its intensity curve was parallel to that of Ti. This behavior indicated the formation of a Ti-Si phase.

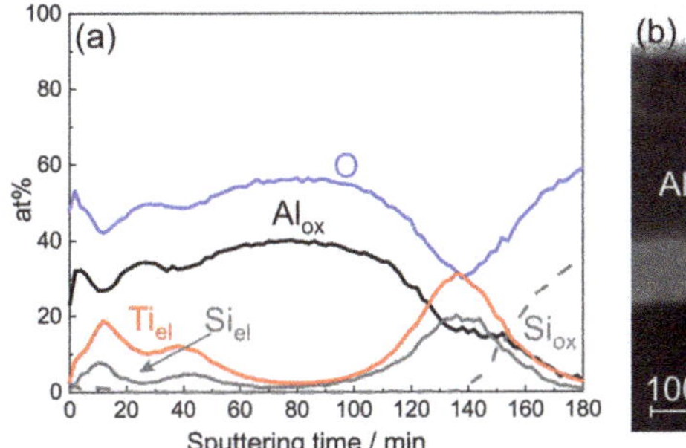

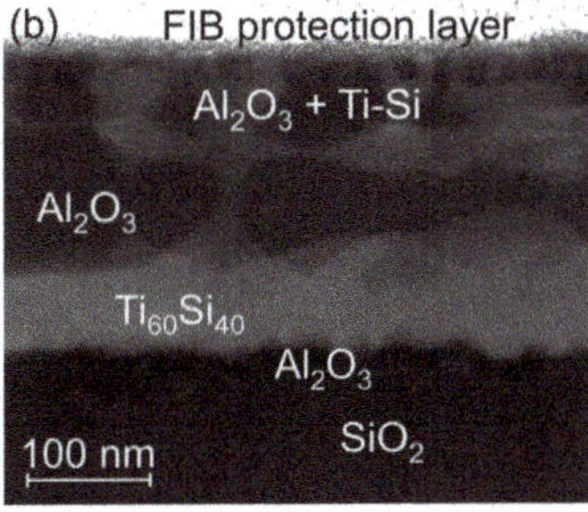

Figure 4. (**a**) Results of the XPS measurement of the ML sample with 20 nm individual layer thickness; and (**b**) STEM image with dominant chemical contrast, both after annealing at 800 °C in HV. In (**a**), the subscript "el" refers to the elemental and "ox" to the oxidized state of the respective element.

In contrast to the ML samples, in the co-sputtered films, the O signal was not parallel to that of Al (see Figure 3d). Around the sputtering time of 30 min, the film without the cover layer showed a strong decrease in the Al, a slightly increasing O, and a maximum in the Ti signal. An analysis of the AES spectral shape revealed that in this region of the film a part of the Ti was oxidized. In the co-sputtered film with SiO_2 cover as well, at the sputtering time around 70 min, a stronger O signal as compared to Al was visible. Again, a part of the Ti was oxidized, however, to a much reduced amount as compared to the sample without cover layer.

The reaction between Ti and SiO_2 is widely described in the literature (see, e.g., [27,28]). There is a chemical reaction between Ti and SiO_2 at their interface, leading to the formation of a Ti silicide. The free O diffuses through the Ti layer, which is possible because of the high O solubility of 34 at% in this material [29]. This solution of O in the bottom Ti layer was already visible for the ML systems after annealing at 400 °C (see Figure 3b). During annealing at 600 °C, the interface reaction became stronger, and the O which diffused through the Ti layer then reacted with the Al on top of it to Al_2O_3. Ti not only reacted with the SiO_2 of the substrate, but also with the covering SiO_2 layer, so that at the film surface Ti-Si was also formed.

In the upper region of the film, a small carbon signal of less than 5 at% was measured by XPS, indicating C which was present in a carbide state. However, most likely the carbide was formed by the Ar sputtering and did not originate from the sample. (More details on the XPS measurement of the Ti-Al samples were published by Oswald et al. [30].)

To get more information on the morphology of this sample, a STEM analysis was performed. A STEM image predominantly showing the chemical contrast is presented in Figure 4b. The EDX analysis confirmed the formation of Al_2O_3 and the presence of Ti-Si in the upper and lower region of the film. The chemical composition of the Ti-Si grains in the lower region of the film was about $Ti_{60}Al_{40}$. Close to this composition, there are the two Ti-Si phases: Ti_5Si_4 and Ti_5Si_3 [31]. There are reports describing the formation of the Ti_5Si_3 phase [27,28], and for our samples this phase was confirmed by the small Ti_5Si_3 (002) peak in the XRD data at $2\theta = 40.6°$ in Figure 1c,d.

Figure 5 summarizes SEM images of the FIB cross sections of the different samples in the as-prepared state and after the annealing procedures. In the as-prepared state (Figure 5a), the co-sputtered films showed a homogeneous structure. It was challenging to resolve the 10 nm multilayers; however, the 20-nm Ti and Al layers were clearly visible (dark layers: Ti; brighter layers: Al). After annealing at 400 °C, the co-sputtered samples as well as the 10-nm ML film appeared homogeneous. In the 20-nm ML sample, it was still possible to identify the layer structure (Figure 5b).

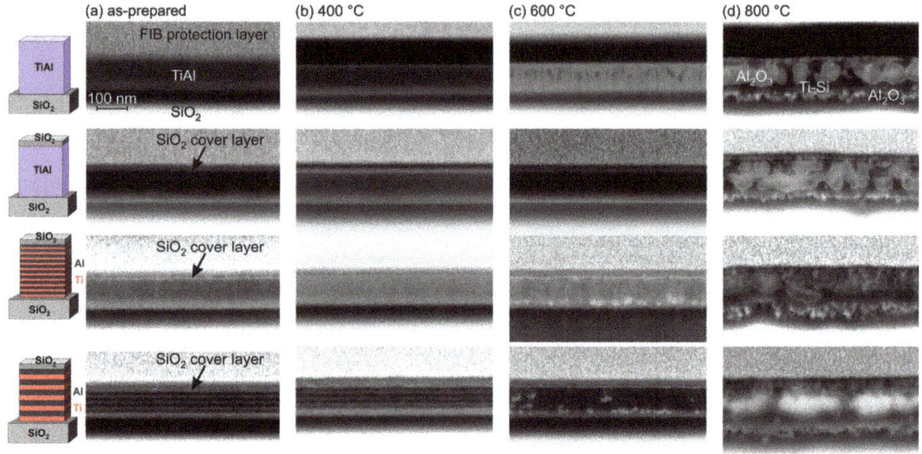

Figure 5. SEM images (inLens, 3 kV) of the FIB cross sections of the co-sputtered Ti-Al film, co-sputtered Ti-Al film with SiO_2 cover layer, and Ti/Al ML film with 10- and 20-nm individual layer thickness, both with SiO_2 cover layer in: (**a**) the as-prepared state; (**b**) after annealing at 400 °C; (**c**) after annealing at 600 °C; and (**d**) after annealing at 800 °C.

Changes in the microstructure became visible in the co-sputtered film without cover layer after annealing at 600 °C. The SEM image showed that darker grains were present in the upper region of the film, and that there are no visible inhomogeneities in the lower region. The dark structures were most likely Ti rich grains. This is in agreement with the AES measurements (Figure 3c), which showed besides Al and O a strong Ti signal in the upper region of the film. In contrast to the uncovered film, the co-sputtered film with SiO_2 layer did not possess such inhomogeneities as was expected from the AES measurements, which revealed a homogeneous composition. After annealing at 600 °C, in the images of both ML samples, small bright structures were present—in the case of the 10-nm ML in the lower region of the film and in the case of the 20-nm ML additionally in the center region of the TiAl layer. These bright structures were Al_2O_3 grains on top of the Ti_5Si_3 layer, which is in agreement with the AES results. In the 20-nm ML films, the AES showed a small O signal during the sputtering time between 40 and 55 min, which was not present in the 10-nm ML sample, caused by the Al_2O_3 grains that were distributed in the center region of the TiAl layer.

For all samples, strong changes in the morphology appeared after annealing at 800 °C (Figure 5d). Irregular formed structures were visible and the film thickness increased significantly. From the evaluation of the EDX data shown in Figure 4 in combination with the AES measurements, it can be derived that the large bright structures were Al_2O_3 grains. In the case of the uncovered film, these grains were embedded in a Ti and Ti-O matrix, and, in the case of the films with the SiO_2 cover layer, the matrix consisted of Ti-Si. In all samples, the interface between the film and substrate was no longer clearly defined and showed a large roughness due to the reaction between Ti and SiO_2.

4. Conclusions

This paper reports on the high temperature stability of very thin Ti-Al films. Co-sputtered or multilayered Ti-Al films with a thickness of 200 nm were deposited on thermally oxidized Si substrates, and annealed up to 800 °C in HV. Subsequently, they were analyzed regarding their phase formation and degradation processes up to 800 °C. In addition, the influence of a SiO_2 cover layer was evaluated.

The results show that already after annealing at 400 °C co-sputtered Ti-Al films without cover layer started to oxidize. In contrast to this, films with a 20-nm-thick SiO_2 cover layer were stable up to 600 °C. XRD measurements revealed that in the ML films the γ-TiAl phase formed during annealing at 600 °C. For the co-sputtered films, no phase formation was observed at all.

During annealing at 600 °C, in the ML samples, a chemical reaction between the Ti and SiO_2 of the substrate, but also with the SiO_2 cover layer, took place. This reaction led to the formation of Ti_5Si_3. The free O was solved in the TiAl layer and locally led to the formation of Al_2O_3. All films were destroyed during annealing at 800 °C. Ti silicide was formed to a large extent, and all Al, as well as in the film without cover layer some Ti, was oxidized.

In summary, the results reveal that the deposition of Ti/Al MLs is a prerequisite to achieve the γ-TiAl phase in the very thin films. In contrast to the μm thick TiAl films, which are described in literature, for the thin films, the chemical reaction between the Ti and SiO_2 plays a crucial role. On the one hand, the part of the Ti that is chemically bonded to the Si is not available for the TiAl phase formation. On the other hand, the O that is emitted by this reaction leads to the formation of Al_2O_3 at 600 °C within the 10 h of annealing. This effect will be more pronounced for longer annealing times and higher temperatures, and with this contributes to the degradation of the film. For thick films, these processes are negligible, since they only affect a very small part of the total layer. The results also show that a SiO_2 cover layer improves the high temperature stability of the TiAl films as compared to samples without cover layer despite the chemical reaction between Ti and SiO_2.

To realize a high temperature stability of γ-TiAl thin films, a contact to SiO_2 needs to be avoided. This could be achieved by substituting the barrier with another material, e.g., AlN, or by using an additional protection layer. For the application in SAW devices, a high temperature stability of TiAl thin films on other piezoelectric oxide materials such as CTGS substrates also has to be realized.

Author Contributions: Conceptualization, M.S., E.L., and S.B.M.; Data curation, M.S., E.L., and S.O.; Funding acquisition, S.B.M.; Investigation, M.S., E.L., and S.O.; Project administration, M.S.; Resources, S.B.M., S.O., and T.G.; Supervision, M.S.; Validation, M.S., S.O., and T.G.; Visualization, M.S.; Writing—original draft, M.S.; and Writing—review and editing, M.S., S.B.M., and T.G. All authors have read and agreed to the published version of the manuscript.

Funding: The work was supported by German BMBF under grant InnoProfile-Transfer 03IPT610Y and BMWI under grant 03ET1589 A.

Acknowledgments: The authors gratefully acknowledge Thomas Wiek as well as Dina Bieberstein for FIB cuts and TEM lamella preparation and Steffi Kaschube for AES and XPS measurements. Dina Bieberstein is also acknowledged for the support of the texture measurements.

Conflicts of Interest: The authors declare no conflict of interest. The funders had no role in the design of the study; in the collection, analyses, or interpretation of data; in the writing of the manuscript, or in the decision to publish the results.

Abbreviations

The following abbreviations are used in this manuscript:

AES	Auger electron spectroscopy
EDX	energy dispersive X-ray spectroscopy
FIB	focussed ion beam
HV	high vacuum
ML	multilayer
SAW	surface acoustic wave
SEM	scanning electron microscopy
STEM	scanning electron transmission microscopy
XPS	X-ray photoelectron spectroscopy
XRD	X-ray diffraction

References

1. Bewlay, B.P.; Nag, S.; Suzuki, A.; Weimer, M.J. TiAl alloys in commercial aircraft engines. *Mater. High Temp.* **2016**, *33*, 549–559. [CrossRef]
2. Rugg, D.; Dixon, M.; Burrows, J. High-temperature application of titanium alloys in gas turbines. Material life cycle opportunities and threats—An industrial perspective. *Mater. High Temp.* **2016**, *33*, 536–541. [CrossRef]
3. Ramos, A.S.; Cavaleiro, A.J.; Vieira, M.T.; Morgiel, J.; Safran, G. Thermal stability of nanoscale metallic multilayers. *Thin Solid Films* **2014**, *571*, 268–274. [CrossRef]
4. Banerjee, R.; Swaminathan, S.; Wheeler, R.; Fraser, H.L. Phase evolution during crystallization of sputter-deposited amorphous titanium–aluminium alloy thin films: Dimensional and solute effects. *Philos. Mag.* **2000**, *80*, 1715–1727. [CrossRef]
5. Illekova, E.; Gachon, J.C.; Rogachev, A.; Grigoryan, H.; Schuster, J.C.; Nosyrev, A.; Tsygankov, T. Kinetics of intermetallic phase formation in the Ti/Al multilayers. *Thermochim. Acta* **2008**, *469*, 77–85. [CrossRef]
6. Aubert, T.; Elmazira, O.; Assouar, B.; Bouvot, L.; Hehn, M.; Weber, S.; Oudich, M.; Geneve, D. Behavior of platinum/tantalum as interdigital transducers for SAW devices in high-temperature environments. *IEEE Trans. Ultrason. Ferroelect. Freq. Contr.* **2011**, *58*, 603–610. [CrossRef]
7. Thiele, J.A.; Pereira da Cunha, M. Platinum and palladium high-temperature transducers on langasite. *IEEE Trans. Ultrason. Ferroelect. Freq. Contr.* **2005**, *52*, 545–549. [CrossRef]
8. Aubert, T.; Bardong, J.; Elmazria, O.; Bruckner, G.; Assouar, B. Iridium interdigital transducers for high-temperature surface acoustic wave applications. *IEEE Trans. Ultrason. Ferroelectr. Freq. Control* **2012**, *59*, 194–197. [CrossRef]
9. Taguett, A.; Aubert, T.; Elmazria, O.; Bartoli, F.; Lomello, M.; Hehn, M.; Ghanbaja, J.; Boulet, P.; Mangin, S.; Xu, Y. Comparison between Ir, $Ir_{0.85}Rh_{0.15}$ and $Ir_{0.7}Rh_{0.3}$ thin films as electrodes for surface acoustic waves applications above 800 °C in air atmosphere. *Sens. Actuators A Phys.* **2017**, *266*, 211–218. [CrossRef]
10. Moulzolf, S.C.; Frankel, D.J.; Pereira da Cunha, M.; Lad, R. High temperature stability of electrically conductive Pt-Rh/ZrO_2 and Pt-Rh/HfO_2 nanocomposite thin film electrodes. *Microsyst. Technol.* **2014**, *20*, 523–531. [CrossRef]
11. Menzel, S.B.; Seifert, M.; Priyadarshi, A.; Rane, G.K.; Park, E.; Oswald, S.; Gemming, T. Mo-La_2O_3 Multilayer Metallization Systems for High Temperature Surface Acoustic Wave Sensor Devices. *Materials* **2019**, *12*, 2651. [CrossRef] [PubMed]
12. Okamoto, H. Al-Ru (aluminum-ruthenium). *J. Phase Equilib.* **1997**, *18*, 105. [CrossRef]
13. Mücklich, F.; Ilić, N.; Woll, K. RuAl and its alloys, Part II: Mechanical properties, environmental resistance and applications. *Intermetallics* **2008**, *16*, 593–608. [CrossRef]
14. Seifert, M. High temperature stable RuAl thin films on piezoelectric CTGS and LGS substrates. *Materials* **2020**, *13*, 1605. [CrossRef] [PubMed]
15. Mouritz, A.P. (Ed.) Titanium alloys for aerospace structures and engines. In *Introduction to Aerospace Materials*; Woodhead Publishing: Cambridge, UK, 2012; pp. 202–223.
16. Seifert, M.; Rane, G.K.; Menzel, S.B.; Oswald, S.; Gemming, T. Improving the oxidation resistance of RuAl thin films with Al_2O_3 or SiO_2 cover layers. *J. Alloys Compd.* **2019** *776*, 819–825. [CrossRef]
17. *MultiPak, Software Package, V. 9.5*; ULVAC-PHI: Kanagawa, Japan, 1994–2014.

18. Ramos, A.S.; Vieira, M.T. Kinetics of the thin films transformation Ti/Al multilayer → γ-TiAl. *Surf. Coat. Technol.* **2005**, *200*, 326–329. [CrossRef]
19. Hofmann, M.; Gemming, T.; Menzel, S.; Wetzig, K. Microstructure of Al/Ti Metallization Layers. *Z. Metallkd.* **2003**, *94*, 317–322. [CrossRef]
20. Hofmann, M.; Gemming, T.; Wetzig, K. Microstructure and composition of annealed Al/Ti-metallization layers. *Anal. Bioanal. Chem.* **2004**, *379*, 547–553. [CrossRef]
21. Senkov, O.N.; Uchic, M.D. Microstructure evolution during annealing of an amorphous TiAl sheet. *Mater. Sci. Eng. A* **2003**, *340*, 216–224. [CrossRef]
22. Kim, H.C.; Theodore, N.D.; Gadre, K.S.; Mayer, J.W.; Alford, T.L. Investigation of thermal stability, phase formation, electrical, and microstructural properties of sputter-deposited titanium aluminide thin films. *Thin Solid Films* **2004**, *460*, 17–24. [CrossRef]
23. Senkov, O.N.; Uchic, M.D.; Menon, S.; Miracle, D.B. Crystallization kinetics of an amorphous TiAl sheet produced by PVD. *Scr. Mater.* **2002**, *46*, 187–192. [CrossRef]
24. Gaskell, D.R. *Introduction to Metallurgical Thermodynamics*; McGraw-Hill Book Company: New York, NY, USA, 1973; p. 269.
25. Rahmel, A.; Spencer, P.J. Thermodynamic aspects of TiAl and $TiSi_2$ oxidation: The Al-Ti-O and Si-Ti-O Phase diagrams. *Oxid. Met.* **1991**, *35*, 53–68. [CrossRef]
26. Dadabhai, F.; Gaspari, F.; Zukotynski, S.; Bland, C. Reduction of silicon dioxide by aluminum in metal–oxide–semiconductor structures. *J. Appl. Phys.* **1996**, *80*, 6505–6509. [CrossRef]
27. Ting, C.Y.; Wittme, M.; Iyer, S.S.; Brodsky, S.B. Interaction Between Ti and SiO_2. *J. Electrochem. Soc.* **1984**, *131*, 2934–2938. [CrossRef]
28. Russell, S.W.; Strane, J.W.; Mayer, J.W.; Wang, S.Q. Reaction kinetics in the Ti/SiO_2 system and Ti thickness dependence on reaction rate. *J. Appl. Phys.* **1994**, *76*, 257–263. [CrossRef]
29. Okamoto, H. O-Ti (Oxygen-Titanium). *J. Phase Equilib.* **2001**, *22*, 515–517. [CrossRef]
30. Oswald, S.; Lattner, E.; Seifert, M. XPS chemical state analysis of sputter depth profiling measurements for annealed TiAl-SiO_2 and TiAl-W layer stacks. *Surf. Interface Anal.* **2019**, submitted.
31. Seifert, H.J.; Lukas, H.L.; Petzow, G. Thermodynamic Optimization of the Ti-Si System. *Z. Metallkd.* **1996**, *87*, 2–13.

Article

Microstructure and Optical Properties of E-Beam Evaporated Zinc Oxide Films—Effects of Decomposition and Surface Desorption

Lukasz Skowronski [1,*,†], Arkadiusz Ciesielski [2,†], Aleksandra Olszewska [1], Robert Szczesny [3], Mieczyslaw Naparty [1], Marek Trzcinski [1] and Antoni Bukaluk [1]

1 Institute of Mathematics and Physics, UTP University of Science and Technology, Kaliskiego 7, 85-796 Bydgoszcz, Poland; aleksandra.lewandowska@utp.edu.pl (A.O.); mieczyslaw.naparty@utp.edu.pl (M.N.); marek.trzcinski@utp.edu.pl (M.T.); antoni.bukaluk@utp.edu.pl (A.B.)

2 Faculty of Physics, University of Warsaw, Pasteura 5, 02-093 Warsaw, Poland; Arkadiusz.Ciesielski@fuw.edu.pl

3 Faculty of Chemistry, Nicolaus Copernicus University in Torun, Gagarina 7, 87-100 Torun, Poland; robert.szczesny@umk.pl

* Correspondence: lukasz.skowronski@utp.edu.pl

† These authors contributed equally to this work.

Received: 18 July 2020; Accepted: 6 August 2020 ; Published: 9 August 2020

Abstract: Zinc oxide films have been fabricated by the electron beam physical vapour deposition (PVD) technique. The effect of substrate temperature during fabrication and annealing temperature (carried out in ultra high vacuum conditions) has been investigated by means of atomic force microscopy, scanning electron microscopy, powder X-ray diffraction, X-ray photoelectron spectroscopy and spectroscopic ellipsometry. It was found that the layer deposited at room temperature is composed of Zn and ZnO crystallites with a number of orientations, whereas those grown at 100 and 200 °C consist of ZnO grains and exhibit privileged growth direction. Presented results clearly show the influence of ZnO decomposition and segregation of Zn atoms during evaporation and post-deposition annealing on microstructure and optical properties of zinc oxide films.

Keywords: ZnO; decomposition; segregation; annealing; optical constants; microstructure; e-beam

1. Introduction

Zinc oxide exhibits high thermal and chemical stability and has been studied for decades [1,2]. ZnO is a semiconductor with a direct band-gap of $E_g \sim 3.3$ eV [3,4] similar to the value reported to gallium nitride ($E_g \sim 3.4$ eV) [5], however the binding energy of zinc oxide exciton (60 meV) [4] is 2.4 times larger than that of GaN (25 meV) [5]. These features indicate that ZnO is a promising material for short-wavelength optoelectronic devices such as light-emitting and laser diodes [6]. Due to good optical transmittance in the visible spectral range zinc oxide is used as a transparent electrode in organic and hybrid solar cells [7,8]. Moreover, ZnO is used in thin-film transistors [9] and gas sensors [10]. Zinc oxide is a very good candidate for space applications due to its stability to high energy radiation [1]. On the other hand, ZnO can be easily etched in acids and alkali [11]. This feature is commonly used to produce ZnO in form of nanosheeds, nanoshells, multipods, nanorods or nanowires typicaly implemented as periodic structures using a number of methods [2,12].

The structural and optical properties of ZnO films strongly depend on the fabrication conditions that affect the optimal performance of the device. It has been shown that optical constants (complex refractive index or complex dielectric function) of zinc oxide demonstrate significant film-thickness relationship, especially for thicknesses below 20 nm [3,13–15]. Moreover, Pal et al. [3] revealed the influence of the substrate on the optical properties of ZnO films. Therefore, systematic study of the properties of zinc oxide layers fabricated under various growing conditions is desirable. This analysis was performed for ZnO produced by different techniques such as atomic layer deposition [3,4,9], electron beam evaporation [16–19], magnetron sputtering [10,19,20], chemical vapour deposition [6], spin coating [21,22] as well as molecular beam epitaxy [23]. In addition, the existence of oxygen or zinc vacancies and their concentration in ZnO, significantly affects luminescence and photocatalytic properties [22], depending on the method of production.

Electron beam physical vapour deposition (e-beam PVD) technique is commonly used to produce both metallic [24–26] and oxide coatings [27] including zinc oxide films [16,19]. The influence of microstructure on the optical response of zinc oxide films fabricated under various conditions, such as the substrate temperature (and/or annealing after deposition) as well as deposition rate, has been investigated in recent years [16,18,19]. It has been reported that increasing the substrate temperature and/or annealing of ZnO after deposition significantly improves their stoichiometry, resistivity, free carrier mobility and optical transparency [16,18]. The post-evaporation annealing process in oxygen atmosphere also serves to ensure that there is no deficiency of oxygen atoms in the deposited films [19], due to decomposition and release oxygen into the vacuum [28–31]. This process was also observed for ZnO films [19,20].

In this paper we show the influence of substrate temperature during deposition by means of the e-beam technique and annealing in ultra high vacuum (UHV) conditions on the microstructure and optical properties of zinc oxide films. The above relationship was studied based on the results of atomic force microscopy (AFM), scanning electron microscopy (SEM), X-ray diffraction (XRD), X-ray photoelectron spectroscopy (XPS) and spectroscopic ellipsometry (SE) measurements. The above-mentioned methods were used to explain the influence of ZnO decomposition and segregation of Zn atoms on the optical response of the deposited films.

2. Materials and Methods

Fabrication of the samples has been conducted using the PVD75 e-beam evaporation system from Lesker (St. Leonards-on-Sea, UK). ZnO films have been deposited from 4 N ZnO pieces in graphite crucible onto polished Si substrates (100). Silicon is widely used in fabrication of a variety of optical devices (e.g., [32,33]). Substrates were located 40 cm away from the crucible. Three sets of samples have been fabricated—with the substrate at room temperature (RT), at 100 and 200 °C. The base pressure was 2×10^{-5} Torr, however, due to the decomposition of ZnO, during the deposition process the pressure approached 10^{-4} Torr. The deposition rate and total film thickness were monitored by two quartz weights inside the deposition chamber. The nominal deposition rate was 0.3 Å/s, however, due to the decomposition of ZnO, it is difficult to estimate the true value (although it is expected to be within 30% of the nominal value). After the deposition, the samples were annealed for two hours at 300, 500 and 800 °C. The annealing was performed in a UHV preparation chamber (base pressure $\leq 5 \times 10^{-10}$ mbar). The list of prepared samples is summarized in Table 1.

After cooling to room temperature, the samples were transferred without venting to the analysis chamber (base pressure $\leq 2 \times 10^{-10}$ mbar) for XPS measurements. The incident radiation was produced by Al K_α source (1486.6 eV) at 55 degrees with respect to the normal of the sample. The energy of photoelectrons was analyzed by the VG-Scienta R3000 (Uppsala, Sweden) spectrometer (the energy step was set at ΔE = 0.1 eV. For quantitative analysis, the experimental data was fitted to Gauss–Lorentz shapes by using CasaXPS software (version 2.3.16, Casa Software Ltd., Teignmouth, UK).

The surface topography of the films was examined by means of the AFM Innova device from Bruker (Billerica, MA, USA) equipped with the standard Si tips for a tapping mode. Roughness parameters R_a (the arithmetical mean deviation):

$$R_a = \frac{1}{N}\sum_{j=1}^{N}|Z_j| \tag{1}$$

and R_q (the root mean squared roughness):

$$R_q = \sqrt{\frac{1}{N}\sum_{j=1}^{N}Z_j^2} \tag{2}$$

where Z_j and N are the current surface height value and the number of points measured, respectively. They were calculated based on the images, recorded for an area 2 µm × 2 µm, using the NanoScope Analysis software (version 1.40). Scanning electron microscopy (SEM) investigations were performed with a Quanta 3D FEG (FEI, Hillsboro, OR, USA) (EHT = 30 kV) device.

The Phillips X'Pert (Malvern Panalytical Ltd., Malvern, UK) device with Cu K_α radiation (λ = 1.5418 Å) and X'Celerator Scientific (Malvern Panalytical Ltd., Malvern, UK) detector was used to record the XRD patters of the investigated samples. These measurements were performed in the range from 2θ = 20 ° to 60 °.

Table 1. Temperatures of deposition (T_d) and annealing (T_a) for the investigated samples.

Sample id.	T_d (°C)	T_a (°C)
S_RT	RT	–
S_RT_300C	RT	300
S_RT_500C	RT	500
S_RT_800C	RT	800
S_100C	100	–
S_100C_300C	100	300
S_100C_500C	100	500
S_100C_800C	100	800
S_200C	200	–
S_200C_300C	200	300
S_200C_500C	200	500
S_200C_800C	200	800

The thickness and optical constants of the prepared thin films were investigated by means of the V-VASE device from J.A. Woollam Co., Inc., Lincoln, NE, USA. The ellipsometric azimuths Ψ and Δ were measured for three angles of incidence (65°, 70° and 75°) in the NIR-vis-UV spectral range (193–2000 nm; 0.6–6.5 eV). The analysis of ellipsometric data was performed using the WVASE32 software (J.A. Woollam Co., Inc.).

3. Results

Spectroscopic ellipsometry (SE) was used to obtain the thickness of the zinc oxide film as well as the optical constants of the deposited coatings. The measured Ψ and Δ ellipsometric azimuths for S_ RT, S_ RT_ 800C and S_ 200C_ 800C are presented in Figure 1. The Ψ and Δ curves vary significantly for films fabricated in different conditions. Spectra shown in Figure 1a exhibit metallic-like behaviour, while data presented in Figure 1b,c are typical for a semiconducting material. Oscillations in the ellipsometric Ψ and Δ angles for wavelengths longer than about 390 nm for the S_ RT_ 800C sample (Figure 1b) indicate that the thickness of the layer is significantly greater compared to the thickness of S_ 200C_ 800C film. The thickness of the layer and its optical constants were determined by means of the five-medium optical model of a sample (from bottom to top: substrate-Si\native SiO_2\ZnO film\rough layer\ambient). The complex refractive index of the ZnO film ($\tilde{n} = n + ik$, where n and k

are the real part of $\tilde{n}$ and the extinction coefficient, respectively) was parametrised using a model containing the following components/oscillators:

- a Sellemeier-type dispersion relation (a Pole oscillator; ε_P),
- a Drude term ($\widetilde{\varepsilon}_D$; for the S_RT and S_ RT_ 300 samples) or a Tauc–Lorentz oscillator ($\widetilde{\varepsilon}_{T_L}$; for the other specimens) and
- a sum of Gaussian oscillators ($\widetilde{\varepsilon}_G$).

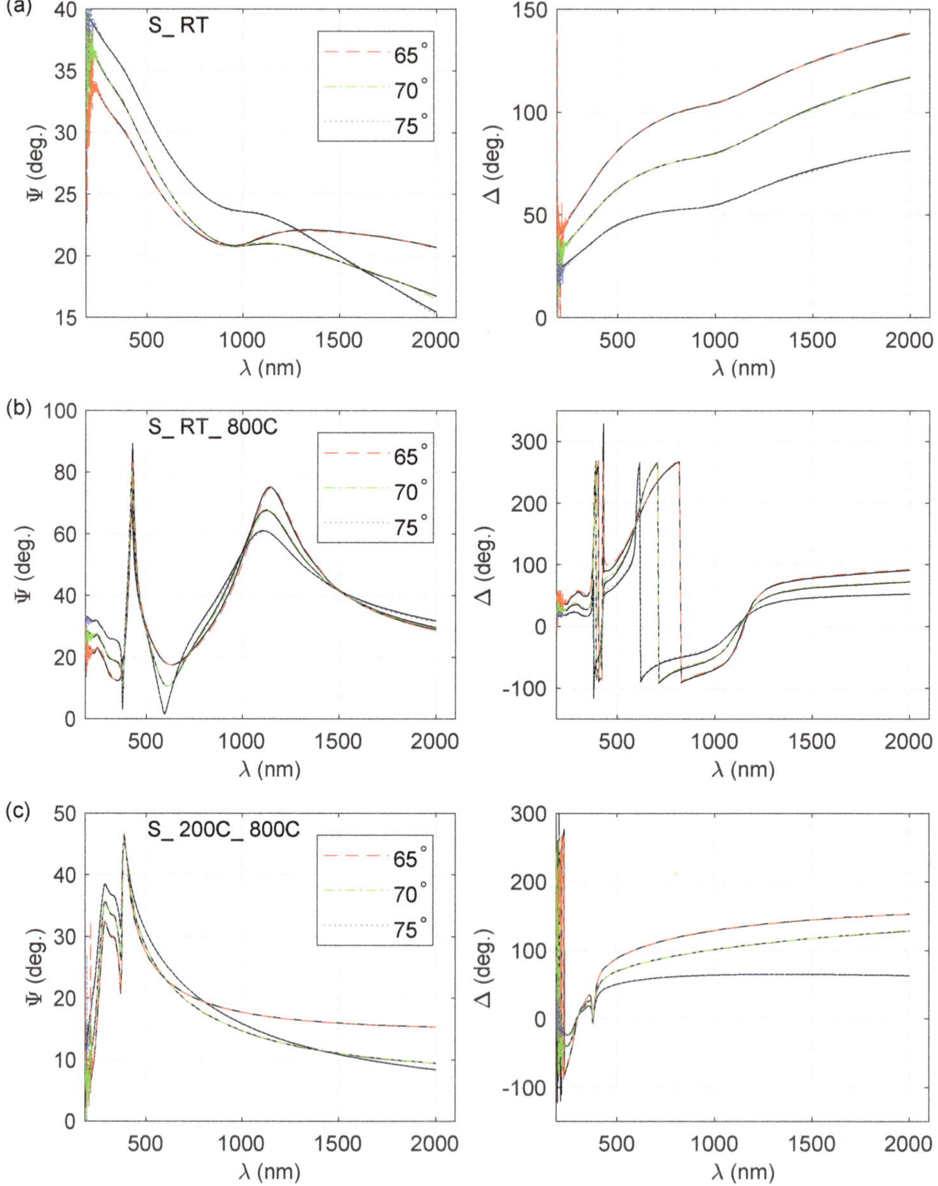

Figure 1. Experimental and calculated ellipsometric azimuths Ψ and Δ for (**a**) S_ RT (χ^2 = 2.233), (**b**) S_ RT_ 800 (χ^2 = 2.242) and (**c**) S_ 200C_ 800C (χ^2 = 2.57) samples.

The selection of these oscillators was performed based on the shape of the measured ellipsometric azimuths (see Figure 1). The complex refractive index of the deposited film can be written in the following form:

$$\widetilde{n}^2 = \varepsilon_\infty + \varepsilon_P\left(A_0, E_0\right) + \widetilde{\varepsilon}_{T_L}\left(A_{T-L}, E_g, E_n, Br_{T_L}\right) + \sum_j \widetilde{\varepsilon}_G\left(A_j, E_j, Br_j\right) \tag{3}$$

or

$$\widetilde{n}^2 = \varepsilon_\infty + \varepsilon_P\left(A_0, E_0\right) + \widetilde{\varepsilon}_D\left(\hbar\omega_P, \hbar\Gamma\right) + \sum_j \widetilde{\varepsilon}_G\left(A_j, E_j, Br_j\right). \tag{4}$$

In Equations (3) and (4), ε_∞ is the high-frequency dielectric constant, E_0 and A_0 are the pole oscillator position and magnitude, respectively, while $\hbar\omega_P$ is the unscreened plasma energy and $\hbar\Gamma$ is the free-carrier dumping. The quantities A, E and Br (with adequate subscripts) represent amplitude, energy and broadening of the line shape, respectively, while E_g is the band gap energy. Mathematical formulas for the particular line shapes can be found elsewhere [34,35]. The form of $\widetilde{n}$ (Equation (3) or Equation (4)) used to determine optical constants of the produced films are summarized in Table 2.

The complex refractive index of Si and SiO_2 were employed from the database of optical constants [35]. The thickness of native oxide was set to be 2.3 nm and was established in the separate experiment [36]. The imperfect surface of the deposited film (the rough layer) was described based on the Bruggeman Effective Medium Approximation (EMA) model [34,35]. In this approach, the optical constants of the rough film are represented as a combination of optical properties of ZnO film and ambient (the assumed volume fraction of each medium was set to be 1/2). The model quantities were minimized to reduce the standard mean squared error (χ^2) [35]:

$$\chi^2 = \frac{1}{N-P}\sum_j \left(\frac{\left(\Psi_j^{mod} - \Psi_j^{exp}\right)^2}{\sigma_{\Psi\, j}^2} + \frac{\left(\Delta_j^{mod} - \Delta_j^{exp}\right)^2}{\sigma_{\Delta\, j}^2} \right). \tag{5}$$

In Equation (5) N and P are the total number of data points and the number of fitted model parameters respectively. The Ψ_j and Δ_j represent measured (with the superscript 'exp') or calculated (with the superscript 'mod') ellipsometric azimuths. Quantities $\sigma_{\Psi\, j}$ and $\sigma_{\Delta\, j}$ are standard deviations for measured Ψ and Δ angles. Example of the fits are presented in Figure 1. It should be noted that the reduced mean squared error (χ^2) was below 2.57 for all the samples produced, which means that the calculated Ψ and Δ values are in a very good agreement with the experimental data.

The thicknesses of the prepared films (d_{ZnO}) are in the range from about 37 to 325 nm, while thicknesses of the rough layer (d_r) were found to be 5–41 nm (see Table 2), whereas for the samples deposited at room temperature these values are one order of magnitude higher, than those obtained for the specimens produced at higher temperatures. The real part (n) of the complex refractive index ($\widetilde{n}$) and the extinction coefficient (k) determined by the spectroscopic ellipsometry method are presented in Figure 2. The spectra shown in Figure 2a (for S_ RT and S_ RT_ 300C samples) exhibit metallic-like behaviour (noticeable extinction coefficient in the UV-vis-IR), however without the strong Drude contribution characteristic for conductors. Plasma energy ($\hbar\omega_P$), found during the analysis of the SE data, established to be 1.7 and 1.8 eV for S_ RT_ 300C and S_ RT samples, respectively, is much lower than that reported for pure zinc (10 eV) [37]. The n and k spectra presented in Figure 2b–d are typical for semiconducting materials, i.e., a normal dispersion for longer wavelengths and interband transitions in the high-energy region.

Table 2. The form of the complex refractive index used to parametrise optical constants of the produced films, thicknesses of ZnO (d_{ZnO}) and rough (d_r) layers as well as roughness parameters (R_a, R_q and R_{max}) determined for the investigated samples.

Sample id.	Form of $\tilde{n}$	d_{ZnO} (nm)	d_r (nm)	R_a (nm)	R_q (nm)	R_{max} (nm)
S_RT	Equation (3)	325 ± 10	37.2 ± 1.3	20.7	16.6	150
S_RT_300C	Equation (3)	299 ± 20	41.3 ± 1.5	19.0	14.9	151
S_RT_500C	Equation (4)	266.0 ± 0.3	40.6 ± 0.2	19.4	15.4	147
S_RT_800C	Equation (4)	199.5 ± 0.5	28.3 ± 0.4	19.5	15.2	161
S_100C	Equation (4)	37.6 ± 0.1	7.5 ± 0.2	1.5	1.2	11.5
S_100C_300C	Equation (4)	38.5 ± 0.2	6.9 ± 0.3	2.3	1.7	18.7
S_100C_500C	Equation (4)	38.3 ± 0.2	7.5 ± 0.3	1.7	1.2	26.9
S_100C_800C	Equation (4)	38.2 ± 0.2	5.4 ± 0.3	2.6	1.9	32.4
S_200C	Equation (4)	36.5 ± 0.2	6.7 ± 0.2	1.5	1.2	11.3
S_200C_300C	Equation (4)	36.7 ± 0.3	5.9 ± 0.3	1.9	1.5	16.0
S_200C_500C	Equation (4)	37.1 ± 0.3	6.5 ± 0.4	2.1	1.7	15.6
S_200C_800C	Equation (4)	36.6 ± 0.2	5.8 ± 0.3	2.8	2.2	21.9

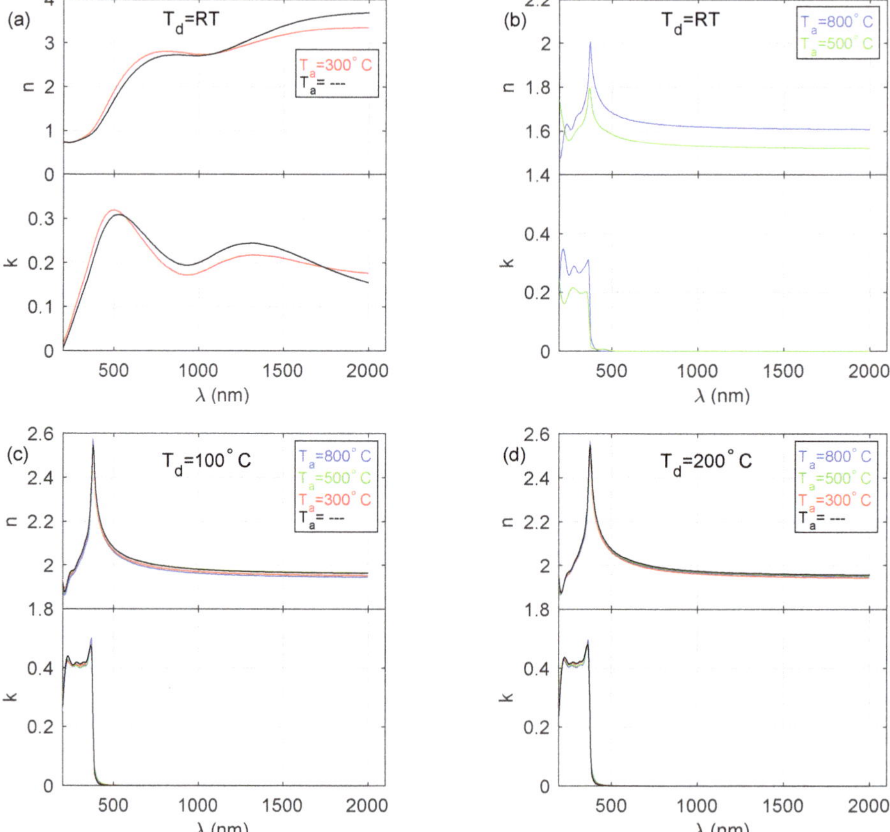

Figure 2. The index of refraction (n) and the extinction coefficient (k) for the layers deposited at (**a**) and (**b**) room temperature, (**c**) 100 °C and (**d**) 200 °C and annealed at 300, 500 and 800 °C.

The band-gap energy (E_g) for the produced ZnO films was determined during the parametrisation of their optical constants (a Tauc–Lorentz oscillator; see Equation (3)) and are summarized in Table 3. Values of E_g were established to be 3.2 eV for the layers deposited at 100 and 200 °C as well as at RT (however only for films annealed at 500 and 800 °C) and are in a perfect agreement with published data [3,4,13,15,38]. For metallic-like samples (S_RT and S_RT_300C) the band-gap energy could not be determined.

Table 3. The band-gap energy (E_g) estimated for the deposited films.

Sample id.	E_g (eV)	E_g (nm)
S_RT	- [a]	- [a]
S_RT_300C	- [a]	- [a]
S_RT_500C	3.239 ± 0.002	382.8 ± 0.3
S_RT_800C	3.238 ± 0.003	382.9 ± 0.4
S_100C	3.191 ± 0.001	388.5 ± 0.2
S_100C_300C	3.194 ± 0.001	388.2 ± 0.2
S_100C_500C	3.206 ± 0.001	386.7 ± 0.2
S_100C_800C	3.219 ± 0.001	385.2 ± 0.2
S_200C	3.191 ± 0.002	388.5 ± 0.3
S_200C_300C	3.191 ± 0.002	388.5 ± 0.3
S_200C_500C	3.201 ± 0.002	387.2 ± 0.3
S_200C_800C	3.214 ± 0.002	385.8 ± 0.3

[a] the band-gap energy has not been determined.

AFM images of the ZnO films deposited at RT, at 100 °C and at 200 °C (both non-annealed and annealed after deposition) are presented in Figure 3. Surfaces of all samples exhibit the nanogranular structure with the lateral size of grains smaller than 100 nm (for specimens produced at RT; Figure 3a) or 50 nm (for the other surfaces; Figure 3b,c). It should be taken into account that the samples deposited at RT exhibit significantly expanded surface compared to the surface of the other ZnO layers (see profiles in Figure 3). The surface morphology of the deposited films has been confirmed in the SEM measurements (see Figure A1 in Appendix A). For ZnO films deposited at RT the average roughness (R_a) is about 20 nm (see Table 2) and the maximum roughness (R_{max}) 150–160 nm, whereas for the zinc oxide layers grown at higher temperatures (100 and 200 °C) these values are significantly lower (R_a = 1.5–2.9 nm; R_{max} = 11–32 nm).

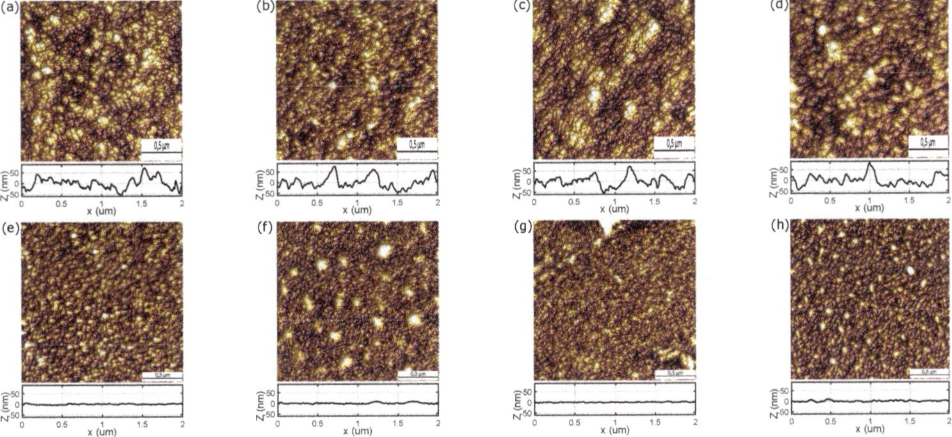

Figure 3. *Cont.*

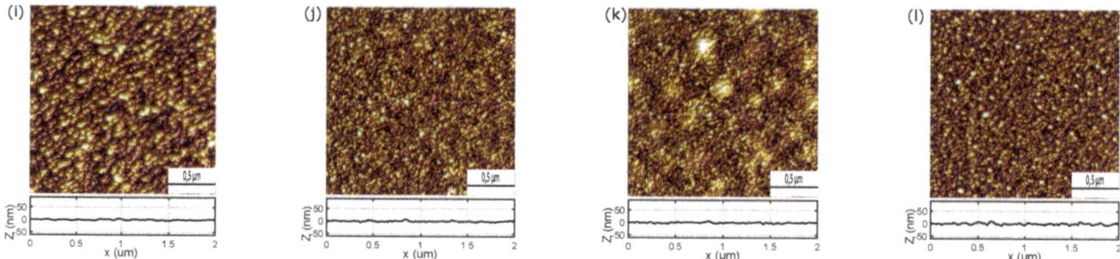

Figure 3. AFM images of ZnO layers deposited at RT: (**a**) not annealed and annealed at (**b**) 300 °C, (**c**) 500 °C and (**d**) 800 °C and deposited at 100 °C: (**e**) not annealed and annealed at (**f**) 300 °C, (**g**) 500 °C and (**h**) 800 °C as well as deposited at 200 °C: (**i**) not annealed and annealed at (**j**) 300 °C, (**k**) 500 °C and (**l**) 800 °C.

The XRD patterns recorded for the deposited coatings are presented in Figure 4. Diffraction peaks related to ZnO are observed at $2\theta \approx 32°$ (100), 34° (002), 36° (101), 48° (102) and 57° (110), while the signal recorded at $2\theta \approx 39°$ (100), 43° (101) and 55° (102) can be referred to metallic zinc. The strong and narrow peak at $2\theta \approx 33°$ can be assigned to the substrate-Si (200). The average crystallite size ($< D >$) determined for the identified Zn and ZnO phases, summarized in Table 4, has been estimated based on the Scherrer formula:

$$< D >= \frac{0.94\lambda}{\beta \cos(2\theta)} \quad (6)$$

where λ is the X-ray wavelength and β is the full-width at half-maximum (FWHM) of the Bragg diffraction peak at angle 2θ.

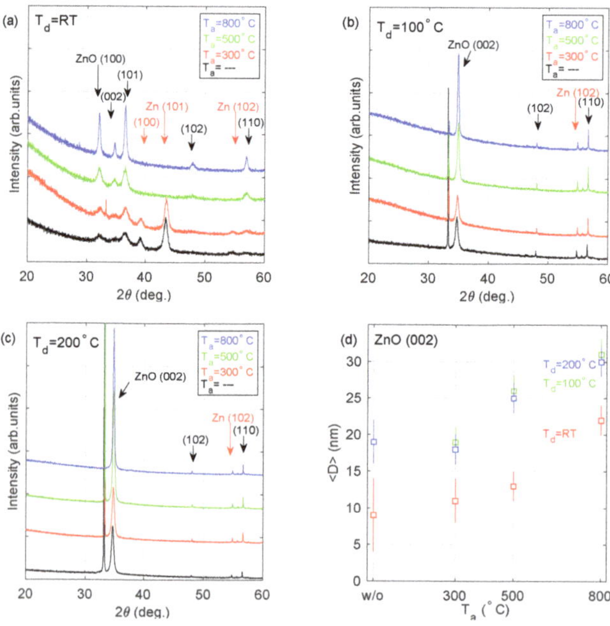

Figure 4. XRD patterns for the samples deposited at (**a**) room temperature, (**b**) 100 °C and (**c**) 200 °C and annealed at 300, 500 and 800 °C. (**d**) The average size ($< D >$) of the ZnO(002) crystallite. Used ICDD cards: Zn-00-001-1238 and ZnO-00-001-1136.

The layer deposited at RT (a sample S_ RT; see Figure 4a) consists of both ZnO and Zn crystallites, which are relatively small ($< D >$ = 6–13 nm) and show a number of orientations. Annealing of the deposited film at 300 °C (a sample S_ RT_ 300C) does not change the phase composition of the coating. The layers annealed at higher temperatures ($T_a \geq 500$ °C; samples S_ RT_ 500C and S_ RT_ 800C) contain only ZnO grains. The most prominent XRD peaks (2θ = 30°–40°) for these two samples are narrower than that recorded for S_ RT and S_ RT_ 300C—the average crystallite size is in the range 11–15 nm and 12–25 nm for S_ RT_ 500C and S_ RT_ 800C films, respectively. It should be noted that no growth direction is privileged. Completely different XRD patterns have been registered for samples deposited at 100 and 200 °C (see Figure 4b,c). Those films are composed of only ZnO (200) crystallites (the other XRD peaks show negligible intensity). The registered (200) direction of growth (as provileged orientation) was observed earlier for ZnO layers produced using various methods [14,17,19]. The average crystallite size increases with the increase in the annealing temperature from about 19 to about 30 nm (see Figure 4d).

Table 4. The average Zn and ZnO crystallite size ($< D >$) estimated based on the Scherrer formula.

	$< D >$ (nm)						
Sample id.	**Zn (100)**	**Zn (101)**	**Zn (102)**	**ZnO (100)**	**ZnO (002)**	**ZnO (101)**	**ZnO (110)**
S_RT	13 ± 2	13 ± 2	9 ± 3	6 ± 4	9 ± 5	9 ± 5	9 ± 3
S_RT_300C	13 ± 2	10 ± 3	9 ± 3	9 ± 5	11 ± 3	9 ± 5	9 ± 4
S_RT_500C	- [a]	- [a]	- [a]	15 ± 3	13 ± 2	11 ± 2	12 ± 4
S_RT_800C	- [a]	- [a]	- [a]	25 ± 3	22 ± 2	21 ± 2	12 ± 4
S_100C	- [a]	- [a]	- [b]	- [a]	19 ± 2	- [b]	- [b]
S_100C_300C	- [a]	- [a]	- [b]	- [a]	19 ± 2	- [b]	- [b]
S_100C_500C	- [a]	- [a]	- [b]	- [a]	26 ± 2	- [b]	- [b]
S_100C_800C	- [a]	- [a]	- [b]	- [a]	31 ± 2	- [b]	- [b]
S_200C	- [a]	- [a]	- [b]	- [a]	19 ± 3	- [b]	- [b]
S_200C_300C	- [a]	- [a]	- [b]	- [a]	18 ± 2	- [b]	- [b]
S_200C_500C	- [a]	- [a]	- [b]	- [a]	25 ± 2	- [b]	- [b]
S_200C_800C	- [a]	- [a]	- [b]	- [a]	30 ± 2	- [b]	- [b]

[a] not detected; [b] negligible intensity.

To investigate the chemical state of the surface of the produced samples the Zn 2p, O 1s and C 1s peaks have been recorded. Figure 5 shows Zn 2p and O 1s XPS signals for samples deposited at room temperature (both non-annealed and annealed at 300, 500 and 800 °C). The spectrum of zinc exhibits a characteristic doublet at 1022 eV (2p 3/2) and 1045 eV (2p 1/2) [14,20,39]. The O 1s signal consists of three components. The most prominent (with maximum at 530 eV; O(Zn)) can be assigned to oxygen atoms in the ZnO compound [14,20,21,40,41]. The second one (at 531 eV; O(vac.)) is associated with the O^{2-} in the oxygen-deficient region (vacancies) within the matrix of zinc oxide [42,43]. The last one (at 532 eV; O(ads.))—is connected to the adsorbed oxygen at the surface of the specimen [41]. It should be noted that the intensity (and thus concentration) of the second and the third components of the O 1s peak significantly decreases with the increase of the annealing temperature. Concentrations of particular elements are summarized in Table 5. The concentration of Zn (C_{Zn}) and O bounded with Zn ($C_{O(Zn)}$) atoms increases with the increase in the annealing temperature (for all series of samples). The opposite can be observed for the adsorbed oxygen ($C_{O(ads.)}$) and oxygen ions (O^{2-}) in the oxygen-deficient region as well as carbon (related to both sp^3 hybridization and adsorbed at the surface as CO_2) [39]. Despite the fact that the concentration of carbon ($C_{C(tot.)}$) for the non-annealed sample is 12–15% (see Table 5), it decreases rapidly with increasing annealing temperature and is about 1% for samples annealed at 800 °C.

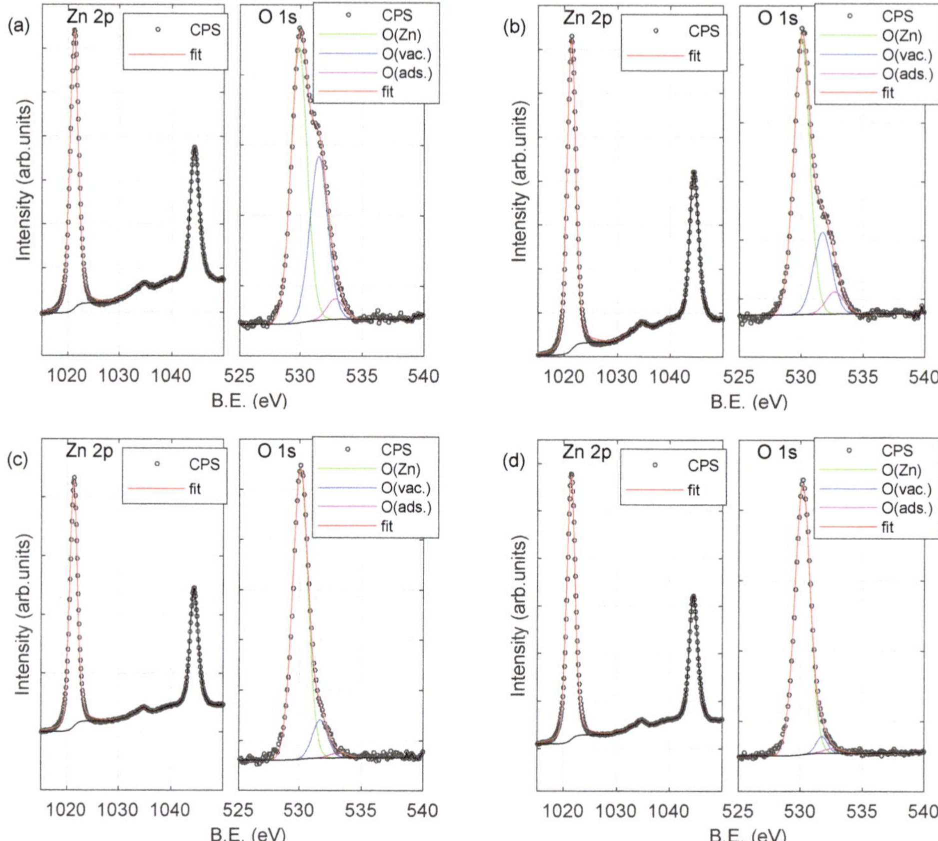

Figure 5. X-ray photoelectron spectroscopy (XPS) spectra (Zn 2p and O 1s peaks) of the sample deposited at room temperature (RT): (**a**) non-annealed and annealed at (**b**) 300 °C, (**c**) 500 °C and (**d**) 800 °C.

Deconvolution of the O 1s signal allowed us to determine the ratio of Zn and O atoms at the surface of the prepared films taking into account total concentration of oxygen ($C_{O(tot.)}$) or the part related only to O ($C_{O(Zn)}$) atoms bounded with Zn atoms. These estimated ratios are presented in Table 6. The ratio between zinc and oxygen atoms (total) for the non-annealed samples (S_ RT, S_ 100C and S_ 200C) is about 1 and increases to about 1.6 with the increase of the annealing temperature. Significantly higher values have been determined taking into account only the part of O 1s peak related to oxygen atoms bounded with zinc atoms. These values are in the range from about 1.5 to 1.9 (see Table 6).

Table 5. Concentration of detected elements (Zn, O and C) at the surface of the produced layers as well as the ratio of Zn to O atoms.

Sample id.	C_{Zn} (%)	$C_{O(Zn)}$ (%)	$C_{O(vac.)}$ (%)	$C_{O(ads.)}$ (%)	$C_{O(tot.)}$ (%)	$C_{C(tot.)}$ (%)
S_RT	44.23	25.28	16.16	1.84	43.28	12.49
S_RT_300C	48.19	29.15	9.67	2.65	41.47	10.34
S_RT_500C	51.78	33.89	4.35	0.70	38.94	9.28
S_RT_800C	59.43	37.64	1.72	0.79	40.15	0.42
S_100C	43.23	25.99	12.00	3.95	41.94	14.83
S_100C_300C	43.82	29.46	5.80	6.34	41.60	14.58
S_100C_500C	55.60	35.99	1.64	1.25	38.88	5.52
S_100C_800C	58.50	36.19	2.90	1.28	40.37	1.13
S_200C	46.98	28.09	9.93	3.16	41.18	11.84
S_200C_300C	48.08	30.30	8.26	1.70	40.26	11.66
S_200C_500C	56.41	34.67	2.04	0.24	36.95	6.64
S_200C_800C	60.45	33.64	4.82	0.48	38.94	0.61

Table 6. The estimated ratio of Zn to O atoms.

Sample id.	Zn:O (tot.)	Zn:O (Zn)
S_RT	1.07	1.75
S_RT_300C	1.24	1.65
S_RT_500C	1.35	1.53
S_RT_800C	1.51	1.58
S_100C	1.14	1.66
S_100C_300C	1.24	1.49
S_100C_500C	1.48	1.54
S_100C_800C	1.50	1.62
S_200C	1.24	1.67
S_200C_300C	1.25	1.59
S_200C_500C	1.54	1.63
S_200C_800C	1.57	1.80

4. Discussion

Both AFM (see Figure 3 and Table 2) and XRD (see Figure 4 and Table 4) results show that the grain size perpendicular to the surface increases with annealing temperature as the grains coalesce, regardless of the sample deposition temperature. The trend is similar for the lateral grain size, however in the case of ZnO deposited at RT the lateral grain size is already so large that no annealing-induced coalescence is expected.

Further results presented in the previous section unambiguously show that the main changes during annealing were observed for the samples deposited at room temperature. The changes were much less prominent for samples deposited at elevated temperatures. Therefore in this section, we focus our attention on the explanation of this phenomenon and on the effect of annealing on microstructure and optical properties of the fabricated films.

First, it was noticed that the thicknesses of ZnO films deposited onto substrates at elevated temperatures during the deposition process (100 and 200 °C) as determined form ellipsometric measurements are significantly lower (by an order of magnitude) than in the case of films deposited onto substrates at room temperature (see Table 2). Second, as described in the Results section, the permittivity of these films exhibits a more oxide-like behaviour (Figure 2b–d), whereas films deposited onto substrates at room temperature have more metal-like characteristics (Figure 2a). This might be a consequence of two phenomena—ZnO decomposition during the deposition process (which will result in an excess of Zn atoms in the deposited film) as well as subsequent excess Zn atom surface segregation as desorption in elevated temperatures, which would improve stoichiometry. The details are described below.

During the e-beam evaporation process, many oxide materials are known to decompose [28–31] releasing oxygen into the vacuum. The decomposition of ZnO has also been observed [19,20]. Thus, to ensure there is no deficiency of oxygen atoms in the evaporated film, it has been proposed to perform the deposition process in an oxygen atmosphere [19]. In our case, the ZnO films have been deposited in a high vacuum. This approach results in the evaporated films containing an excess of Zn atoms with respect to typical ZnO stoichiometry. Since samples fabricated using e-beam technology are typically polycrystalline, the excess Zn atoms are expected to segregate to the ZnO grain boundaries as well as to the film's free surface, as this process would result in the release of the elastic strain energy accumulated in the non-stoichiometric ZnO crystallites [44,45]. This is due to the fact that within the crystallites there is a limited number of lattice vacancies which enable the excess Zn atoms to occupy substitutional lattice sites. Most of the excess Zn atoms are therefore expected to occupy the interstitial lattice sites. This would induce much higher local distortions (strain) to the crystal lattice. Because of numerous lattice vacancies being present on the free surface and at the grain boundaries, there are more substitutional lattice sites in those regions for the excess Zn atoms to occupy. Moreover, the excess Zn atoms present in the interstitial sites on the surface and at the grain boundaries induce strain in a lower volume of the crystal lattice than the excess Zn atoms present in the interstitial lattice sites inside the ZnO grains [44,45]. Although elastic strain energy can properly be calculated only for mixtures of monoatomic substances, its contribution to the enthalpy of segregation is always negative [26], indicating that the segregation of the excess Zn atoms towards the surface or the grain boundaries of the ZnO films might be spontaneous.

In the famous model of segregation proposed by Wynblatt and Ku [46], one of the contributions to the enthalpy of segregation—besides the elastic strain energy–is also the difference between the surface energies of the minority substance (in our case Zn) and the majority substance (in our case ZnO), multiplied by the surface area per atom. As the surface energy of Zn γ_{Zn} is below 0.573 $\mathrm{J/m^2}$ [47–49] and the surface energy of ZnO γ_{ZnO} is within 0.94 and 4.05 $\mathrm{J/m^2}$ [50–52], then the value of $\gamma_{Zn} - \gamma_{ZnO}$ will always be negative. Thus we might assume that segregation of Zn atoms towards the surface or the grain boundaries of the ZnO films is occurring spontaneously.

In the case of samples deposited at room temperature, this process would result in a non-uniform distribution of Zn atoms throughout a sample, with the possibility of Zn precipitations at the ZnO grain boundaries and at the film's surface. The presence of diffraction peaks at 39° and 43° corresponding to 100 and 101 surfaces of Zn crystallites (see Figure 4), confirms the presence of such precipitates. Moreover, the Zn-to-O atomic ratio estimated from the XPS spectra recorded from the surface is 1.89 (see Table 6) which indicates a high excess of Zn atoms with respect to normal ZnO stoichiometry.

The situation is, however, drastically different in the case of samples deposited on substrates at elevated temperatures. The higher the substrate temperature, the higher the probability that the deposited material will grow in an island (Volmer–Weber) mode. A metric that helps determine this is the homologous temperature T_h defined as [53,54]:

$$T_h = T_s / T_m \quad (7)$$

where T_s is the substrate temperature and T_m is the bulk melting temperature of the deposited material. As a threshold, 0.3 is generally accepted as a value above which an island mode is observed [53]—and the higher the value, the longer the island mode persists during the deposition process. As in the case of e-beam evaporated ZnO there is a large excess of Zn atoms due to the decomposition of the zinc oxide, the T_m is probably closer to the bulk melting temperature of Zn (693 K) than the bulk melting temperature of ZnO (2247 K). For pure Zn, even the case of the substrate at room temperature, the value of homologous temperature is already as high as 0.425 and for higher temperatures, it will be even greater. Thus, we can assume, that the ZnO films during our deposition process grow in an island mode.

If that is the case, then the surface-to-volume ratio at the initial stages of the growth is very large, and for elevated substrate temperatures it may remain so even in the late stages of the growth. Thus, the excess of Zn atoms might quickly segregate towards the surface of these ZnO islands and form Zn surface precipitations even during deposition. This should progress at a faster rate for elevated substrate temperatures—for instance, it has been demonstrated that elevating the substrate temperature from room temperature to 100 °C results in faster segregation of Ge atoms within Ag films [55]. The Zn surface precipitations are expected to be very thin, not exceeding the thickness of several nm. It has been shown previously that for such thin metal layers, the melting temperature (and thus also the boiling temperature) is much lower than in the case of bulk materials [56]. This suggests that in the case of elevated substrate temperatures, the excess Zn precipitations boil or sublimate from the surface and leave the system thus forming a much thinner, though a more stoichiometric ZnO film. The Zn-to-O atomic ratio, in this case, is closer to the stoichiometric value of 1 (see Table 6) than in the case of samples deposited at room temperature, but still an excess of Zn atoms in XPS measurements is observed. This is probably due to the fact, that the XPS spectra (see Figure 5) are collected from the very surface, to which Zn atoms segregate. The phenomenon described above certainly contributes to the fact, that the thickness of the samples deposited at elevated temperatures are lower than in the case of room temperature (see Table 2), but is probably not the only cause. We believe that, at elevated substrate temperatures, the adsorption energy of Zn and O adatoms might be smaller than their thermal energy, which may result in whole ZnO clusters desorbing from the surface of the substrate, which reduced the overall film thickness. Since the deposition rate of the fabricated samples was very low (0.3 Å/s), the movement of adatoms is not restricted by the stream of subsequent adatoms, which, in a typical case, would prevent such desorption.

The shape of the optical constants of the deposited films (see Figure 2) can be explained taking into account the above-mentioned reason. The metallic-like layers (S_ RT and S_ RT_ 300) consist of both ZnO and Zn crystallites (see Figure 4a). Their optical constants exhibit metallic-like behaviour, however without strong absorption caused by the interaction of electromagnetic radiation with free-carriers. This is due to negligible electrical continuity between zinc crystallites since they are separated by ZnO. The layers deposited at RT and annealed at 500 and 800 °C contain only ZnO crystallites. General shapes of the refractive index and the extinction coefficient curves (see Figure 2b) are close to those determined for ZnO [38], however, the values are significantly lower. In the IR spectral range, n equals only 1.5–1.6. Although the band-gap energy was established to be 3.24 eV (value reported for stoichiometric ZnO [4,13]), the value of k in the absorbing spectral range (wavelengths below ~380 nm) is about 0.2 or 0.3 for the S_ RT_ 500C and S_ RT_ 800C specimens, respectively. These discrepancies most probably result from the fact that the ZnO crystallites in these samples grow in different orientations, which also influences the nanoporosity of the fabricated material. The films deposited at 100 °C and 200 °C contain the ZnO crystallites with almost exclusively the 002 direction (see Figure 5b,c) and their optical constants, as well as band-gap energy (3.19 eV), are close to the values reported for the stoichiometric zinc oxide [4,13]. The Zn-rich extremely thin surface film does not affect the optical constants of the deposited ZnO layers. It should be noted that annealing of ZnO films deposited at 100 and 200 °C leads to the increase in the mean crystallite size (see Table 4 and Figure 4d) for the 002 direction from 18–19 nm to 30–31 nm. This effect causes the slight blue-shift of band-gap energy from 3.19 to 3.21–3.22 eV (see Table 3).

5. Conclusions

Zinc oxide films were produced by means of the electron beam evaporation method and investigated using atomic force microscopy, scanning electron microscopy, powder X-ray diffraction, X-ray photoelectron spectroscopy and spectroscopic ellipsometry techniques. Influence of substrate and annealing temperatures on the film composition as well as microstructure and optical properties was examined. We have shown that decomposition of ZnO during evaporation and segregation of Zn atoms during evaporation and post-deposition annealing significantly affect the real part of the complex refractive index and the extinction coefficient. We have shown that zinc oxide film produced at room temperature consists of ZnO and Zn crystallites and its optical constants exhibit a metallic-like behaviour. Annealing at 500 and 800 °C leads to desorption of Zn atoms form the surface during deposition. As a result, we have obtained a layer with only ZnO crystallites which is characterised by a complex refractive index typical for semiconducting materials. However, because of a variety of orientations in which the crystallites have grown there is an increased volume of grain boundary voids. Thus, the values of the refractive index are significantly lower (1.5–1.7 in the visible spectral range) than those reported for high-quality zinc oxide films. In the case of the ZnO layers deposited at 100 and 200 °C are characterised by an order of magnitude smaller thickness than that obtained for films evaporated at RT. The real part of the complex refractive index exhibits high value (1.9) in the non-absorbing spectral range and annealing does not lead to significant changes in the optical constants of the fabricated zinc oxide layers. The value of the band-gap energy for all the films exhibiting semiconducting behaviour was established to be ∼3.2 eV.

Author Contributions: Conceptualization, L.S. and A.C.; methodology, L.S. and A.C.; software, L.S.; validation, L.S., A.C. and A.O.; formal analysis, L.S., A.C. and A.O; investigation, L.S., A.O., M.T., R.S. and M.N.; resources, A.O. and L.S.; writing—original draft preparation, L.S. and A.C.; writing—review and editing, L.S. and A.C.; visualization, L.S.; supervision, A.B. All authors have read and agreed to the published version of the manuscript.

Funding: This research received no external funding.

Conflicts of Interest: The authors declare no conflict of interest.

Appendix A. SEM Images

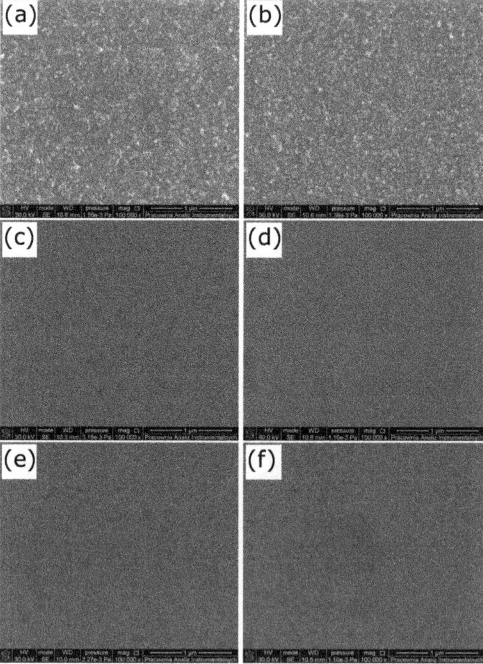

Figure A1. Scanning electron microscopy (SEM) images of ZnO layers deposited at RT: (**a**) not annealed and (**b**) annealed at 500 °C, deposited at 100 °C: (**c**) not annealed and (**d**) annealed at 500 °C as well as deposited at 200 °C: (**e**) not annealed and (**f**) annealed at 500 °C.

References

1. Özgür, Ü.; Alivov, Y.I.; Liu, C.; Teke, A.; Reshchikov, M.A.; Doğan, S.; Avrutin, V.; Cho, S.-J.; MorkoÇ, K. A comprehensive review of ZnO materials and devices. *J. Appl. Phys.* **2005**, *98*, 041301. [CrossRef]
2. Djurišić, A.B.; Ng, A.M.C.; Chen, X.Y. ZnO nanostructures for optoelectronics: Material properties and device applications. *Prog. Quantum Electron.* **2010**, *34*, 191–259. [CrossRef]
3. Pal, D.; Singhal, J.; Mathur, A.; Singh, A.; Dutta, S.; Zollner, S.; Chattopadhyay, S. Effect of substrates and thickness on optical properties in atomic layer deposition grown ZnO thin films. *Appl. Surf. Sci.* **2017**, *421*, 341–348. [CrossRef]
4. Guziewicz, E.; Kowalik, I.A.; Godlewski, M.; Kopalko, K.; Osinniy, V.; Wójcik, A.; Yatsunenko, S.; Łusakowska, E.; Paszkowicz, W.; Guziewicz, M. Extremely low temperature growth of ZnO by atomic layer deposition. *J. Appl. Phys.* **2008**, *103*, 033515. [CrossRef]
5. Muth, J.F.; Lee, J.H.; Shmagin, I.K.; Kolbas, R.M.; Casey, H.C., Jr.; Keller, B.P.; Mishra, U.K.; DenBaars, S.P. Absorption coefficient, energy gap, exciton binding energy, and recombination lifetime of GaN obtained from transmission measurements. *Appl. Phys. Lett.* **1997**, *71*, 2572–2574. [CrossRef]
6. Tan, S.T.; Chen, B.J.; Sun, X.W.; Fan, W.J.; Kwok, H.S.; Zhang, X.H.; Chua, S.J. Blueshift of optical band gap in ZnO thin films grown by metal-organic chemical-vapor deposition. *J. Appl. Phys.* **2005**, *98*, 013505. [CrossRef]
7. Peiró, A.M.; Ravirajan, P.; Govender, K.; Boyle, D.S.; O'Brien, P.; Bradley, D.D.; Nelson, J.; Durrant, J.R. Hybrid polymer/metal oxide solar cells based on ZnO columnar structures. *J. Mater. Chem.* **2006**, *16*, 2088–2096. [CrossRef]
8. Huang, J.; Yin, Z.; Zheng, Q. Applications of ZnO in organic and hybrid solar cells. *Energy Environ. Sci.* **2011**, *4*, 3861–3877. [CrossRef]
9. Carcia, P.F.; McLean, R.S.; Reilly, M.H. High-performance ZnO thin-film transistors on gate dielectrics grown by atomic layer deposition. *Appl. Phys. Lett.* **2006**, *88*, 123509. [CrossRef]
10. Suchea, M.; Christoulakis, S.; Moschovis, K.; Katsarakis, N.; Kiriakidis, G. ZnO transparent thin films for gas sensor applications. *Thin Solid Films* **2006**, *515*, 551–554. [CrossRef]
11. Kim, J.; Lee, J. Wet chemical etching of ZnO films using NH_x-based $(NH_4)_2CO_3$ and NH_4OH alkaline solution. *J. Mater. Sci.* **2017**, *52*, 13054–13063. [CrossRef]
12. Velayi, E.; Norouzbeigi, R. Synthesis of hierarchical superhydrophobic zinc oxide nano-structures for oil/water separation. *Ceram. Int.* **2018**, *44*, 14202–14208. [CrossRef]
13. Nie, J.C.; Yang, J.Y.; Piao, Y.; Li, H.; Sun, Y.; Xue, Q.M.; Xiong, C.M.; Dou, R.F.; Tu, Q.Y. Quantum confinement effect in ZnO thin films grown by pulsed laser deposition. *Appl. Phys. Lett.* **2008**, *93*, 173104. [CrossRef]
14. Mosquera, A.A.; Horwat, D.; Rashkovskiy, A.; Kovalev, A.; Miska, P.; Wainstein, D.; Albella, J.M.; Endrino, J.L. Exciton and core-level electron confinement effects in transparent ZnO thin films. *Sci. Rep.* **2013**, *3*, 1714. [CrossRef]
15. Li, X.D.; Chen, T.P.; Liu, P.; Liu, Y.; Liu, Z.; Leong, K.C. A study on the evolution of dielectric function of ZnO thin films with decreasing film thickness. *J. Appl. Phys.* **2014**, *115*, 103512. [CrossRef]
16. Agarwal, D.C.; Chauhan, R.S.; Kumar, A.; Kabiraj, D.; Singh, F.; Khan, S.A.; Avasthi, D.K.; Pivin, J.C.; Kumar, M.; Ghatak, J.; et al. Synthesis and characterization of ZnO thin film grown by electron beam evaporation. *J. Appl. Phys.* **2006**, *99*, 123105. [CrossRef]
17. Wu, H.Z.; He, K.M.; Qiu, D.J.; Huang, D.M. Low-temperature epitaxy of ZnO films on Si(001) and silica by reactive e-beam evaporation. *J. Cryst. Growth* **2017**, *217*, 131–137. [CrossRef]
18. Varnamkhasti, M.G.; Fallah, H.R.; Zadsar, M. Effect of heat treatment on characteristics of nanocrystalline ZnO films by electron beam evaporation. *Vacuum* **2012**, *86*, 871–875. [CrossRef]
19. Asmar, R.D.; Zaouk, D.; Bahouth, P.H.; Podleki, J.; Foucaran, A. Characterization of electron beam evaporated ZnO thin films and stacking ZnO fabricated by e-beam evaporation and rf magnetron sputtering for the realization of resonators. *Microelectron. Eng.* **2006**, *83*, 393–398. [CrossRef]
20. Yun, E.-J.; Jung, J.W.; Han, Y.H.; Kim, M.-W.; Lee, B.C. Effect of high-energy electron beam irradiation on the properties of ZnO thin films prepared by magnetron sputtering. *J. Appl. Phys.* **2009**, *105*, 123509. [CrossRef]
21. Chen, Y.; Nayak, J.; Ko, H.-U.; Kim, J. Effect of annealing temperature on the characteristics of ZnO thin films. *J. Phys. Chem. Solids* **2012**, *73*, 1259–1263. [CrossRef]

22. Kaviyarasu, K.; Magdalane, C.M.; Kanimozhi, K.; Kennedy, J.; Siddhardha, B.; Reddy E.S.; Rotte, N.K.; Sharma, C.S.; Thema, F.T.; Letsholathebe, D.; et al. Elucidation of photocatalysis, photoluminescence and antibacterial studies of ZnO thin films by spin coating method. *J. Photochem. Photobiol. B* **2017**, *173*, 466–475. [CrossRef] [PubMed]
23. Kato, H.; Sano, M.; Miyamoto, K.; Yao, T. Effect of O/Zn flux ratio on crystalline quality of ZnO films grown by plasma-assisted molecular beam epitaxy. *Jpn. J. Appl. Phys.* **2003**, *42*, 2241–2244. [CrossRef]
24. Ciesielski, A.; Skowronski, L.; Trzcinski, M.; Górecka, E.; Trautman, P.; Szoplik, T. Evidence of germanium segregation in gold thin films. *Surf. Sci.* **2018**, *674*, 73–78. [CrossRef]
25. Ciesielski, A.; Skowronski, L.; Trzcinski, M.; Szoplik, T. Controlling the optical parameters of self-assembled silver films with wetting layers and annealing. *Appl. Surf. Sci.* **2017**, *421*, 349–356. [CrossRef]
26. Ciesielski, A.; Trzcinski, M.; Szoplik, T. Inhibiting the segregation of germanium in silver nanolayers. *Crystals* **2020**, *10*, 262. [CrossRef]
27. Garlisi, C.; Scandura, G.; Szlachetko, J.; Ahmadi, S.; Sa, J.; Palmisano, G. E-beam evaporated TiO_2 and Cu-TiO_2 on glass: Performance in the discoloration of methylene blue and 2-propanol oxidation. *Appl. Catal. A* **2016**, *526*, 191–199. [CrossRef]
28. Material Deposition Chart. Available online: https://www.lesker.com/newweb/deposition_materials/materialdepositionchart.cfm?pgid=0 (accessed on 8 August 2020).
29. Jiang, N. Electron beam damage in oxides: A review. *Rep. Prog. Phys.* **2016**, *79*, 016501. [CrossRef]
30. Watanabe, H.; Fujita, S.; Maruno, S.; Fujita, K.; Ichikawa, M. Electron-beam-induced selective thermal decomposition of ultrathin SiO_2 layers used in nanofabrication. *Jpn. J. Appl. Phys.* **1997**, *36*, 7777–7781. [CrossRef]
31. Mens, A.J.M.; Gijzeman, O.L.J. AES study of electron beam induced damage on TiO_2 surfaces. *Appl. Surf. Sci.* **1996**, *99*, 133–143. [CrossRef]
32. Pinhas, H.; Malka, D.; Danan, Y.; Sinvani, M.; Zalevsky, Z. Design of fiber-integrated tunable thermo-optic C-band filter based on coated silicon slab. *J. Eur. Opt. Soc. Rapid Publ.* **2017**, *13*, 32. [CrossRef]
33. Samoi, E.; Benezra, Y.; Malka, D. An ultracompact 3×1 MMI power-combiner based on Si slot-waveguide structures. *Photonic. Nanostruct.* **2020**, *39*, 100780. [CrossRef]
34. Fujiwara, H. Spectroscopic Ellipsometry. In *Principles and Applications*; John Wiley & Sons Ltd.: Chichester, UK, 2009.
35. J.A. Woollam Co., Inc. *Guide to Using WVASE32®*; Wextech Systems Inc.: New York, NY, USA, 2010.
36. Rerek, T.; Skowronski, L.; Szczesny, R.; Naparty, M.K.; Derkowska-Zielinska, B. The effect of the deposition rate on microstructural and opto-electronic properties of β-Sn layers. *Thin Solid Films* **2019**, *670*, 86–92. [CrossRef]
37. Mosteller, L.P.; Wooten, F. Optical Properties of Zn. *Phys. Rev.* **1968**, *171*, 743–749. [CrossRef]
38. Yoshikawa, H.; Adachi, S. Optical Constants of ZnO. *Jpn. J. Appl. Phys.* **1997**, *36*, 6237–6243. [CrossRef]
39. Hüfner, S. *Photoelectron Spectroscopy: Principles and Applications*; Springer: Berlin/Heidelberg, Germany, 2003.
40. Nefedov, V.I. *XPS of Solid Surfaces*; VSP: Utrecht, The Netherlands, 1988.
41. Mass, J.; Bhattacharya, P.; Katiyar, R.S. Effect of high substrate temperature on Al-doped ZnO thin films grown by pulsed laser deposition. *Mater. Sci. Eng. B* **2003**, *103*, 9–15. [CrossRef]
42. Chen, T.; Liu, S.-Y.; Xie, Q.; Detavernier, C.; Van Meirhaeghe, R.L.; Qu, X.-P. The effects of deposition temperature and ambient on the physical and electrical performance of DC-sputtered n-ZnO/p-Si heterojunction. *Appl. Phys. A* **2010**, *98*, 357–365. [CrossRef]
43. Lai, L.-W.; Lee, C.-T. Investigation of optical and electrical properties of ZnO thin films. *Mater. Chem. Phys.* **2008**, *110*, 393–396. [CrossRef]
44. Paul, A.; Laurila, T.; Vuorinen, V.; Divinski, S.V. *Thermodynamics, Diffusion and the Kirkendall Effect in Solids*; Springer: Berlin/Heidelberg, Germany, 2014.
45. Lejček, P. *Grain Boundary Segregation in Metals*; Springer: Berlin/Heidelberg, Germany, 2010.
46. Wynblatt, P.; Ku, R.C. Surface energy and solute strain energy effects in surface segregation. *Surf. Sci.* **1977**, *65*, 511–531. [CrossRef]
47. Rudawska, A.; Kuczmaszewski, J. Surface free energy of zinc coating after finishing treatment. *MSP* **2006**, *24*, 975–981.
48. Maitland, A.H.; Chadwick, G.A. The cleavage surface energy of zinc. *Philos. Mag.* **1969**, *19*, 645–651. [CrossRef]

49. Bilello, J.C.; DewHughes, D.; Pucino, A.T. The surface energy of zinc. *J. Appl. Phys.* **1983**, *54*, 1821–1826. [CrossRef]
50. Mora Fonz, D.P. A Theoretical Study on the Surfaces of Zinc Oxide. Ph.D. Thesis, University College London, London, UK, 2016.
51. Tang, C.; Spencer, M.J.S.; Barnard, A.S. Activity of ZnO polar surfaces: An insight from surface energies. *Phys. Chem. Chem. Phys.* **2014**, *16*, 22139–22144. [CrossRef] [PubMed]
52. Wang, S.; Fan, Z.; Koster, R.S.; Fang, C.; van Huis, M.A.; Yalcin, A.O.; Tichelaar, F.D.; Zandbergen, H.W.; Vlugt, T.J.H. New Ab initio based pair potential for accurate simulation of phase transitions in ZnO. *J. Phys. Chem. C* **2014**, *118*, 11050–11061. [CrossRef]
53. Flötotto, D.; Wang, Z.M.; Jeurgens, L.P.H.; Bischoff, E.; Mittemeijer, E.J. Effect of adatom surface diffusivity on microstructure and intrinsic stress evolutions during Ag film growth. *J. Appl. Phys.* **2012**, *112*, 043503. [CrossRef]
54. Mahieu, S.; Ghekiere, P.; Depla, D.; De Gryse, R. Biaxial alignment in sputter deposited thin films. *Thin Solid Films* **2006**, *515*, 1229–1249. [CrossRef]
55. Ciesielski, A.; Skowroński, Ł.; Górecka, E.; Kierdaszuk, J.; Szoplik, T. Growth model and structure evolution of Ag films deposited on Ge. *Beilstein J. Nanotechnol.* **2018**, *9*, 66–76. [CrossRef]
56. Kitsyuk, E.P.; Gromov, D.G.; Redichev, E.N.; Sagunova, I.V. Specifics of low-temperature melting and disintegration in to drops of silver thin films. *Prot. Metals Phys. Chem.* **2012**, *48*, 304–309. [CrossRef]

Article

Effect of Ni on the Suppression of Sn Whisker Formation in Sn-0.7Cu Solder Joint

Aimi Noorliyana Hashim [1,2], Mohd Arif Anuar Mohd Salleh [1,2,*], Andrei Victor Sandu [1,3,4,*], Muhammad Mahyiddin Ramli [5], Khor Chu Yee [6], Noor Zaimah Mohd Mokhtar [1,2] and Jitrin Chaiprapa [7]

1 Centre of Excellence Geopolymer and Green Technology, Universiti Malaysia Perlis (UniMAP), Taman Muhibah, Jejawi, Arau, Perlis 02600, Malaysia; aimiliyana@unimap.edu.my (A.N.H.); noorzaimah3287@gmail.com (N.Z.M.M.)
2 Faculty of Chemical Engineering Technology, Universiti Malaysia Perlis (UniMAP), Taman Muhibah, Jejawi, Arau, Perlis 02600, Malaysia
3 Faculty of Materials Science and Engineering, Gheorghe Asachi Technical University, D. Mangeron 41, 700050 Iasi, Romania
4 Romanian Inventors Forum, Str. Sf. P. Movila 3, L11, 700089 Iasi, Romania
5 School of Microelectronic Engineering, Pauh Putra Campus, University Malaysia Perlis (UniMAP), Arau, Perlis 02600, Malaysia; mmahyiddin@unimap.edu.my
6 Faculty of Engineering Technology (FETech), Universiti Malaysia Perlis (UniMAP), Level 1, Block S2, UniCITI Alam Campus, Sungai Chuchuh, Padang Besar, Perlis 02100, Malaysia; cykhor@unimap.edu.my
7 Synchrotron Light Research Institute (SLRI), 111 University Avenue, Muang District, Nakhon Ratchasima 30000, Thailand; jitrin@slri.or.th
* Correspondence: arifanuar@unimap.edu.my (M.A.A.M.S.); sav@tuiasi.ro (A.V.S.)

Citation: Hashim, A.N.; Salleh, M.A.A.M.; Sandu, A.V.; Ramli, M.M.; Yee, K.C.; Mohd Mokhtar, N.Z.; Chaiprapa, J. Effect of Ni on the Suppression of Sn Whisker Formation in Sn-0.7Cu Solder Joint. *Materials* **2021**, *14*, 738. https://doi.org/10.3390/ma14040738

Received: 8 December 2020
Accepted: 13 January 2021
Published: 5 February 2021

Abstract: The evolution of internal compressive stress from the intermetallic compound (IMC) Cu_6Sn_5 growth is commonly acknowledged as the key inducement initiating the nucleation and growth of tin (Sn) whisker. This study investigates the effect of Sn-0.7Cu-0.05Ni on the nucleation and growth of Sn whisker under continuous mechanical stress induced. The Sn-0.7Cu-0.05Ni solder joint has a noticeable effect of suppression by diminishing the susceptibility of nucleation and growth of Sn whisker. By using a synchrotron micro X-ray fluorescence (μ-XRF) spectroscopy, it was found that a small amount of Ni alters the microstructure of Cu_6Sn_5 to form a $(Cu,Ni)_6Sn_5$ intermetallic layer. The morphology structure of the $(Cu,Ni)_6Sn_5$ interfacial intermetallic layer and Sn whisker growth were investigated by scanning electron microscope (SEM) in secondary and backscattered electron imaging mode, which showed that there is a strong correlation between the formation of Sn whisker and the composition of solder alloy. The thickness of the $(Cu,Ni)_6Sn_5$ IMC interfacial layer was relatively thinner and more refined, with a continuous fine scallop-shaped IMC interfacial layer, and consequently enhanced a greater incubation period for the nucleation and growth of the Sn whisker. These verification outcomes proposes a scientifically foundation to mitigate Sn whisker growth in lead-free solder joint.

Keywords: Sn whisker; reliability; lead-free solder; Sn-0.7Cu-0.05Ni; whisker growth; mitigation; interconnects; soldering

1. Introduction

Presently, the lead (Pb)-free solder has become a top priority in electronic industries. This is concurrent with the enactment of the EU Directive 2002/95/EC that states the use of lead (Pb) must either be eliminated or limited in a very low concentration in electrical and electronic equipment (EEE) [1–8]. The regulation to limit the use of lead in solder has brought back the Sn whiskers phenomenon and triggered numerous system failures [1,3,8–13]. Furthermore, the trends of electronics miniaturization in microelectronics packages have resulted in a serious reliability issue of lead-free solder in electronic devices [6,8,14,15]. Numerous studies have already been carried out, and researchers agree that the main inducement factor for Sn whisker nucleation and growth due to the stress progression from the development of a SnCu intermetallic compound (IMC) layer at room temperature [2,3,8,11,13,16–19]. A Cu_6Sn_5 IMC interfacial layer begins to form at the solder/substrate interface as the copper (Cu) atoms diffuse into the tin (Sn) layer. The growth of intermetallic particles into the tin (Sn) layer to reach the solution saturation limit can lead to a significant effect on thermal expansion mismatch, and thus create a residual stress gradient at the solder/substrate interface [20].

Illes et al. [21] surmise that the main source for whiskers growth is the non-uniform shape of IMC interfacial layers and expansion of the IMC interfacial layer at the solder/substrate interface layers [8,21,22]. As explained by Kim et al. [22], the Sn whisker prone to grow with irregularly shaped due to high residual compressive stresses concentrated at IMC interfacial layer. The natural surroundings of the Sn whisker development are spontaneous and slow-moving, and consequently, an assessment of the Sn whisker kinetic growth can often result in an unreliable and unpredictable growth rate [8]. It is most beneficial to have access to a fundamental study of the Sn whiskers phenomenon in a well-ordered stimulus condition [8]. Multiple studies have been carried out to study the acceleration behavior of Sn whisker formation and growth through mechanical stress induced [13,23], but only few works have shown continuous mechanical stress induced to accelerate Sn whisker nucleation and growth [2,8]. From previous studies, both nucleation and growth of Sn whisker are capable to form, even though there is no external induced stress applied. However, as determined by Lin et al. [2], the Sn whisker growth is longer and denser, with continuous mechanical stress induced [2]. In addition, Jagtap et al. [23] also claimed that the stress gradient in the solder/substrate interface layers accelerates the nucleation and growth of Sn whiskers with time evolvements under continuous mechanical stress induced by indentation test fixture [23].

Many existing studies have verified that lead (Pb) noticeably has a great suppression significance on the Sn whisker nucleation and formation by enhancing the stress relaxation behavior in the solder layer [3,9,18,20]. Much effort has been aimed at reliability improvement techniques, such as adding a third element [10,24,25], but a spot-on lead-free solder replacement has not yet been found to suppress whiskers formation as effectively as the addition of Pb in solder [14,22,26]. Many studies found similar results that specify the mechanism for the stress relaxation enhancement in solder layers with the addition of micro-alloying elements in the solder joint [11,15]. Considering the effects of adding micro-alloying elements in lead-free solder can certainly lead to the formation of new compounds in the IMC interfacial layer in addition to Cu_6Sn_5, and can significantly modify the properties of the IMC interfacial layer [7,27].

Micro-alloying with Ni indicates noticeably high solubility in Cu_6Sn_5 and pointedly influences the microstructure of the lead-free solder joint. Sn-0.7Cu-0.05Ni enables the nucleation of Cu_6Sn_5 and is possible to solidify as a eutectic, similar to tin lead solder [28]. It has been reported that Ni is able to reduce both magnitude [29] and expansion discontinuity of thermal expansion of Cu_6Sn_5 with the polymorphic phase transformation, mainly along the a-axis [27]. The distinctive properties of Sn-0.7Cu-0.05Ni could diminish the internal stress residual in solder joints, particularly during the thermal cyclic condition [27]. It has also been discovered that Ni micro-alloying can generate stabilization at high-temperature (around 250 °C) phase transition of the $(Cu,Ni)_6Sn_5$ IMC hexagonal phase [30] and is able to increase the stress relaxation in the layer with the formation of fine, needle-like $(Cu,Ni)_6Sn_5$ at the Sn–Cu interface [8]. In addition, Sn-0.7Cu-0.05Ni has good fluidity and oxidation resistance condition compared to the Sn–0.7Cu alloy, and also, $(Cu,Ni)_6Sn_5$ grows as an initial phase during solidification [27]. Therefore, it is remarkable to determine the correlation between the $(Cu,Ni)_6Sn_5$ IMC interfacial layer and Sn whisker formation with regard to the IMC interfacial layer suppression and stress relaxation enhancement for both the nucleation and growth of the Sn whisker.

This study investigated the effects of Sn-0.7Cu-0.05Ni solder joints on the nucleation and growth of the Sn whisker under continuous mechanical stress induced using scanning electron microscope (SEM), synchrotron microbeam X-ray fluorescence (μ-XRF) spectroscopy and atomic force microscopy (AFM). The findings resulted an in-depth understanding of the suppression effects of the $(Cu,Ni)_6Sn_5$ IMC interfacial layer on the nucleation and growth of the Sn whisker in the Sn-0.7Cu-0.05Ni solder joint.

2. Experimental Procedures

2.1. Materials

The raw materials of Sn-0.7Cu and Sn-0.7Cu-0.05 Ni solder alloy were provided by Nihon Superior (M) Sdn. Bhd., Ipoh, Perak, Malaysia. The solder alloys were prepared by melting ingots in a solder pot at a temperature of 265 °C. The industrial high-purity Cu above 99.9% were used as substrates with dimensions of 1.5×1.5 cm^2 and 1.0 cm in thickness. Table 1 depicts the elemental composition analysis by arc spark spectrometer used throughout the study.

Table 1. The chemical composition analysis of lead (Pb)-free solder alloy and substrate (wt.%).

Composition	Sn-0.7Cu (wt.%)	Sn-0.7Cu-0.05Ni (wt.%)	Cu Substrate (wt.%)
Sn	99.2276	99.2184	–
Cu	0.7683	0.6563	99.9426
Ni	–	0.0528	–
Sb	0.0024	0.0064	–
Pb	0.0013	0.0162	0.0053
Zn	<0.0002	<0.0002	0.0142
Fe	0.0042	0.0041	–
Al	0.0003	0.0004	0.0148
In	0.0012	0.0034	–

2.2. Sample Preparation

The hot-dipping approach was used to prepare Sn-0.7Cu and Sn-0.7Cu-0.05Ni solder joints. The Cu substrate was manually dipped into acid liquid solution comprised of 5% hydrochloric acid (HCl) and deionized water at ambient temperature for 5 min to get rid of surface oxides film and then rinsed with acetone (C_3H_6O) and distilled water, and subsequently air-dried to remove oil contamination. To enhance the soldering and for the removal of oxidation, substrate was immersed in a flux solution. A standard B-type solution flux based on Japanese Industrial Standards (JIS Z3198-4) was applied which contains of mixtures of resin, isopropyl alcohol and diethylamine hydrochloride.

The Cu substrate was then positioned on the pneumatic angular gripper of an automated solder-dip machine and then dipped in the molten solder bath with an immersion dwell time of 2 s and speeds of withdrawal of 10 mm/s. The temperature of the molten solder bath was held at 265 °C. During withdrawal, the surface of the solder coating was blown off using a compressed air knives blower with controllable angle and speed of pressure to remove the excess solder coating and to improve smoothness of the solder coating layer. Then, the samples were air-cooled and water-rinsed using an ultrasonic bath. Around 10–20 μm thickness of solder coating layer was applied on the substrate. Figure 1 shows the schematic diagram of the solder hot-dipping process. The solder hot-dipping temperature profile process is presented in Table 2. After the hot-dip coating, samples were aged at room temperature for 48 h to saturate the internal residue stress induced by IMC formation.

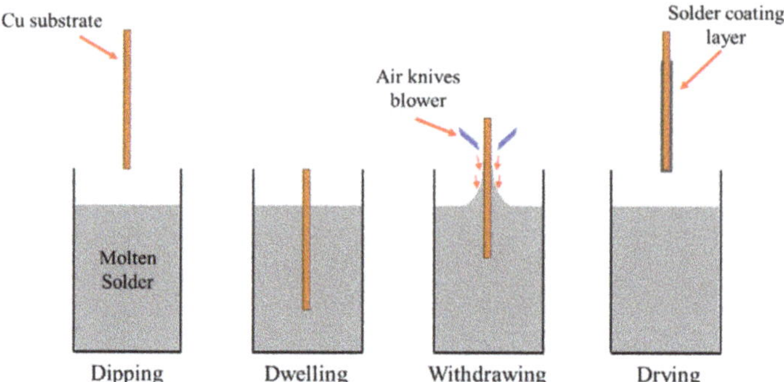

Figure 1. Schematic diagram of the hot-dipping process.

Table 2. Solder hot-dipping process parameters.

Hot-Dip Soldering Parameters	Value
Preheat Temperature (°C)	200
Preheat Time (s)	60
Preheat Rate (°C/s)	2.5
Peak Temperature (°C)	265
Immersion Withdrawal Speed (mm/s)	10
Immersion Dwell Time (s)	2

2.3. Testing and Characterization Method

The micro-indentation test was employed to stimulate the Sn whiskers nucleation and growth through continuous mechanical stress induced with a constant load. This accelerated testing method increases the distribution of compressive stress that is transmitted to the weaker grain boundaries. Subsequently, the stress could be accumulated and promotes the nucleation and growth of Sn whiskers from surface flaws, such as from voids and cracks. Figure 2 illustrates the schematic of micro-indentation apparatus used in this study. The test apparatus was located in the clean room at ambient temperature. Periodic observations of the samples under continuous mechanically induced stress on the nucleation and growth of Sn whiskers were employed for each sample throughout 20 days using a JEOL JSM 6460LA scanning electron microscope (SEM), JEOL Ltd. Tokyo, Japan in secondary and backscattered electron imaging mode at an accelerating voltage at 20 KV. The Sn whisker morphology was observed with a 30° tilt angle. This investigation outcomes presented as a parameters to assess the trends of the Sn whisker nucleation and growth in Sn-0.7Cu-0.05Ni.

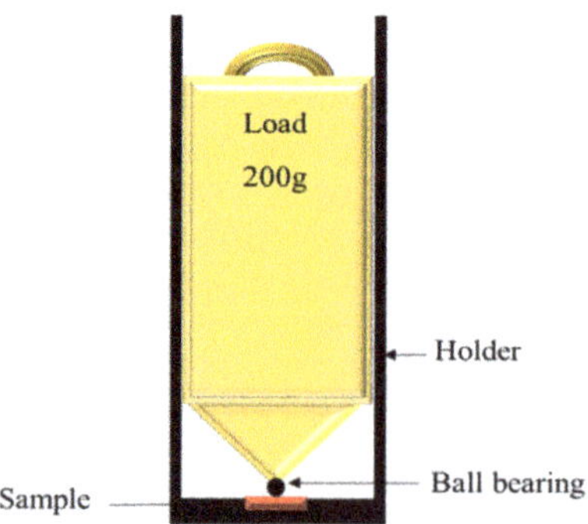

Figure 2. Schematic diagram of micro-indentation test apparatus.

The length and density of the Sn whisker distribution were considered based on Whisker Standards (JESD22-A121A) of the Joint Electron Device Engineering Council (JEDEC) [8,15] by picking five equivalent sampling areas for each sample, using a Java image processing program, image-J Software with an average value of five interpretations per analysis. The measurement of axial Sn whisker length is calculated from the bottom of the Sn whisker growth to the whisker tip, whereas for bend, the Sn whisker is composed of the straight section of the whisker, as shown in Figure 3. The precision for the Sn whisker length and Sn whisker density calculation was about ±5%. It is essentially reliant on the magnification ratio of the particular image micrograph. The density of the Sn whisker is defined as the total number of Sn whiskers' growth per unit area of the observed area and considered in a unit of pcs/μm^2. The error bars were measured from the standard deviation (SD) of the Sn whiskers' distribution.

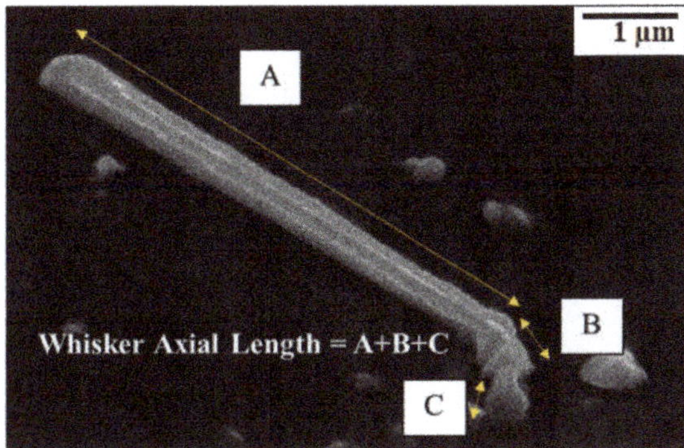

Figure 3. A method to calculate total axial Sn whisker length by adding all straight section of Sn whisker based on the JEDEC whisker standards.

The samples were then mounted using a mixture of resin and hardener to cross-section, followed by a grinding and polishing process in order to measure an average IMC interfacial thickness of the solder joint. The IMC interfacial layer thickness (T), is equal to total IMC area (A), divided by total IMC length (L), and the unit is expressed in µm. With the purpose of observing the top view of the three-dimensional (3D) morphology of the IMC interfacial layer, the solder joint was then entirely deep-etched, consuming a mixed solution of 2% ortho-nitrophenol, 5% sodium hydroxide and 93% of distilled water at a temperature 70 °C to selectively eliminate the tin (Sn) element of the solder alloys. The shape, thickness and root mean square (RMS) roughness of the IMC interfacial layer were analyzed by JEOL JSM 6460LA scanning electron microscope (SEM) by JEOL Ltd. Tokyo, Japan, high-resolution of atomic force microscope (AFM) by Bruker Corporation, Billerica, MA, Statele Unite and Profilm online web based software program by Kla corporation, Milpitas, CA, Statele Unite. The procedure of AFM comprises scanning a sharp tip incorporated in a flexible cantilever across the sample surface, monitoring the tip–sample interaction to produce a 3D topographic micrograph figure of the surface. This technique permits to compute the surface properties, such as RMS roughness, particle size, step height, etc. Surface roughness provides the quantitative data of the surface features based on the statistical deviation from average height. Besides, AFM data able to generate a data containing information of surface roughness in a graphic arrangement to show the growth features of the IMC interfacial layer.

A methodology observation through synchrotron microbeam X-ray fluorescence (µ-XRF) spectroscopy was accomplished for elemental mapping to detect small trace metal element distribution, such as Cu (70 ppm) and Ni (500 ppm), in solder alloys. This analysis was performed at the Synchrotron Light Research Institute (SLRI), Kanchanaburi, Thailand. The BL6b beamline exploited the continuous synchrotron radiation that is emitted from the bending magnet of the 1.2 GeV storage ring with a micro-focused beam of $30 \times 30\ \mu m^2$. Imaging analysis was processed using the PyMca 5.6.3-win64.exe software. In order to observe the elemental distribution mapping of the lead-free solder joint, samples of micro-XRF were mounted using a mixture of resin and hardener and polished in the cross-sectioned condition. The thicknesses of all solder coating layers were synthesized in the range from 200 to 300 μm due to the limits of spatial resolution level, with X-ray beam spot size in the order of 50 μm and above.

3. Results and Discussion

The surface morphologies of Sn-0.7Cu and Sn-0.7Cu-0.05Ni solder joints were periodically observed from day 1 up to day 20 under continuous mechanically induced stress. The observation of Sn whisker growth behavior resulted from two types of solder joint, Sn-0.7Cu and Sn-0.7Cu-0.05Ni, as shown in Figure 4. The observed distribution of Sn whisker growth was increasing over storage time. In Figure 4a,b, hillock- or nodule-shaped Sn whiskers were noticed on both of the solder joints on day 1. The Sn whiskers appear to progressively extrude from the surface of Sn-0.7Cu and Sn-0.7Cu-0.05Ni. It was evidently seen that the initial process of whiskering is a nucleation that protruded as a hillock or nodule shape on the surface. In response to the stress gradient of the IMC interfacial layer, hillock could probably to grow as a risky long filament-shaped Sn whisker [31]. With an increase of storage time from day 5 to day 20 (Figure 4c–f), the hillock-shaped Sn whisker rapidly grew and became larger on the surface of the solder joint. In contrast, Sn whiskers' growth were found greater on the Sn-0.7Cu solder with the growth of filament-shaped and bent-shaped Sn whiskers, while on the surface of the Sn-0.7Cu-0.05Ni solder joint, the hillock shape was observed.

It can be observed that the growth trend of Sn whiskers' length and density on Sn-0.7Cu and Sn-0.7Cu-0.05Ni solder joints increased as the storage time increased. As shown in Figure 5, the Sn whiskers' length on Sn-0.7Cu was observed to be higher than on the Sn-0.7Cu-0.05Ni solder joint. The size of the average Sn whisker length on Sn-0.7Cu increased more rapidly at day 3 until day 20. This occurrence tends to contribute a high tendency to short-circuit fine pitch components in electronic packages and is considerably dangerous to microelectronic reliability.

It can also be noticed that Sn whisker tendency, with respect to Sn whisker density, significantly increased with storage time and finally reached a saturated condition at day 20 for both Sn-0.7Cu and Sn-0.7Cu-0.05Ni solder joints. From this investigation, as presented in Figure 6, the average density of Sn-0.7Cu was obviously higher than the Sn-0.7Cu-0.05Ni solder joint, where the Sn-0.7Cu-0.05Ni solder joint has been found to be less prone to Sn whiskers. This observation also indicates that the density of Sn whiskers was greater as storage time progressed. The trend of Sn whisker nucleation and growth corresponds to the IMC interfacial growth and residual stress evolution that increases over time to reach the solution saturation [32]. Therefore, it could be concluded that there is a significant relationship between the Sn whisker growth and composition of the lead-free solder joint. The Sn-0.7Cu-0.05Ni solder joint has an obvious suppression effect by diminishing the susceptibility for Sn whisker growth and increase the incubation time for nucleation of Sn whiskers.

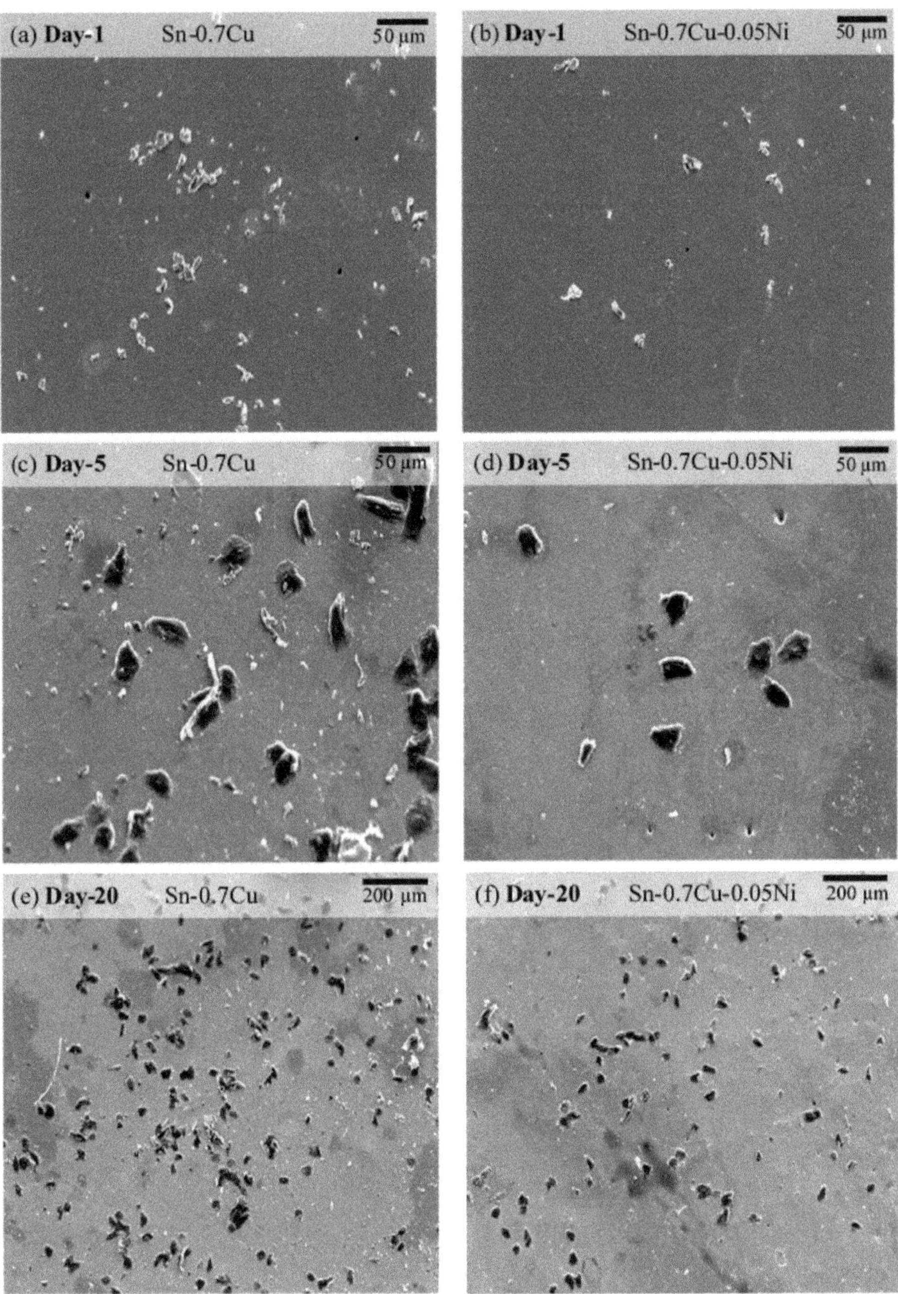

Figure 4. Surface morphologies of Sn whiskers' distribution from Sn-0.7Cu and Sn-0.7Cu-0.05Ni solder coatings over a period of (**a**,**d**) 1 day, (**b**,**e**) 5 days and (**c**,**f**) 20 days.

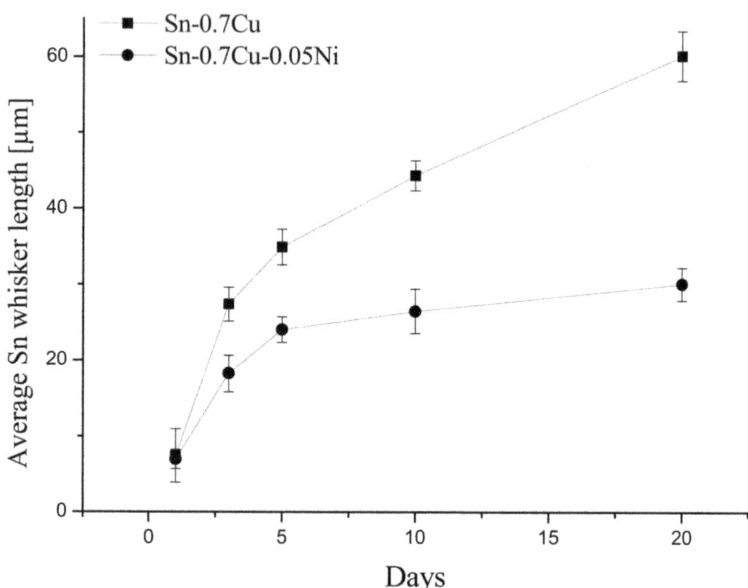

Figure 5. The average length of Sn whiskers in Sn-0.7Cu and Sn-0.7Cu-0.05Ni solder joints.

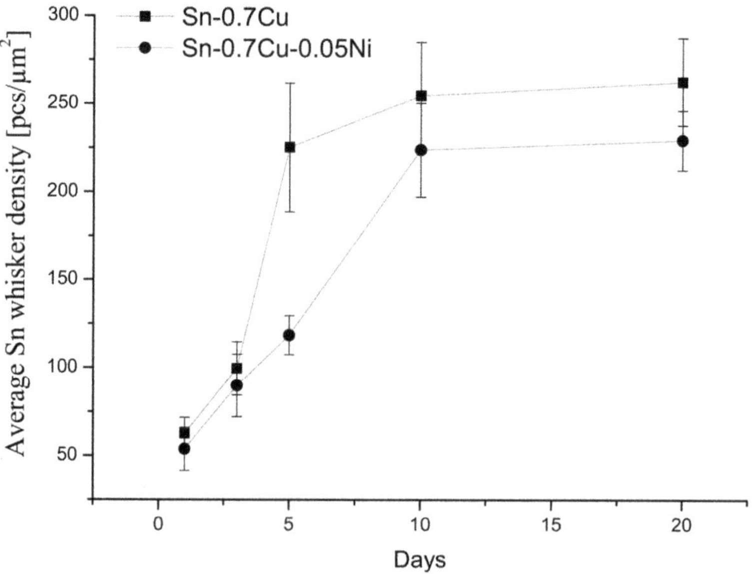

Figure 6. The average density of Sn whiskers in Sn-0.7Cu and Sn-0.7Cu-0.05Ni solder joints.

Since the residual stress and IMC interfacial reaction cannot be regulated, a methodical observation of Sn whisker nucleation and growth is very challenging. Apparently, the propensity for Sn whisker nucleation and growth is increased with the implementation of both internal residual stress and mechanical compressive stress induced in the solder layer. To generate well-regulated inducement of Sn whisker growth, a continuous mechanical compressive stress was induced in all samples of the Sn–Cu solder joint to accelerate the Sn whiskers' growth by micro-indentation. Figure 7 show SEM images with a 30°

tilt angle of the surface of the Sn-0.7Cu-0.05Ni solder joint that have been indented for 12 h. The indentation area indicates a localized effect and has a diameter of approximately 83.94 μm. It was observed that the hillock- or nodule-shaped Sn whisker growth was most abundant nearby the indentation area. The length of the hillock shape was found to vary from 0.5 to 3 μm in the Sn-0.7Cu-0.05Ni solder joint.

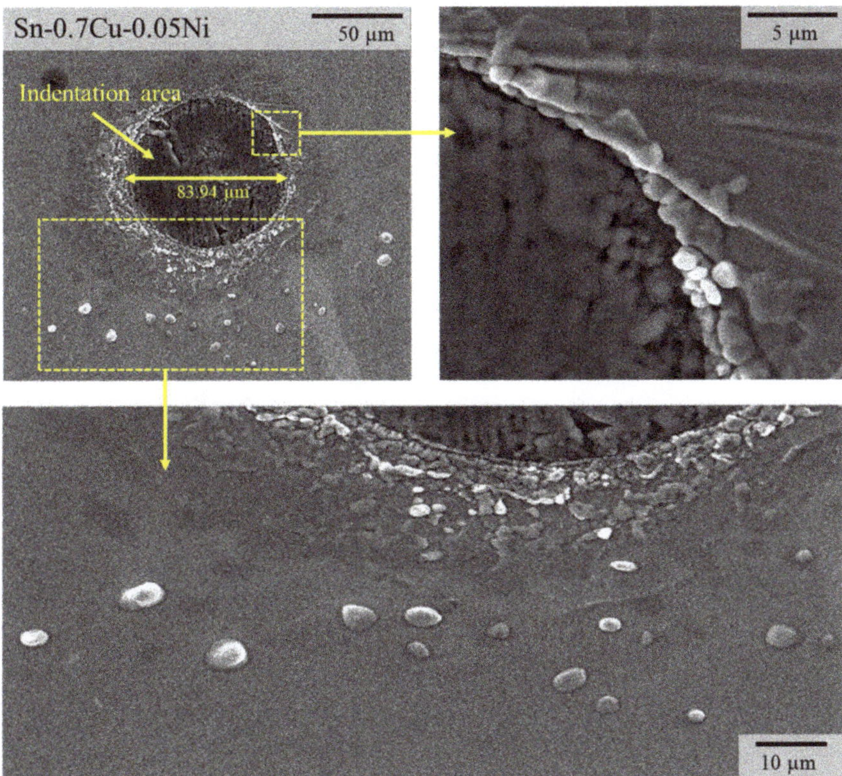

Figure 7. Scanning electron microscope (SEM) images of Sn whisker nucleation growth in the Sn-0.7Cu-0.05Ni solder joint after indentation for 12 h.

Regarding the morphology of Sn whisker nucleation growth, the primary nucleation takes place with the growth of a hillock or nodule shape, followed by protrusion of Sn whiskers. The nucleation kinetics of the Sn whisker revealed here are consistent with previous work [23] under a constant continuous mechanical stress using a clamping fixture, wherein the compressive stress extends into the region away from the punch, organized by the relaxation processes in the solder layer with the growth of the Sn whisker. The results revealed that the initiation of Sn whiskers' growth is also consistent with Fei et al.'s [19] study. They claimed that a primary stage in the Sn whisker formation is nucleation, which is a Sn whisker extruding position. In addition, according to Skwarek et al. [33], the Sn whiskers grew out from the whole grains of the solder/substrate interface, not from the grain boundaries.

The composition and distribution analysis of Ni, Cu and Sn alloying elements in the Sn-0.7Cu-0.05Ni solder joint using synchrotron micro-XRF mapping are presented in Figure 8. The micro-XRF mapping area is ~450 × 1750 μm. The mapping of the blue area indicates low concentration or no element in the solder joint. The mapping of green and yellow areas indicates intermediate concentration, and the mapping of the red area indicates high concentration of specific composition. The small, well-distributed Ni and Cu have been found at primary IMC of the Sn-0.7Cu-0.05Ni solder joint. It is direct confirmation from the mapping result that a small amount of Ni (500 ppm) alters the microstructure of Cu_6Sn_5 to form a $(Cu, Ni)_6Sn_5$ intermetallic layer. The formation of new compounds in IMC interfacial layers is very significant to the phase equilibria of the system, and in addition, is able to modify the properties of the IMC interfacial layer. In the presence of Ni in Sn-0.7Cu, the Cu_6Sn_5 interfacial IMC layer takes the form of the $(Cu, Ni)_6Sn_5$ interfacial IMC layer, where Ni atoms occupy copper lattice sites [34]. A previous study documented that Ni is able to enhance the IMC interfacial formation with a fine scallop shape during liquid–solid interfacial reactions [27].

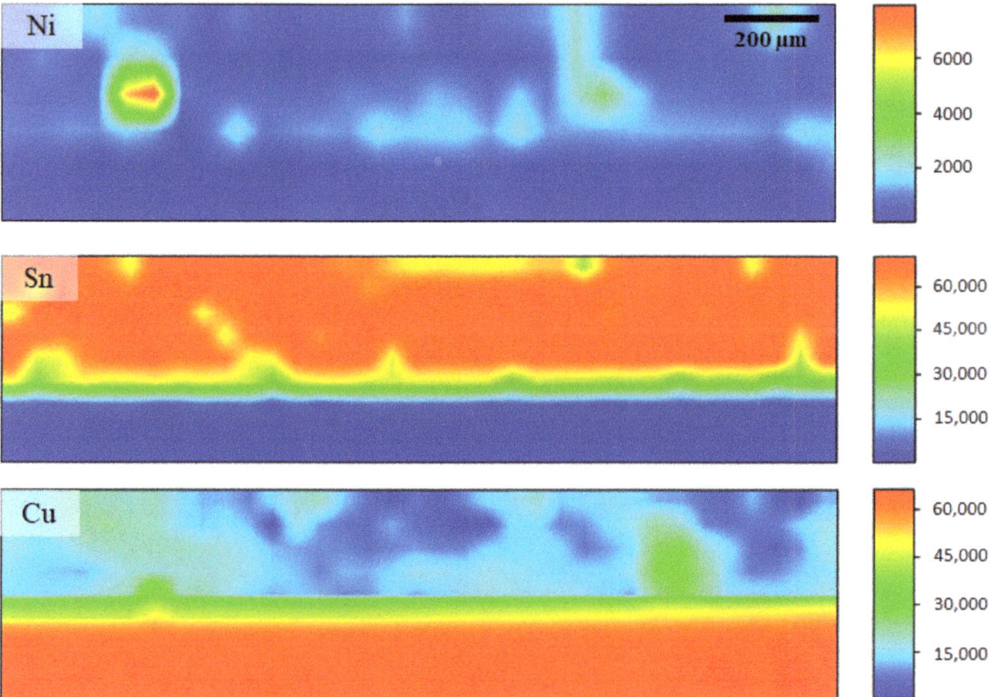

Figure 8. Elemental distribution of Sn-0.7Cu-0.05Ni on Cu substrate.

According to the literature [35], Ni is dispersed in a relatively homogeneous manner through the Cu_6Sn_5 IMC interfacial layer. However, in this study, Ni distribution was not fitting at the IMC interfacial layer due to the limits of spatial resolution level, with X-ray beam spot size at the order of 50 μm and above. It is believed that homogeneous concentration of Ni in Cu_6Sn_5 IMC Ni sustains the phase stability in the hexagonal form of the $(Cu, Ni)_6Sn_5$ IMC interfacial layer and prevents the transformation of polymorphic phase (hexagonal to monoclinic) of Cu_6Sn_5 [35]. This phase stability will minimize the stress concentrations in the IMC layer, and subsequently reduce the Sn whiskering potential on the surface of the solder joint.

The morphology of Sn whiskers was intensely inclined to the shape and thickness of the interfacial intermetallic layer (IMC) [3,25]. A quantitative analysis was done to examine the correlation of the Sn whiskers' formation and the IMC interfacial layer using SEM in secondary and backscattered electron imaging mode at an accelerating voltage of 20 kV. An image structure of the IMC interfacial layer that formed in Sn-0.7Cu and Sn-0.7Cu-0.05Ni solder joints is presented in Figure 9. It can be observed that from the cross-sectioned view of the morphology, the Cu_6Sn_5 IMC interfacial layer on Sn-0.7Cu (Figure 9a) is rougher than the (Cu, Ni)$_6$Sn$_5$ IMC layer on Sn-0.7Cu-0.05Ni (Figure 9b). The (Cu, Ni)$_6$Sn$_5$ IMC interfacial layer results in a finer grained, more continuous layer with stable shape of the IMC interfacial layer. Moreover, regarding the thickness of the IMC interfacial layer, the 1.76 μm thick scalloped Cu_6Sn_5 IMC layer was higher compared to the 1.28 μm thick (Cu, Ni)$_6$Sn$_5$ IMC layer, whereas Ni micro-alloying changed the thickness of the IMC interfacial layers significantly and revealed an approximately uniform height of the IMC grains.

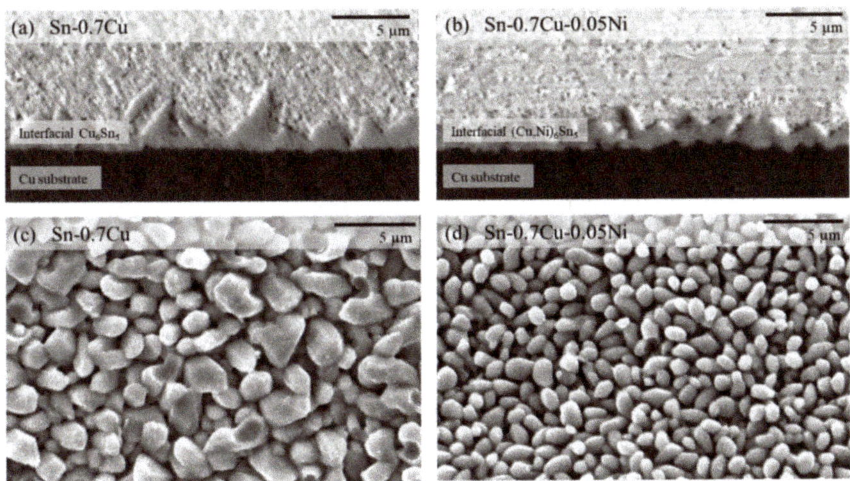

Figure 9. The morphology of the intermetallic compound (IMC) interfacial layer of Sn-0.7Cu and Sn-0.7Cu-0.05Ni solder joints. (**a**) Cross-section view of the interfacial Cu_6Sn_5, (**b**) cross-section view of the interfacial (Cu,Ni)$_6$Sn$_5$, (**c**) top view of the interfacial Cu_6Sn_5 and (**d**) top view of the interfacial (Cu,Ni)$_6$Sn$_5$.

Sn-0.7Cu-0.05Ni has a significant implication on the Sn whisker formation and growth behavior. The outcomes of this work appear to validate that the Sn whisker susceptibility is associated with the solder composition (refer to Figure 4) as well as the IMC interface topography. From the top view in Figure 9c and d, it can be noticeably observed that there is strong correlation between morphology and size of IMC interfacial intermetallic grains and compositions of the solder alloys. At the interface of solder/substrate in the Sn-0.7Cu-0.05Ni solder joint, significantly fine and almost uniform IMC grains formed compared to the interface of the solder/substrate in the Sn-0.7Cu solder joint. Remarkably, the grain size of (Cu,Ni)$_6$Sn$_5$ is significantly smaller than the Cu_6Sn_5 IMC interfacial layer. This means that the Ni micro-alloying decreased the IMC grain size. The uniform and smaller average size of IMC grains may possibly result in a lower residual stress and suppress the propensity of Sn whisker formation [33].

The significance of the IMC interfacial layer growth directly creates internal stress in the grain boundary of the solder layer, which can induce Sn whisker growth, as illustrated schematically in Figure 10. The rapid dissolution and growth of the thick scallop shape of the IMC interfacial layer with pyramid-like IMC grains significantly produced higher cumulative stress in the IMC interfacial layer and can lead to the formation of Sn whiskers [3]. In contrast, the thinner scallop shape of IMC interfacial layers can lead to lower stress of the IMC interfacial layer and thus suppress the growth of Sn whiskers [4]. This is also consistent with the work of Chason et al. [11], who showed that the stress relaxation processes are significantly more rapid and greater for thicker Sn solder. Therefore, the morphology structure of the (CuNi)$_6$Sn$_5$ IMC interfacial layer was relatively thinner and more refined, with a fine scallop shape thus suppressed the thickness of the IMC interfacial layer and enhanced longer incubation period for the formation and growth of Sn whiskers. Consistent with Illes et al. [21], the mitigation of Sn whisker formation and growth can be achieved by reducing the Sn whisker incubation time, which produces thicker layers of lead-free solder, since this formation improves the stress relaxation properties.

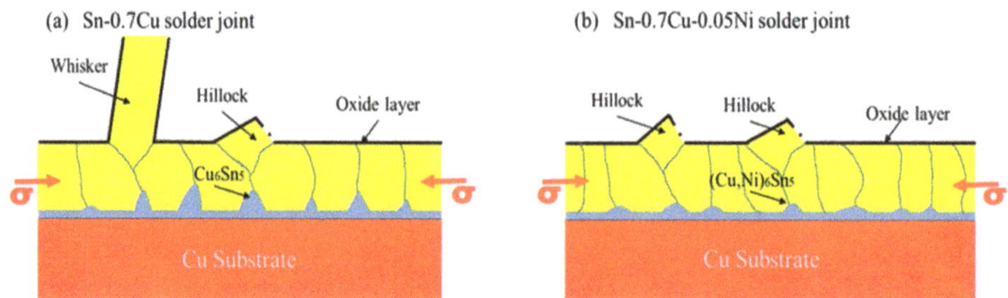

Figure 10. A schematic diagram illustrating the correlation between IMC interfacial growth and Sn whisker growth on (**a**) Sn-0.7Cu and (**b**) Sn-0.7Cu-0.05Ni.

Figure 11 indicates the topographies of IMC interfacial properties of deep-etched Sn-0.7Cu and Sn-0.7Cu-0.05Ni solder joints by AFM. The scale of contrast of Cu_6Sn_5 and $(Cu,Ni)_6Sn_5$ IMC interfacial layers is 2.5 μm and 2.0 μm. The higher contrast enhanced the number of thicker and higher peaks of IMC grain that formed. It was observed that the three-dimensional shape morphology of the scallop-shaped Cu_6Sn_5 IMC interfacial layer on Sn-0.7Cu (Figure 10a) was significantly thicker and coarser compared to the $(Cu,Ni)_6Sn_5$ IMC interfacial layer on Sn-0.7Cu-0.05Ni (Figure 10b). These results corresponded with a quantitative analysis of the SEM, as shown in Figure 9. Comparing the 0.05 wt.% Ni addition to Sn-0.7Cu, the formation of the $(Cu,Ni)_6Sn_5$ IMC layer was relatively thinned and had a finer scallop shape, which has been found to be less prone to whiskers.

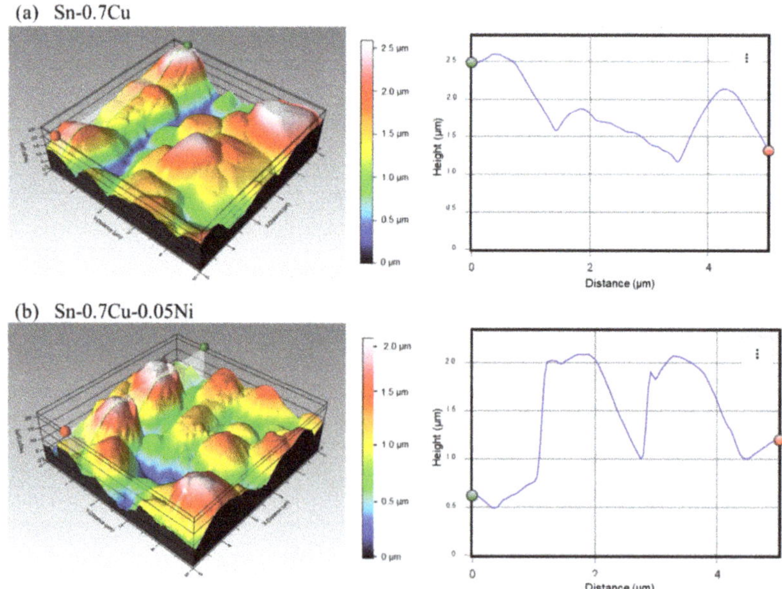

Figure 11. Atomic force microscope (AFM) three-dimensional (3D) images (5 × 5 μm) of interfacial IMC formed from (**a**) Sn-0.7Cu and (**b**) Sn-0.7Cu-0.05Ni.

Investigation of the Cu_6Sn_5 and $(Cu,Ni)_6Sn_5$ IMC interfacial layer was further carried out by measuring the surface roughness values and the highest peak of the IMC interfacial layer. Surface roughness is an observation parameter for describing the surface features and is commonly signified in terms of the statistical deviation from average height [8]. It was established that the surface roughness values of the IMC interfacial layers Cu_6Sn_5 and $(Cu,Ni)_6Sn_5$ were 1.83 and 1.26 μm, respectively. Moreover, it was observed that the highest peak value of the Sn-0.7Cu IMC interfacial layer was ~2.641 μm, whereas the highest peak value of the Sn-0.7Cu-0.05Ni IMC interfacial layer was lower, at ~2.126 μm. This implies that a more uneven, higher peak of the IMC interfacial layer was observed to be more susceptible to Sn whisker nucleation and growth. This outcome indicates that Ni micro-alloying plays an important role in the IMC interfacial layer. The thinner thickness and fine scallop-shaped formation of the $(Cu,Ni)_6Sn_5$ IMC layer enhances stress relaxation and suppress the nucleation and growth of the Sn whisker.

Similar results were found in studies by Chason et al. [11], that validate that the stress relaxation is suppressed with small grain size of IMC interfacial layers. The growth of the IMC interfacial layer can have a significant impact on the residual stress of the solder joint as the stress gradient in the solder joints is induced by volume expansion of polymorphic phase transformation. It has been stated that $(Cu,Ni)_6Sn_5$ greatly reduced the volume change associated with the polymorphic phase transformation of Cu_6Sn_5 [27]. This is also consistent with a finding by Nogita et al. [29], who suggest that the phase stabilization of the hexagonal $(Cu,Ni)_6Sn_5$ may possibly prevent volume changes of the IMC interfacial layer that could provide the stress relaxation in the solder layer [8,36].

4. Conclusions

The effect of the Sn-0.7Cu-0.05Ni solder joint on the nucleation and growth of Sn whiskers under continuous mechanically induced stress was investigated. The correlation of the Sn whiskers' growth and the IMC interfacial layer was also established. The following conclusions can be drawn:

1. There was a noticeable correlation between the Sn whisker growth and the composition of the lead-free solder alloy. The Sn-0.7Cu-0.05Ni solder joint had a great suppression effect on the nucleation and growth of Sn whiskers of the Pb-free solder joint.
2. The small amount of Ni addition (~500 ppm) was able to alter the microstructure of Cu_6Sn_5 to form a $(Cu,Ni)_6Sn_5$ IMC intermetallic layer, and it is very significant to the nucleation and growth of Sn whiskers.
3. The methodic structure of the $(Cu,Ni)_6Sn_5$ IMC interfacial layer was relatively thinner and more refined, with a continuous fine scallop-shaped IMC interfacial layer consequently enhanced a greater incubation period for the nucleation and growth of the Sn whisker.

Author Contributions: Conceptualization, A.N.H. and M.A.A.M.S.; methodology, A.N.H., N.Z.M.M. and K.C.Y.; software, J.C. and M.M.R.; validation, M.A.A.M.S., A.V.S.; formal analysis, J.C. and M.M.R.; investigation, M.A.A.M.S., J.C. and N.Z.M.M.; resources, M.A.A.M.S. and A.V.S.; data curation, A.N.H. and N.Z.M.M.; writing—original draft preparation, A.N.H.; writing—review and editing, M.A.A.M.S., A.V.S. and K.C.Y.; visualization, A.N.H., J.C. and M.M.R.; supervision, M.A.A.M.S.; project administration, M.A.A.M.S.; funding acquisition, M.A.A.M.S. and A.V.S. All authors have read and agreed to the published version of the manuscript.

Funding: This study was financed by Nihon Superior (Grant No. 2016/10/0001) and Tin Solder Technology Research Group (Grant No. 9002-00084). The work was monetarily supported by Collaborative Research in Engineering, Science and Technology (CREST) under grant No. P14C1-17/001 with Nihon Superior.

Data Availability Statement: The data presented in this study are available in this article.

Acknowledgments: The authors appreciatively recognize the Centre of Excellent Geopolymer and Green Technology (CeGeoGTech), UniMAP and Synchrotron Light Research Institute (SLRI), Thailand, under Beamtime ID: 3774 and 3774-7. The authors would like to acknowledge University Malaysia Perlis and Nihon Superior.

Conflicts of Interest: The authors declare no conflict of interest

References

1. Kim, K.-S.; Yu, C.-H.; Yang, J.-M. Behavior of tin whisker formation and growth on lead-free solder finish. *Thin Solid Film.* **2006**, *504*, 350–354. [CrossRef]
2. Lin, C.-K.; Lin, T.-H. Effects of continuously applied stress on tin whisker growth. *Microelectron. Reliab.* **2008**, *48*, 1737–1740. [CrossRef]
3. Baated, A.; Kim, K.-S.; Suganuma, K. Effect of intermetallic growth rate on spontaneous whisker growth from a tin coating on copper. *J. Mater. Sci. Mater. Electron.* **2011**, *22*, 1685–1693. [CrossRef]
4. Ashworth, M.A.; Wilcox, G.D.; Higginson, R.L.; Heath, R.J.; Liu, C.; Mortimer, R.J. The effect of electroplating parameters and substrate material on tin whisker formation. *Microelectron. Reliab.* **2015**, *55*, 180–191. [CrossRef]
5. Salleh, M.; al Bakri, A.M.M.; Somidin, F.; Sandu, A.V.; Saud, N.; Kamaruddin, H.; McDonald, S.D.; Nogita, K. A comparative study of solder properties of Sn-0.7 Cu lead-free solder fabricated via the powder metallurgy and casting methods. *Revista Chimie* **2013**, *64*, 725–728.
6. Ibrahim, I.N.A.; Salleh, M.A.A.M. Effect of Zinc Additions on Sn-0.7Cu-0.05Ni LeadFree Solder Alloy. Short review. *IOP Conf. Ser. Mater. Sci. Eng.* **2017**, *238*, 012012. [CrossRef]
7. Mahim, Z.; Salleh, M.; Saud, N. Effect on microstructural and physical properties of Sn-3.0Ag-0.5Cu lead-free solder with the addition of SiC particles. *Eur. J. Mater. Sci. Eng.* **2019**, *4*, 37–43. [CrossRef]
8. Hashim, A.N.; Salleh, M.A.A.M.; Ramli, M.M.; Yee, K.C.; Mokhtar, N.Z.M. Preliminary study on the effect of Ni addition on tin (Sn) whisker growth from lead-free solder coating. *IOP Conf. Ser. Mater. Sci. Eng.* **2020**, *957*, 012062. [CrossRef]
9. Jadhav, N.; Williams, M.; Pei, F.; Stafford, G.; Chason, E. Altering the mechanical properties of Sn films by alloying with Bi: Mimicking the effect of Pb to suppress whiskers. *J. Electron. Mater.* **2013**, *42*, 312–318. [CrossRef]
10. Mokhtar, N.Z.M.; Salleh, M.A.A.M.; Hashim, A.N.; Nazri, S.F. Effects of Gallium addition on the thermal properties and whiskers growth under electrical current stressing. *IOP Conf. Ser. Mater. Sci. Eng.* **2019**, *701*, 012001. [CrossRef]
11. Chason, E.; Jadhav, N.; Pei, F. Effect of layer properties on stress evolution, intermetallic volume, and density during tin whisker formation. *JOM* **2011**, *63*, 62. [CrossRef]
12. Amin, N.L.M.; Yusof, S.Z.; Kahar, T.A.A.; Bakar, A.O.; Fadil, N.A. Tin Whiskers Formation and Growth on Immersion Sn Surface Finish under External Stresses by Bending. *Iop Conf. Ser. Mater. Sci. Eng.* **2017**, *238*, 012001. [CrossRef]
13. Doudrick, K.; Chinn, J.; Williams, J.; Chawla, N.; Rykaczewski, K. Rapid method for testing efficacy of nano-engineered coatings for mitigating tin whisker growth. *Microelectron. Reliab.* **2015**, *55*, 832–837. [CrossRef]
14. Tsukamoto, H.; Dong, Z.; Huang, H.; Nishimura, T.; Nogita, K. Nanoindentation characterization of intermetallic compounds formed between Sn–Cu (–Ni) ball grid arrays and Cu substrates. *Mater. Sci. Eng. B* **2009**, *164*, 44–50. [CrossRef]
15. Hashim, A.N.; Salleh, M.A.A.M.; Mokhtar, N.Z.M.; Idris, S. Tin (Sn) whisker growth from electroplated Sn finished. *IOP Conf. Ser. Mater. Sci. Eng.* **2019**, *701*, 012005. [CrossRef]
16. Cheng, J.; Vianco, P.T.; Zhang, B.; Li, J.C. Nucleation and growth of tin whiskers. *Appl. Phys. Lett.* **2011**, *98*, 241910. [CrossRef]
17. Zhang, W.; Egli, A.; Schwager, F.; Brown, N. Investigation of Sn-Cu intermetallic compounds by AFM: New aspects of the role of intermetallic compounds in whisker formation. *IEEE Trans. Electron. Packag. Manuf.* **2005**, *28*, 85–93. [CrossRef]
18. Illés, B.; Horváth, B. Whiskering behaviour of immersion tin surface coating. *Microelectron. Reliab.* **2013**, *53*, 755–760. [CrossRef]
19. Pei, F.; Briant, C.L.; Kesari, H.; Bower, A.F.; Chason, E. Kinetics of Sn whisker nucleation using thermally induced stress. *Scr. Mater.* **2014**, *93*, 16–19. [CrossRef]
20. Chason, E.; Jadhav, N.; Chan, W.; Reinbold, L.; Kumar, K. Whisker formation in Sn and Pb–Sn coatings: Role of intermetallic growth, stress evolution, and plastic deformation processes. *Appl. Phys. Lett.* **2008**, *92*, 171901. [CrossRef]
21. Illés, B.; Skwarek, A.; Ratajczak, J.; Dušek, K.; Bušek, D. The influence of the crystallographic structure of the intermetallic grains on tin whisker growth. *J. Alloy. Compd.* **2019**, *785*, 774–780. [CrossRef]
22. Kim, K.S.; Yu, C.H.; Han, S.W.; Yang, K.C.; Kim, J.H. Investigation of relation between intermetallic and tin whisker growths under ambient condition. *Microelectron. Reliab.* **2008**, *48*, 111–118. [CrossRef]
23. Jagtap, N.J.P.; Chason, E. Whisker growth under a controlled driving force: Pressure induced whisker nucleation and growth. *Scr. Mater.* **2020**, *182*, 43–47. [CrossRef]
24. Ramli, M.I.I.; Yusuf, M.S.S.; Salleh, M.A.A.M. Effect of bismuth content on microstructure, melting temperature and undercooling of Sn-0.7Cu solder alloy. *Eur. J. Mater. Sci. Eng.* **2018**, *3*, 2013–2017.
25. Somidin, F.; Maeno, H.; Salleh, M.M.; Tran, X.Q.; McDonald, S.D.; Matsumura, S.; Nogita, K. Characterising the polymorphic phase transformation at a localised point on a Cu6Sn5 grain. *Mater. Charact.* **2018**, *138*, 113–119. [CrossRef]
26. Pei, F.; Jadhav, N.; Chason, E. Correlation between surface morphology evolution and grain structure: Whisker/hillock formation in Sn-Cu. *JOM* **2012**, *64*, 1176–1183. [CrossRef]

27. Zeng, G.; McDonald, S.D.; Gu, Q.; Terada, Y.; Uesugi, K.; Yasuda, H.; Nogita, K. The influence of Ni and Zn additions on microstructure and phase transformations in Sn–0.7 Cu/Cu solder joints. *Acta Mater.* **2015**, *83*, 357–371. [CrossRef]
28. Zeng, G.; McDonald, S.D.; Mu, D.; Terada, Y.; Yasuda, H.; Gu, Q.; Salleh, M.M.; Nogita, K. The influence of ageing on the stabilisation of interfacial (Cu, Ni) 6 (Sn, Zn) 5 and (Cu, Au, Ni) 6Sn5 intermetallics in Pb-free Ball Grid Array (BGA) solder joints. *J. Alloy. Compd.* **2016**, *685*, 471–482. [CrossRef]
29. Nogita, K.; Mu, D.; McDonald, S.; Read, J.; Wu, Y. Effect of Ni on phase stability and thermal expansion of Cu6−xNixSn5 (X= 0, 0.5, 1, 1.5 and 2). *Intermetallics* **2012**, *26*, 78–85. [CrossRef]
30. Salleh, M.M.; McDonald, S.; Gourlay, C.; Belyakov, S.; Yasuda, H.; Nogita, K. Effect of Ni on the Formation and Growth of Primary Cu 6 Sn 5 Intermetallics in Sn-0.7 wt.% Cu Solder Pastes on Cu Substrates During the Soldering Process. *J. Electron. Mater.* **2016**, *45*, 154–163. [CrossRef]
31. Jiang, B.; Xian, A.-P. Whisker growth on tin finishes of different electrolytes. *Microelectron. Reliab.* **2008**, *48*, 105–110. [CrossRef]
32. Jadhav, N.; Buchovecky, E.J.; Reinbold, L.; Kumar, S.; Bower, A.F.; Chason, E. Understanding the correlation between intermetallic growth, stress evolution, and Sn whisker nucleation. *IEEE Trans. Electron. Packag. Manuf.* **2010**, *33*, 183–192. [CrossRef]
33. Skwarek, A.; Pluska, M.; Ratajczak, J.; Czerwinski, A.; Witek, K.; Szwagierczak, D. Analysis of tin whisker growth on lead-free alloys with Ni presence under thermal shock stress. *Mater. Sci. Eng. B* **2011**, *176*, 352–357. [CrossRef]
34. Nogita, K.; Gourlay, C.; Nishimura, T. Cracking and phase stability in reaction layers between Sn-Cu-Ni solders and Cu substrates. *JOM* **2009**, *61*, 45–51. [CrossRef]
35. Nogita, K.; Yasuda, H.; Gourlay, C.M.; Suenaga, S.; Tsukamoto, H.; Mcdonald, S.D.; Takeuchi, A.; Uesugi, K.; Suzuki, Y. Synchrotron micro-XRF measurements of trace element distributions in BGA type solders and solder joints. *Trans. Jpn. Inst. Electron. Packag.* **2010**, *3*, 40–46. [CrossRef]
36. Nogita, K. Stabilisation of Cu6Sn5 by Ni in Sn-0.7 Cu-0.05 Ni lead-free solder alloys. *Intermetallics* **2010**, *18*, 145–149. [CrossRef]

 materials

Article

Performance of Sn-3.0Ag-0.5Cu Composite Solder with Kaolin Geopolymer Ceramic Reinforcement on Microstructure and Mechanical Properties under Isothermal Ageing

Nur Syahirah Mohamad Zaimi [1,2], Mohd Arif Anuar Mohd Salleh [1,2,*], Andrei Victor Sandu [1,3,4,*], Mohd Mustafa Al Bakri Abdullah [1,2], Norainiza Saud [1,2], Shayfull Zamree Abd Rahim [1,5], Petrica Vizureanu [1,3,*], Rita Mohd Said [1,2] and Mohd Izrul Izwan Ramli [1,2]

1 Center of Excellence Geopolymer & Green Technology (CeGeoGTech), University Malaysia Perlis (UniMAP), Taman Muhibbah, Perlis 02600, Malaysia; syahirahzaimi25@gmail.com (N.S.M.Z.); mustafa_albakri@unimap.edu.my (M.M.A.B.A.); norainiza@unimap.edu.my (N.S.); shayfull@unimap.edu.my (S.Z.A.R.); rita@unimap.edu.my (R.M.S.); Izrulizwan@unimap.edu.my (M.I.I.R.)
2 Faculty of Chemical Engineering Technology, University Malaysia Perlis (UniMAP), Taman Muhibbah, Perlis 02600, Malaysia
3 Faculty of Materials Science and Engineering, Gheorghe Asachi Technical University of Iasi, 71 D. Mangeron Blv., 700050 Iasi, Romania
4 Romanian Inventors Forum, Str. Sf. P. Movila 3, 700089 Iasi, Romania
5 Faculty of Mechanical Engineering Technology, Pauh Putra Campus, University Malaysia Perlis (UniMAP), Perlis 02600, Malaysia
* Correspondence: arifanuar@unimap.edu.my (M.A.A.M.S.); sav@tuiasi.ro (A.V.S.); peviz@tuiasi.ro (P.V.)

Citation: Zaimi, N.S.M.; Salleh, M.A.A.M.; Sandu, A.V.; Abdullah, M.M.A.B.; Saud, N.; Rahim, S.Z.A.; Vizureanu, P.; Said, R.M.; Ramli, M.I.I. Performance of Sn-3.0Ag-0.5Cu Composite Solder with Kaolin Geopolymer Ceramic Reinforcement on Microstructure and Mechanical Properties under Isothermal Ageing. *Materials* **2021**, *14*, 776. https://doi.org/doi:10.3390/ma14040776

Academic Editor: Prashanth Konda Gokuldoss

Received: 13 January 2021
Accepted: 29 January 2021
Published: 7 February 2021

Abstract: This paper elucidates the effect of isothermal ageing at temperature of 85 °C, 125 °C and 150 °C for 100, 500 and 1000 h on Sn-3.0Ag-0.5Cu (SAC305) lead-free solder with the addition of 1 wt% kaolin geopolymer ceramic (KGC) reinforcement particles. SAC305-KGC composite solders were fabricated through powder metallurgy using a hybrid microwave sintering method and reflowed on copper substrate printed circuit board with an organic solderability preservative surface finish. The results revealed that, the addition of KGC was beneficial in improving the total thickness of interfacial intermetallic compound (IMC) layer. At higher isothermal ageing of 150 °C and 1000 h, the IMC layer in SAC305-KGC composite solder was towards a planar-type morphology. Moreover, the growth of total interfacial IMC layer and Cu_3Sn layer during isothermal ageing was found to be controlled by bulk diffusion and grain-boundary process, respectively. The activation energy possessed by SAC305-KGC composite solder for total interfacial IMC layer and Cu_3Sn IMC was 74 kJ/mol and 104 kJ/mol, respectively. Based on a lap shear test, the shear strength of SAC305-KGC composite solder exhibited higher shear strength than non-reinforced SAC305 solder. Meanwhile, the solder joints failure mode after shear testing was a combination of brittle and ductile modes at higher ageing temperature and time for SAC305-KGC composite solder.

Keywords: composite solder; intermetallics; microstructure; ageing; activation energy; lead-free solder geopolymer; geopolymer ceramic

1. Introduction

Solder alloy is used as an interconnecting material in microelectronic packaging, joining the components to substrate [1]. Development of lead-free solder alloy was ineluctable due to the prohibition of lead usage in the electronic industry implemented by environmental laws. The toxicity of lead has devoted the industry to search for a new generation of solder alloy which is free from lead content. Numerous studies have been carried out in finding a suitable lead-free solder alloy [2–4]. Tin-silver-copper (SAC) solder alloy is identified to be one of promising candidate which could replace the conventional lead solder alloy [4–9]. Other than being widely available in the market industry, SAC solder alloy also has good mechanical properties, and better solderability with a suitable melting temperature [8,10]. Even so, the growth of interfacial intermetallic compound (IMC) phases, such as Cu_6Sn_5 and Cu_3Sn, at the solder/substrate in SAC solder joints is faster as compared to eutectic tin-lead (SnPb) solder joints, resulting in a thicker interfacial IMC layer [11]. A thicker interfacial IMC layer was responsive to the stress and may worsen the reliability of the solder joints as it could promote cracks to be initiated and propagated. Meanwhile, too thin of an interfacial IMC layer may not form a proper solder interconnection. Therefore, controlling the thickness of the interfacial IMC layer is crucial to improve the reliability of the solder joints.

Recently, composite solder approaches have been recognized as one of the potential methods to improve the reliability of the solder joints since it can provide a marked improvement to the microstructure and mechanical strength. Ceramic materials, such as titanium oxide (TiO_2), aluminium oxide (Al_2O_3), silicon carbide (SiC), cerium oxide (CeO_2) and silicon nitride (Si_3N_4), have been chosen as the ones that could be added into solder alloy, forming composite solder. These ceramic materials work by being dispersed in the matrix of Sn solder alloy and distributed homogenously along the grain boundaries of Sn solder alloys. Extensive studies have been reported on how the ceramic materials could improve the performance of lead-free solder alloys by refining the microstructure, inhibiting the growth of interfacial IMC layer and strengthening the solder joints [12–15]. Liu et al. [13] hypothesized that, additions of TiO_2 could inhibit the growth of IMC layer while refining the Cu_6Sn_5 grains in Sn-based solder alloys. Moreover, Wu et al. [13] reported the ability of the Al_2O_3 in reducing the growth rate of interfacial Cu_6Sn_5 in Sn-0.3 Ag-0.7Cu. Meanwhile, Li et al. [16] made an attempt by adding CeO_2 in SAC solder alloys which resulted in lowering the rate of Cu atoms' diffusion at the copper/solder interface through a diffusion-controlled kinetic model.

In our previous works [17], kaolin geopolymer ceramic-reinforced Sn-3.0Ag-0.5Cu (SAC305) composite solders were fabricated with various weight percentages (0–2.0 wt%) and investigated in terms of microstructure, mechanical strength and thermal properties. The results proved that, 1 wt% kaolin geopolymer ceramic was able to yield an optimum result in terms of thickness of the IMC layer, refining the microstructure and shear strength. However, the above-mentioned only focused on the effects of kaolin geopolymer ceramics in SAC305 solder alloys in as-reflowed conditions. As the current development of electronic packaging is continuously downsizing the size of the solder joints, the reliability of the joining becomes very important, especially under actual working conditions. During actual working conditions, solder joints are frequently exposed to solid state ageing conditions, such as continuous use of electronic equipment and switching on-off cycles of the electronic devices. These events could facilitate the growth of interfacial IMC layers as the IMCs are sensitive towards the temperature changes and IMC formations could significantly change during this condition. The reactions during solid state ageing differs significantly under influence of several temperature ranges [18]. For example, Cu_3Sn starts to grow at temperatures below 60 °C and increases at longer duration. Therefore, isothermally age-testing is typically used to simulate the changes of microstructure and performance of the solder joints during the actual working conditions [11,19–21].

Therefore, this paper aims to elucidate the effect of the addition of 1 wt% kaolin geopolymer ceramics on the growth kinetics of Cu_6Sn_5 and Cu_3Sn IMCs in SAC305 solder alloy and examine the suppression effect on the IMC layer and mechanical performance subjected to several ageing times and temperatures.

2. Materials and Methods

Sn-3.0Ag-0.5Cu (SAC305) solder powders with average particle sizes of 25–45 μm with spherical morphology were used as the solder matrix material. The reinforcement, which is kaolin geopolymer ceramic (KGC) powder, having average particle size of ~18 μm, were used in this research. The fabrication of KGC was preceded with the formation of kaolin geopolymer. Formation of kaolin geopolymer took place as the kaolin was activated with alkaline activator solution and underwent curing process at 80 °C for a day. Then, the kaolin geopolymer was pulverized by using a mechanical crusher and uniaxially compressed using a load of 4.5 tons. The sintering process of a compacted pellet was took place at 1200 °C, forming kaolin geopolymer ceramic (KGC). To obtain KGC with an average particle size of ~18 µm, sintered KGC underwent a ball milling process for about 10 h at a speed of 450 rpm. The ball to powder ratio (BPR) used during the ball milling process was 10:1 in a planetary ball milling machine.

To fabricate SAC305-KGC composite solders, powder metallurgy with a hybrid microwave sintering method was used. SAC305 with 1 wt% KGC was weighed and homogeneously mixed in an airtight container by using a planetary mill with a speed of 200 rpm for an hour. Then, the mixture was uniaxially compacted with a load of 4.5 tons in a stainless steel mould. The sintering process of compressed pellets were using a hybrid microwave sintering technique at 185 °C under ambient conditions, with an output power of 800 W, 50 Hz microwave oven for 3 min. For comparison, pure SAC305 solder was compacted and sintered by using the same method. To obtain a thin solder sheet with approximately 50 µm thickness, the sintered pellet was cold rolled using a rolling machine. Fabrication of solder balls with a size of 900 µm took place as a 3.0 mm puncher was used to punch a thin solder sheet. The punched solder sheets were then immersed in rosin mildly-activated flux (RMA) and heated on a Pyrex plate at 250 °C using a reflow oven. Then, the sieving of the solder ball with the use of 1 mm and 0.9 mm sieve was done in order to have a constant size of the solder ball.

A slight amount of RMA flux was applied to the solder ball and was placed on a Cu substrate printed circuit board with an organic solderability perspective surface finish with a ball pitch size of 900 µm. Then, the solder ball was reflowed in a F4N desktop reflow oven at 250 °C with time above liquids of 25 s. To study the growth of interfacial IMC, the samples were isothermally aged at temperatures of 85 °C, 125 °C and 150 °C for 100, 500 and 1000 h, which follows the JEDEC standard temperature (JESD22-A103C) [22] The microstructure analysis was conducted by using optical microscope (OM). The elements in the samples were analysed by using energy dispersive spectroscopy (EDS) on cross-sectioned aged samples. The samples were cold mounted with epoxy resin, mechanically grounded on silicon carbide papers and polished with an alumina suspension of 1.0 and 0.3 µm. To discover the details of the microstructure of the solder joint, oxide polishing suspensions (OPS) were used for final polishing. The thickness of the interfacial IMC was calculated based on optical microscope (OM) images taken by using ImageJ software (developed at National Institutes of Health and Laboratory for Optical and Computational Instrumentation, University of Winconsin, Madison, WI, USA). The thickness of interfacial IMC was measured by divided the total area of IMC with a total length of IMC.

Additionally, the strength of the solder joints after isothermal ageing was determined through single-lap shear solder joint test. The single-lap shear solder joint was performed using Instron machine with reference to ASTM D1002 standard [17]. The average shear strength was calculated from a total of five samples for each compositions and ageing conditions. Then, the fracture mode after single-lap shear test was investigated using a scanning electron microscope (SEM) with an energy dispersive X-ray under secondary imaging mode and an accelerating voltage of 20 kV.

3. Results and Discussions

3.1. Evolutions of IMC of Microstructure

Figure 1 illustrate the microstructure of non-reinforced SAC305 and SAC305-KGC composite solder at the bulk solder prior to reflow soldering. Based on the figure, the microstructure of the solder involves primary IMCs, eutectic and β-Sn area prior to reflow soldering process. Meanwhile, Figures 2–4 present the microstructure formations of non-reinforced SAC305 and SA305-KGC composite solders at the solder bulk subjected to isothermal ageing at 85 °C, 125 °C and 150 °C at several isothermal ageing times. Based on the figure, it can be clearly observed that the increase in ageing temperature and ageing time causes the primary IMCs to become coarser and larger in non-reinforced SAC305 solder. On top of this, the coarsening of IMCs in the eutectic area was also observed in non-reinforced SAC305 especially at higher temperature of 150 °C and longer time (as in Figure 5c). However, for SAC305-KGC composite solder sample, the coarsening of IMCs during solid state isothermal ageing was not obvious. This alludes that addition of KGC particles could controls and suppressed the coarsening of IMCs in eutectic area and formation of primary IMCs even during isothermal ageing. The same phenomenon can be found in a study done by Sobhy et al. [23]. As studied by Gain et al. [24], the incorporation of ceramic particles in the solder may alter the diffusivity and chemical affinity which lead to refining of IMCs particles even during solid state isothermal ageing.

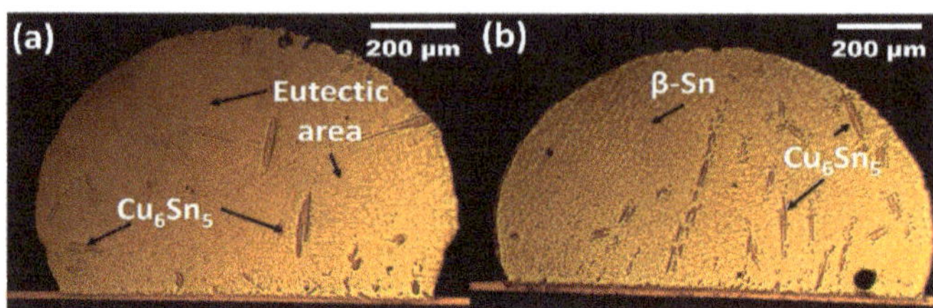

Figure 1. Microstructure of as-reflowed solder bulk for (**a**) non-reinforced SAC305 and (**b**) SAC305-KGC composite solders.

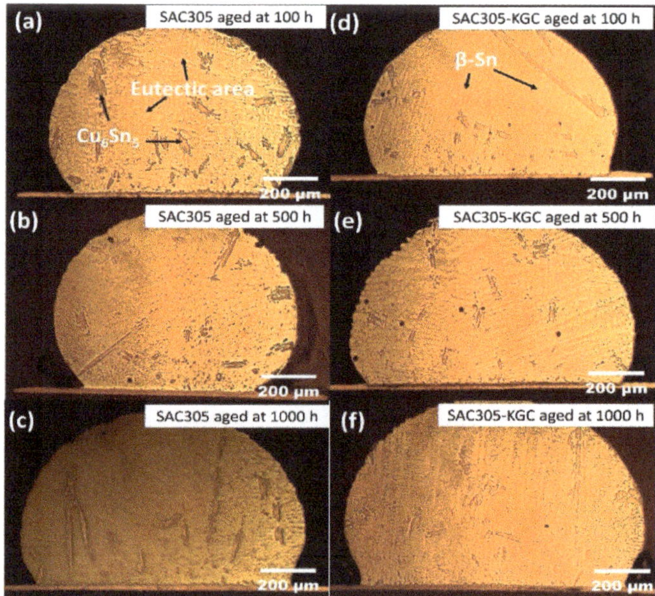

Figure 2. Microstructure formation at the bulk solder after isothermal ageing at 85 °C. Non-reinforced SAC305 solder at (**a**) 100 h, (**b**) 500 h, (**c**) 1000 h and SAC305-KGC composite solder at (**d**) 100 h, (**e**) 500 h, (**f**) 1000 h.

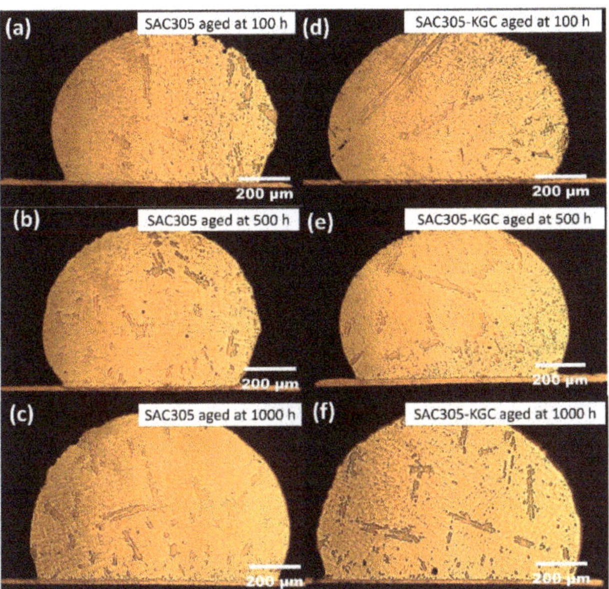

Figure 3. Microstructure formation at the bulk solder after isothermal ageing at 125 °C. Non-reinforced SAC305 solder at (**a**) 100 h, (**b**) 500 h, (**c**) 1000 h and SAC305-KGC composite solder at (**d**) 100 h, (**e**) 500 h, (**f**) 1000 h.

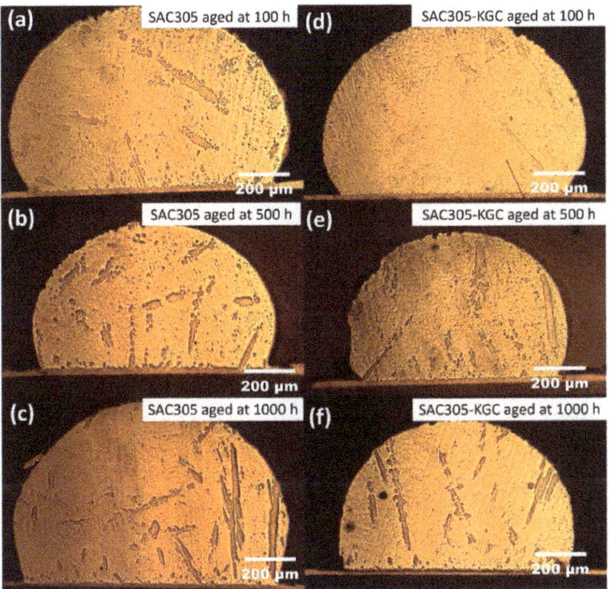

Figure 4. Microstructure formation at the bulk solder after isothermal ageing at 150 °C. Non-reinforced SAC305 solder at (**a**) 100 h, (**b**) 500 h, (**c**) 1000 h and SAC305-KGC composite solder at (**d**) 100 h, (**e**) 500 h, (**f**) 1000 h.

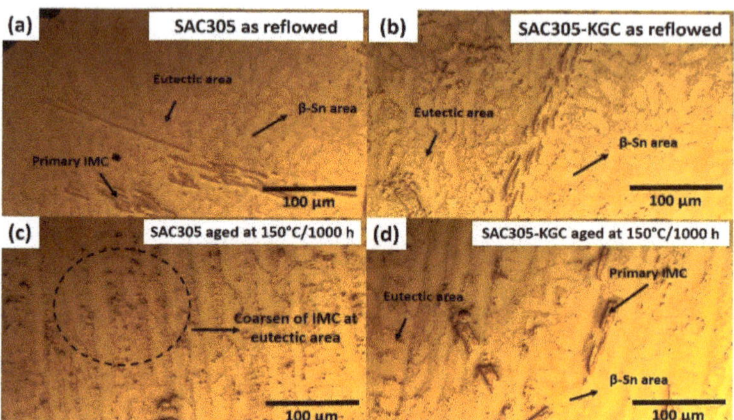

Figure 5. Microstructure at the bulk solder for (**a**) non-reinforced SAC305 as reflowed, (**b**) SAC305-KGC as reflowed, (**c**) SAC305 aged at 150 °C for 1000 h and (**d**) SAC305-KGC aged at 150 °C for 1000 h.

The intermetallic compound (IMC) at the interfacial was investigated based on the average thickness and morphology of IMC layer. Figure 6 shows the cross-sectional images of non-reinforced SAC305 and SAC305-KGC composite solders prior to reflow soldering process. Based on Figure 6a, the morphology of interfacial IMC layer in non-reinforced SAC305 solder had an elongated and rounded scallop of Cu_6Sn_5. Meanwhile, the morphology of the interfacial IMC layer in SAC-KGC composite solder had rounded scallop Cu_6Sn_5 and the formation of elongated scallop Cu_6Sn_5 was not observed. According to our previous study, the elongated scallop was formed due to the increase in concentration of Cu atoms from the substrates, diffusing and reacting with Sn, forming elongated scallop Cu_6Sn_5 [17]. However, in this research Cu_3Sn layer was not be observed on the sample of non-reinforced SAC305 and SAC305-KGC composite solders after reflow soldering due to the significantly thinner layer. The Cu_3Sn phase is known to be formed during the reaction between Sn-Cu/Cu after reflow soldering as reported by Feng et al. [25].

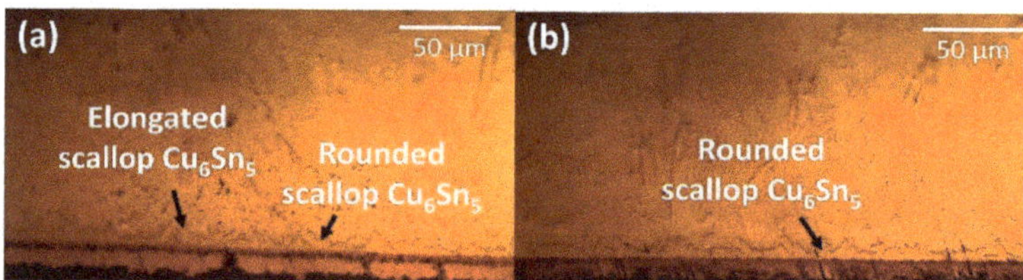

Figure 6. Cross-sectional as-reflowed optical microscope images for (**a**) non-reinforced SAC305 and (**b**) SAC305-KGC composite solder.

Meanwhile, Figures 7–9 show the interfacial IMC layer of non-reinforced SAC305 and SAC305-KGC composite solder subjected to isothermal ageing at temperatures of 85 °C, 125 °C and 150 °C for 100 h, 500 h and 1000 h. Based on the figures, the interfacial IMC layer consists of duplex IMC structures. According to EDX analysis as in Figure 10, the duplex structure consists of light layer (Point 1) which corresponds to Cu_6Sn_5 IMC layer. Then, a dark layer (Point 2) was corresponding to Cu_3Sn IMC layer. In addition, Figure 7 shows the interfacial IMC layer of non-reinforced SAC305 and SAC305-KGC composite solders subjected to a low isothermal ageing temperature of 85 °C for 100 h, 500 h and 1000 h. The samples showed combination of elongated and rounded scallop Cu_6Sn_5 with a very thin Cu_3Sn IMC layer. However, as the isothermal ageing process continues with higher temperatures of 125 °C and 150 °C, the morphology of interfacial

IMC layer in non-reinforced SAC305 was formed with some elongated and small rounded scallop as depicted in Figures 8a–c and 9a–c. Meanwhile, in SAC305-KGC composite solder, the interfacial IMC layer grew towards more planar-type morphology with increasing ageing time and temperature as shown in Figure 9d–f.

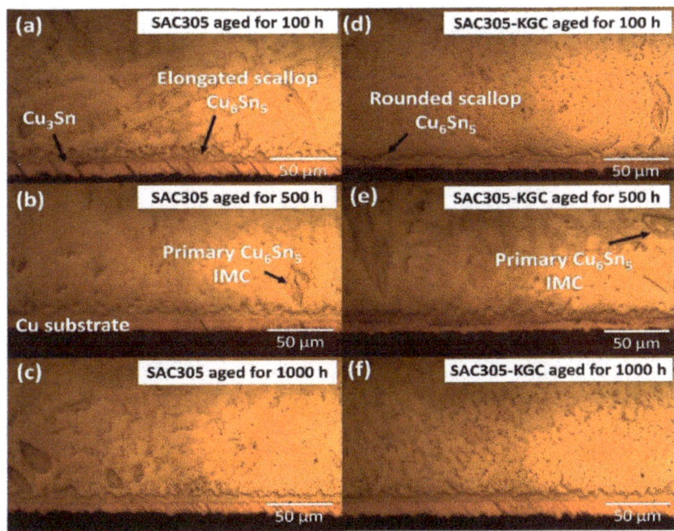

Figure 7. Interfacial IMC layer subjected to isothermal ageing at 85 °C. Non-reinforced SAC305 at (**a**) 100 h, (**b**) 500 h, (**c**) 1000 h. SAC305-KGC composite solder at (**d**) 100 h, (**e**) 500 h and (**f**) 1000 h.

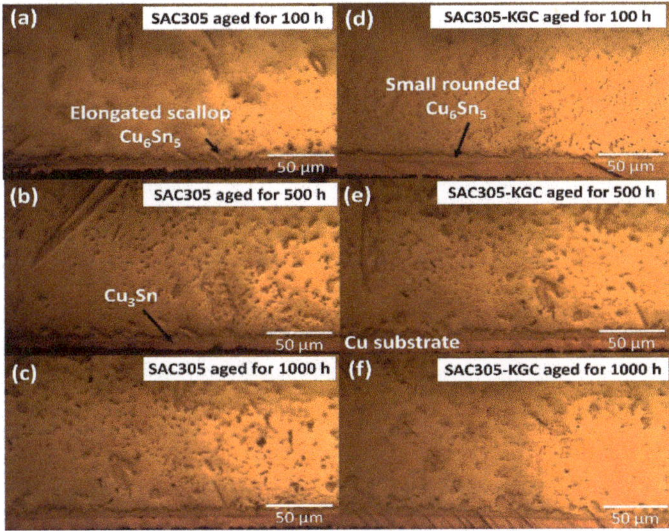

Figure 8. Interfacial IMC layer subjected to isothermal ageing at 125 °C. Non-reinforced SAC305 at (**a**) 100 h, (**b**) 500 h, (**c**) 1000 h. SAC305-KGC composite solder at (**d**) 100 h, (**e**) 500 h and (**f**) 1000 h.

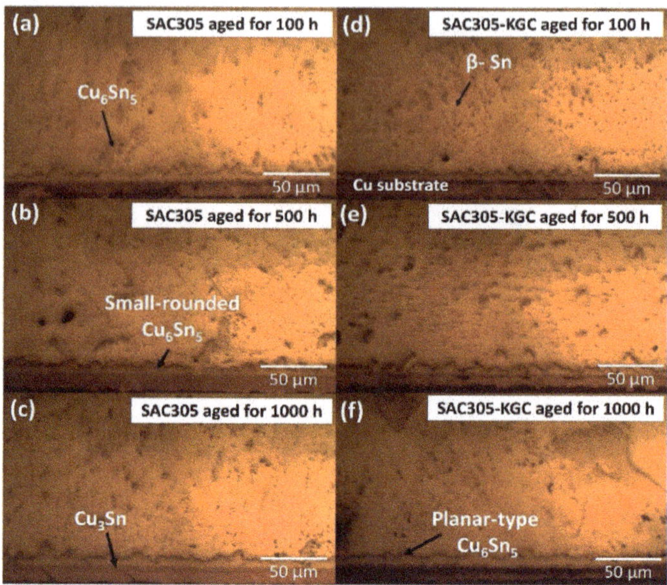

Figure 9. Interfacial IMC layer subjected to isothermal ageing at 150 °C. Non-reinforced SAC305 at (**a**) 100 h, (**b**) 500 h, (**c**) 1000 h. SAC305-KGC composite solder at (**d**) 100 h, (**e**) 500 h and (**f**) 1000 h.

The changes in the morphology of Cu_6Sn_5 IMC in SAC305-KGC composite solder towards more planar-type was due to the distance at scallop valley was closer to the substrates as compared to distance at the scallop of Cu_6Sn_5 peak to the substrates. This occurrence causes a faster Cu diffusion at the scallop valley and, thus, planarizing the morphology of interfacial Cu_6Sn_5 IMC [2,11,22,26]. Furthermore, the changes in the morphology also relates with the differences in the Gibs Free energy. The changes in the IMC structures occurred during the isothermal ageing was aims to lower the surface energy. The surface energy associated with scallop Cu_6Sn_5 IMC was higher compared to the layered type. The heat produced from the isothermal ageing process causes the surface tension of the Cu_6Sn_5 IMC to be unstable. In order to stabilize the surface tension, the excess energy was removed through atoms diffusing, resulting in higher Cu atom diffusion and, thus, producing a layered type of Cu_6Sn_5 IMC [27].

To precisely elucidate the growth of interfacial IMC layer on non-reinforced SAC305 and SAC305-KGC composite solders during isothermal ageing, the average thickness of interfacial IMC layer was calculated by using ImageJ software Figure 11 illustrates the bar graph showing the total (Cu_6Sn_5 + Cu_3Sn) average of interfacial IMC thickness subjected to different isothermal ageing temperatures and time. Based on the figure, it can be seen clearly that the total interfacial IMC thickness increased in non-reinforced SAC305 and SAC305-KGC composite solders as the temperature and time increased. However, the total interfacial IMC thickness of non-reinforced SAC305 was higher compared to SAC305-KGC composite solder. Figure 11c illustrates, that the total interfacial IMC thickness of non-reinforced SAC305 at isothermal ageing of 150 °C/1000 h was ~17 µm. Meanwhile, the total interfacial IMC thickness for SAC305-KGC composite solder under the same condition was ~14 µm. Therefore, this result suggests that the incorporation of KGC particles in SAC3055 solder is beneficial in suppressing the interfacial IMC thickness during solid state ageing with the suppression of approximately 15%.

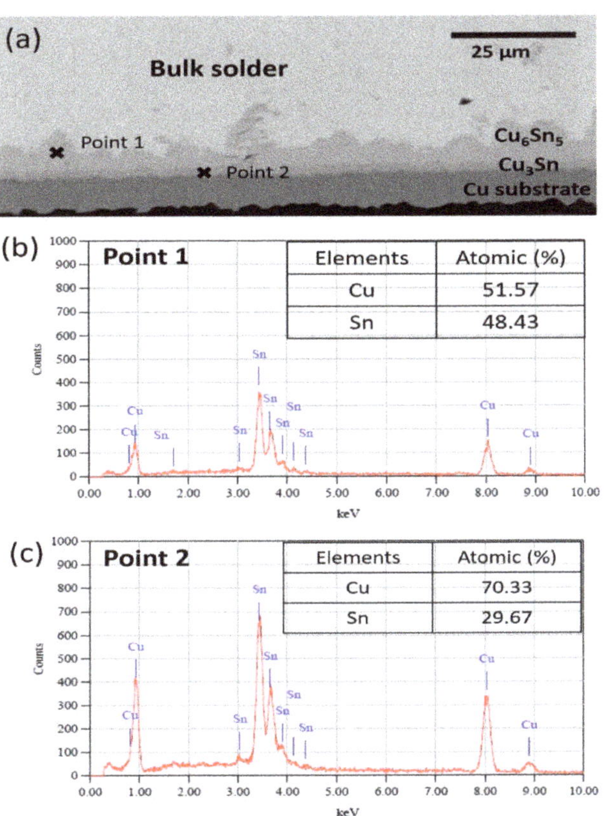

Figure 10. (**a**) Cross-sectional images of EDX point analysis. (**b**) EDX analysis result at point 1 and (**c**) EDX analysis result at point 2.

As reported by Wang et al. [28], Cu_3Sn layer may be formed between Cu_6Sn_5 and Cu substrates during reflow soldering process or after ageing. The formation of Cu_3Sn layer resulted from solid state diffusion. As reported by Mohd Salleh et al. [29], Cu_6Sn_5 were rapidly formed during the early reaction between solder alloy and Cu substrate. However, with continuous diffusion of Cu, a layer of Cu_3Sn will be formed in between Cu_6Sn_5 and Cu substrates. As the Cu atoms reach at the interface between the Cu_6Sn_5/Cu_3Sn, the following reaction will occur [2]:

$$Cu_6Sn_5 + 9Cu \rightarrow 5Cu_3Sn \quad (1)$$

Based on the Equation (1), the reaction of Cu with Cu_6Sn_5 cause Cu_6Sn_5 to be converted to Cu_3Sn. Therefore, the Cu atoms diffusion at Cu_6Sn_5/solder will be reduced. Thus, the growth of Cu_6Sn_5 layer will be interrupted, while the growth of Cu_3Sn layer will be increased with prolong ageing time and temperature. In this research, Cu_3Sn layer was only observed after ageing the sample at 85 °C for 100 h. Figure 12 illustrates the average Cu_3Sn IMC thickness subjected to isothermal ageing at temperature of 85 °C, 125 °C and 150 °C for 100, 500 and 1000 h. From the bar graph, it is clearly shown that, with prolonged ageing time and higher temperature, the thickness of interfacial Cu_3Sn was increased. The average thickness of Cu_3Sn IMC in non-reinforced SAC305 solders during isothermal ageing at 150 °C for 1000 h was 2.20 µm, whereas for SAC305-KGC composite solder, the average thickness of Cu_3Sn IMC was 1.60 µm. The difference about ~24% in the average Cu_3Sn IMC thickness between non-reinforced SAC305 and SAC305-KGC composite solders under isothermal ageing of 150 °C/1000 h prove that the addition of KGC particles were able to suppress the growth of Cu_3Sn IMC layer especially at higher ageing temperature and longer time.

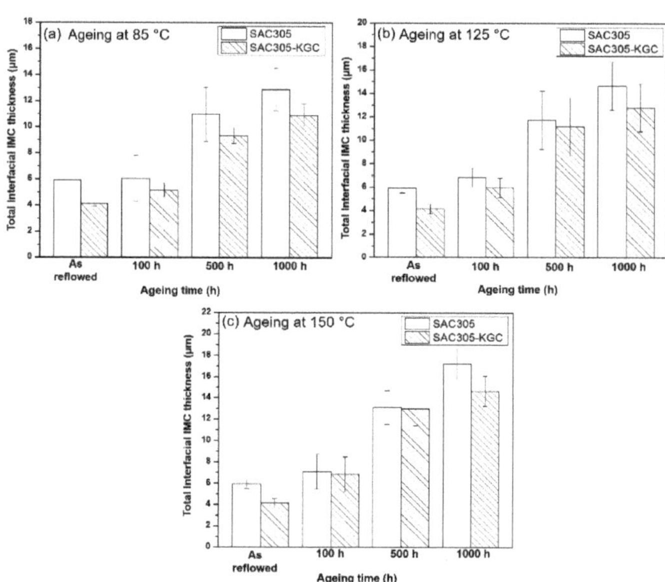

Figure 11. Total interfacial IMC thickness subjected to isothermal ageing at temperature of (**a**) 85 °C, (**b**) 125 °C and (**c**) 150 °C.

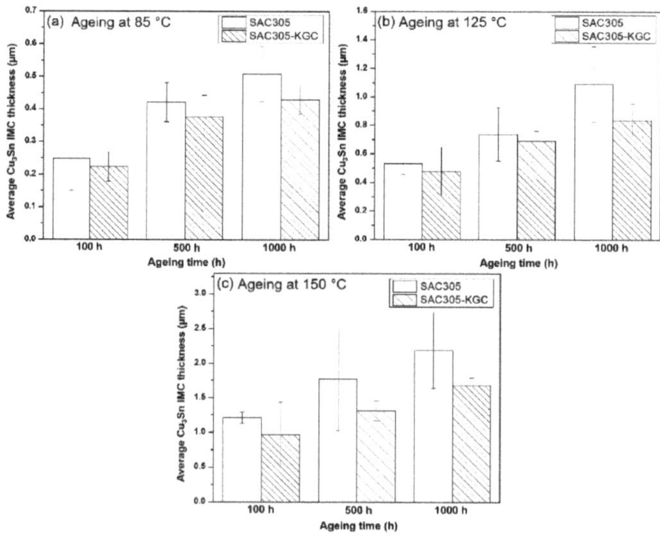

Figure 12. Average Cu_3Sn IMC thickness subjected to isothermal ageing at temperature of (**a**) 85 °C, (**b**) 125 °C and (**c**) 150 °C.

Overall, the thickness of total interfacial IMC layer (Cu_6Sn_5 and Cu_3Sn) and Cu_3Sn in SAC305-KGC composite solder was greatly suppressed under the isothermal ageing process. The suppression in the thickness of interfacial IMC layer was due to the ability of KGC particles across the matrix of solders to hinder the diffusion of Cu at the interfacial IMC layer. The added KGC particles may be absorbed on the surface of IMCs and retarding the further growth of IMC layer. Moreover, as reported by Mohamad Zaimi [17], segregation of KGC particles existed across the molten solder matrix and interfacial layer which supports the ability of KGC particles to alter the growth of interfacial IMC layer.

3.2. Growth Kinetics of IMC Layer

Growth kinetics of IMC layer during the solid-state isothermal ageing can be described according to the empirical power-law relationship as following [30,31]:

$$Xt = Xo + Dt^{n}, \tag{2}$$

where Xt is the IMC thickness (m) at ageing time t (s), Xo is the initial thickness of IMC layer after reflow soldering and D is the diffusion coefficient (m^2/s) and n is the time exponent. The value of time exponent *n* could be considered as an indicator to dictate the controlling mechanism for the growth of IMC layer. If the value of n is closer to n = 0.33, the controlling mechanism for the growth of IMC layer can be described as grain-boundary diffusion. Meanwhile, if the value of n is closer to n = 0.5 and n = 1.0, the controlling mechanism for the growth of IMC layer is described as bulk diffusion-controlled process or an interface reaction rate-controlled process, respectively.

In this research, the value of the time exponent n for total IMC layer (Cu_6Sn_5 and Cu_3Sn) and Cu_3Sn can be attained by using linear regression analysis. The values of time exponent *n* for total IMC layer and Cu_3Sn are presented in Figure 13. Based on Figure 13a,b, the values of time exponent n for total IMC layer in non-reinforced SAC305 and SAC305-KGC composite solders were in the range of 0.45–0.52. The values obtained tend to be closer to 0.5. Thus, it can be considered that the growth of total IMC layer in non-reinforced SAC305 and SAC305-KGC composite solders during solid state isothermal ageing was controlled by the bulk diffusion process. The time exponent n of non- reinforced SAC305 was slightly higher than SAC305-KGC composite solder with an increase in temperature. This phenomenon can be associated with the changes in the morphology of IMC layer. As discussed in Section 3.1, at 150 °C, the SAC305-KGC composite solder had more planar-type morphology in which the channel between the scallop gradually vanishes. Thus, the diffusion process took shorter time and led to small value of time exponent n. This result was aligned with the research by Tang et al. [11]. Meanwhile, for Cu_3Sn IMC layer, the value of time exponent n was in the range of 0.21–0.29. From the values obtained, the growth of Cu_3Sn layer was controlled by grain-boundary diffusion process. Based on the results attained, it was noticed the total interfacial IMC and Cu_3Sn exhibited different IMC growth controlling mechanism. This was due to the difference in the microstructure formation and thickness during the isothermal ageing process [32].

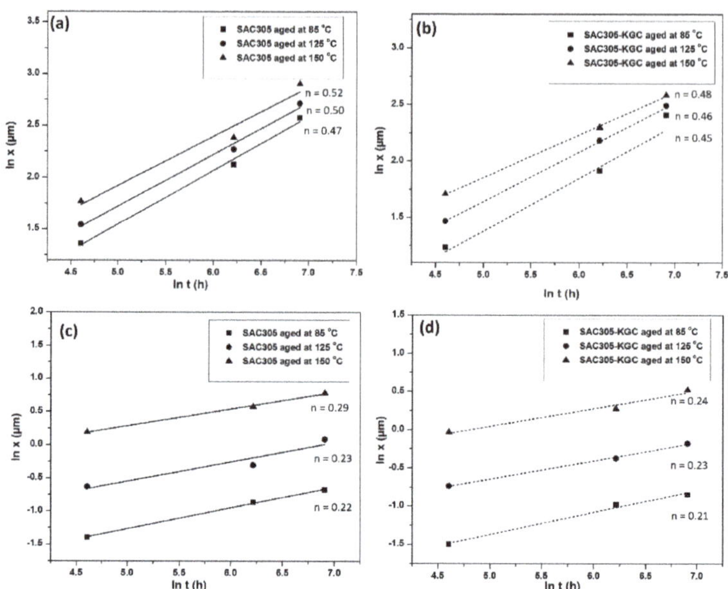

Figure 13. Time exponent of Cu_6Sn_5 for (**a**) non-reinforced SAC305, (**b**) SAC305-KGC composite solder. Cu_3Sn, (**c**) non-reinforced SAC305 and (**d**) SAC305-KGC composite solder.

Moreover, Figure 14 presents the relationship between total interfacial IMC thickness layer (including Cu_6Sn_5 and Cu_3Sn) and Cu_3Sn IMC with several ageing temperatures and times for non-reinforced SAC305 and SAC305-KGC composite solders. Based on the graph plotted in Figure 14, the total interfacial IMC thickness layer and Cu_3Sn IMC increased with an increase ageing time and grew faster at higher ageing temperature. Interestingly, the thickness of IMC layer was suppressed with the incorporation of KGC during the solid-state ageing. Quantitative investigation on the effect of KGC particles towards the interfacial thickness of IMC layer, across non-reinforced SAC305 and SAC305-KGC composite solders during the solid-state ageing, can be described according to Equation (1). The diffusion coefficient D can be determined from a linear regression analysis by plotting the graph presented in Figure 14, where the slope of the graph is equal to D. Thus, the diffusion coefficient, D for the growth of total interfacial IMC layer and Cu_3Sn IMC in non-reinforced SAC and SAC305-KGC composite solders are presented in Table 1. Based on the values of diffusion coefficient obtained, SAC305-KGC composite solder showed lower D value compared to SAC solder at all ageing temperatures. This can be inferred that, KGC particles could hinder the growth of IMC between the solder and substrate. Additionally, at a temperature of 150 °C with ageing time of 1000 h, the values of D for both solders increased corresponding to the thicker IMC at the interfacial layer. This owes to the fact that higher ageing temperature can provide appropriate thermal energy in order to overcome the higher activation energy of diffusion elements [27]. The results obtained was parallel with the studied done by Yin et al. [33].

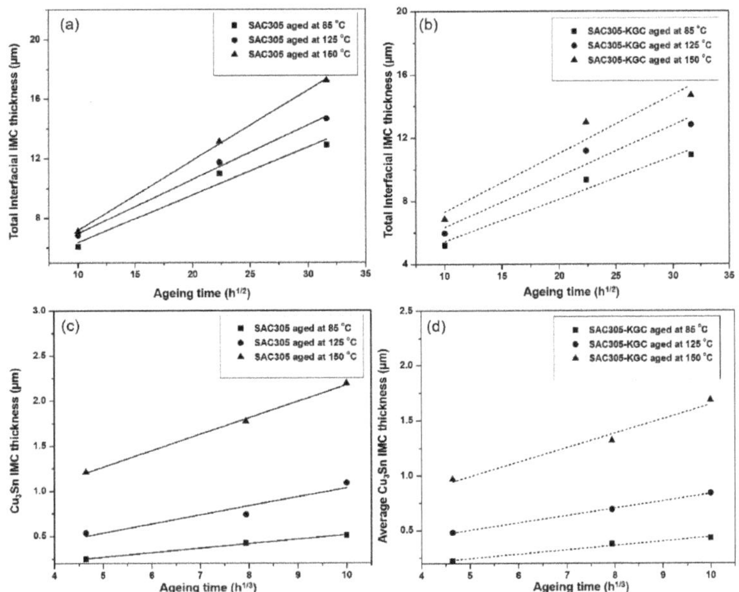

Figure 14. The relationship between the total interfacial IMC thickness curves and ageing temperature and ageing time for (**a**) total interfacial IMC for non-reinforced SAC305, (**b**) total interfacial IMC for SAC305-KGC composite solder (**c**) Cu_3Sn IMC for non-reinforced SAC305 and (**d**) Cu_3Sn IMC for SAC305-KGC composite solder.

Table 1. Diffusion coefficient for total interfacial IMC layer and Cu_3Sn IMC.

Solder Composition	Temperature (°C)	Diffusion Coefficient (Total Interfacial IMC)	Diffusion Coefficient (Cu_3Sn IMC)
Non-reinforced SAC305	85	2.89×10^{-17}	3.29×10^{-26}
	125	3.66×10^{-17}	2.77×10^{-25}
	150	4.25×10^{-17}	1.65×10^{-24}
SAC305-KGC composite solder	85	1.84×10^{-17}	1.68×10^{-26}
	125	2.89×10^{-17}	8.09×10^{-26}
	150	3.79×10^{-17}	6.24×10^{-25}

Meanwhile, the activation energy was determined according to the Arrhenius equation expressed in Equation (3):

$$D = Doe^{-Q/RT} \tag{3}$$

where Do is the temperature-dependent constant, Q is activation energy (kJ/mol), R is universal gas constant and T is the absolute temperature in degrees Kelvin. The activation energy of the interfacial intermetallic compound can be calculated by taking natural logarithm of Equation (3). Thus, the diffusion coefficient D can be expressed as:

$$\ln D = \ln Do - \frac{Q}{R}\left(\frac{1}{T}\right) \tag{4}$$

Equation (4) takes the form of y = mx + C, where the independent variable is (1/T) and dependent variable is ln D. By using Equation (4), the activation energy (Q) can be calculated by plotting the graph of ln D vs. 1/T where the slope of the graph represents the value of Q by using linear regression model. Based on the Equation (4) and the Arrhenius plotted in Figure 15, the activation energy for the total interfacial IMC layer and Cu_3Sn layer can be calculated.

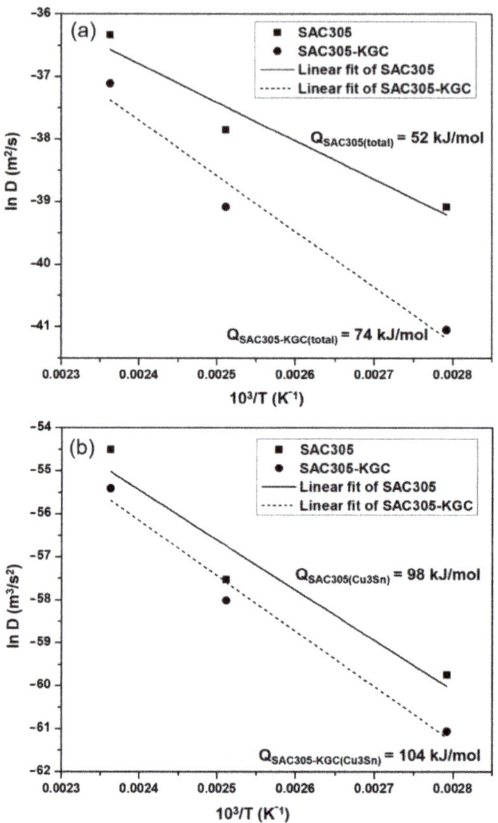

Figure 15. Arrhenius plot of ln D vs. 1/T for (**a**) total interfacial IMC layer and (**b**) Cu_3Sn layer in SAC305 and SAC305-KGC composite solder.

The activation energy for the total interfacial IMC layer in SAC305 solder was 52 kJ/mol, while, for SAC305-KGC composite solder, the activation energy was 74 kJ/mol. As reported by [34–36], the activation energy for SAC solder alloy was in the range of ~44 to 77 kJ/mol. Therefore, the activation energy obtained from this research was comparable with the previous studies. In addition to this, the higher activation energy value in SAC305-KGC composite solder corresponds to the low growth rate at lower temperature and higher growth rate at high temperature, as the value of activation energy was obtained from temperature dependence of diffusion coefficient, D. The higher activation energy experienced by SAC305-KGC composite solders was parallel with the lower total interfacial thickness of IMC layer. Furthermore, the activation energy of Cu_3Sn layer (as plotted in Figure 15b), was 98 kJ/mol and 104 kJ/mol for SAC305 and SAC305-KGC composite solders, respectively. This result indicates that the growth of Cu_3Sn layer in SAC305-KGC composite solder was slightly slower than SAC305 solder. As discussed earlier, the slightly lower activation energy exhibited by SAC305-KGC solders might be due to the effects of KGC particles addition in the solder. Moreover, in a study done by Tang et al. [11], they reported that TiO_2 particles display a little effect in increasing the activation energy of Cu_3Sn layer for Sn-3.0Ag-0.5Cu lead-free solder.

3.3. Shear Joint Strength

Figure 16 depicts the results on average shear strength of non-reinforced SAC305 and SAC305-KGC composite solders prior to reflow soldering and subjected to isothermal ageing of 150 °C at 100 h, 500 h and 1000 h. The average shear strength of non-reinforced SAC305 and SAC305-KGC composite solders after reflow soldering were ~10 MPa and ~13 MPa, respectively. The average shear strength of SAC305-KGC composite solder showed 31% improvement compared to non-reinforced SAC305 after reflow soldering process. However, during the isothermal ageing treatment at higher temperature of 150 °C for 100 h, 500 h and 1000 h, the average shear strength for non-reinforced SAC305 and SAC305-KGC composite solders displayed declining trend. The average shear strength decreased with the increase in temperature and longer ageing time. Nevertheless, the average shear strength of SAC305-KGC composite solder was still higher across all the samples aged at different time and temperature as compared to non-reinforced SAC305 solders. The lowest average shear strength was shown by non-reinforced SAC305 solder aged at 150 °C for 1000 h, which is about ~5 MPa. It was expected that, at higher ageing time and temperature, the shear strength is poor. This can be associated with the increment in the total thickness of interfacial IMC layer as depicted in Figure 11. The decreasing trend of the average shear strength across all the samples were corresponding to the thickness of the IMC layer. The thicker the IMC layer, the lower the average shear strength. This is because, IMC is brittle. Thus, thicker IMC layers were easily exposed to brittle failure and consequently, reducing the strength of solder joints. Nevertheless, the average shear strength in SAC305-KGC composite solder was still higher than non-reinforced SAC305 solders. This suggests that KGC particles play an important role in improving the shear strength during ageing process. In addition to this, KGC particles are also believed to be able to pin the dislocation motions and hinder the grain-boundary sliding across the solder matrix. The results obtained was in good agreement with the study done by Chen et al. [8]. He reported that the improvement in the shear strength of Sn-3.0Ag-0.5Cu during solid-state ageing was attributed to the theory of dispersion strengthening resulting from the addition of reinforcement particles.

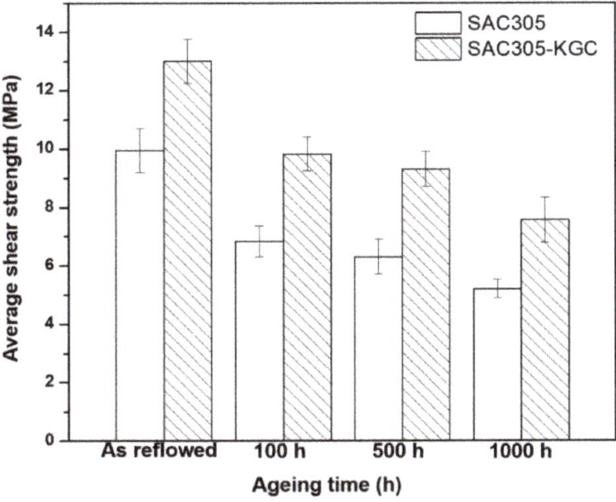

Figure 16. Average shear strength of non-reinforced SAC305 and SAC305-KGC composite prior to reflow soldering and subjected to isothermal ageing at 150 °C.

To ascertain the fracture mode for the samples after the lap shear test, the samples were observed under a scanning electron microscope (SEM) in secondary electron (SEI) mode as presented in Figure 17. Figure 17a,e represent the fracture surfaces of non-reinforced

SAC305 and SAC305-KGC composite solder prior to reflow soldering, respectively. Based on the figure it can be clearly observed that during reflowed condition, non-reinforced SAC305 exhibited a combination of fracture mode which is between ductile and brittle mode, whereas for SAC305-KGC sample, some dimples could be observed which correspond to the ductile fracture mode. By increasing the isothermal ageing temperature and time up to 150 °C for 1000 h, the fracture mode of the samples were slightly changed. Evidently, in SAC305-KGC composite solder, ductile dimple morphology was observed at the samples aged for 100 h and 500 h (as in Figure 17f,g). In addition, the dimple size areas were increased with the increase in ageing time which can be attributed to the coarse planar IMC layer, as suggested by Dele-Afolabi et al. [37]. Additionally, ductile brittle fracture mode was dominant in non-reinforced SAC305 aged at 150 °C for 1000 h, as presented in Figure 17g. The brittle fracture occurred in the region of intermetallics interface [38]. At higher ageing condition, non-reinforced SAC305 had a thicker IMC layer which supports the reason for the occurrence of the brittle fracture mode shown by the sample. Nevertheless, in SAC305-KGC composite solder, the fracture mode was a combination between ductile and brittle fracture modes. The improvement in the fracture mode of SAC305 was partially attributed to suppression of IMC layer in the samples with KGC particles. From this, it can be concluded that the KGC particles are able to alter the fracture mode even though the samples were subjected to higher isothermal ageing temperature for a longer time.

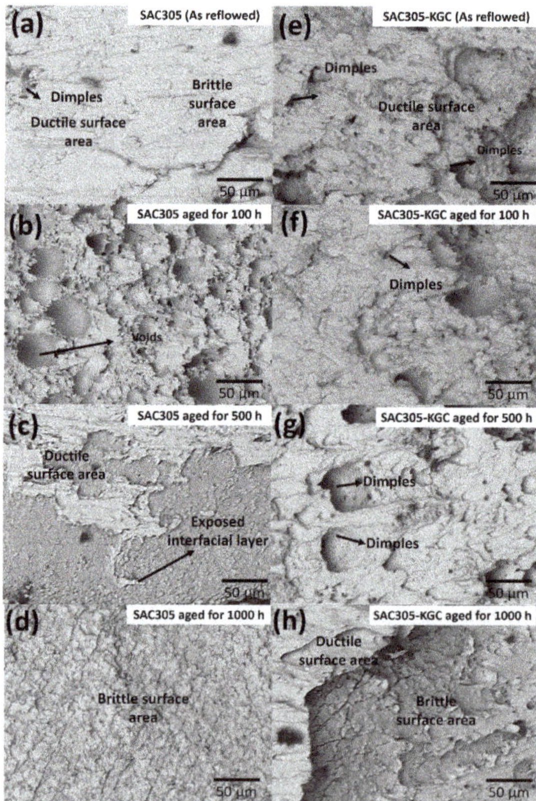

Figure 17. SEM micrographs of fracture surfaces subjected to isothermal ageing at 150 °C for non-reinforced SAC305 (**a**) as-reflowed, (**b**) 100 h, (**c**) 500 h, (**d**) 1000 h and SAC305-KGC composite solder (**e**) as-reflowed, (**f**) 100 h, (**g**) 500 h and (**h**) 1000 h.

4. Conclusions

The effects of KGC particles as the reinforcement in SAC305 solder had been investigated at temperature of 85 °C, 125 °C and 150 °C for 100 h, 500 h and 1000 h. The following conclusions can be gathered:

(a) The morphology of interfacial IMC layer of non-reinforced SAC305 and SAC305-KGC composite solder joints showed a duplex IMC structure comprises of scallop-type Cu_6Sn_5 and layer-type Cu_3Sn. With an increase in ageing time and temperature, the initial scallop Cu_6Sn_5 gradually changes to planar-type in SAC305-KGC composite solder joints. For Cu_3Sn, the morphology consistently maintained as layer-type across all ageing conditions.

(b) The total thickness of interfacial IMC (both Cu_6Sn_5 and Cu_3Sn) layer showed an increasing trend for non-reinforced SAC305 and SAC305-KGC composite solder joints which was about ~6–17 µm. However, with an addition of KGC particles the total thickness of interfacial IMC could be suppressed for about 15% as compared to non-reinforced SAC305 solder joints with the increase in ageing time and temperature. This can be owed to the ability of KGC particles to hinder the diffusion of Cu and resulting in thinner IMC layer. Moreover, the growth of total interfacial IMC layer was controlled by bulk diffusion process with the time exponent, n obtained was towards 0.5. The activation energy for SAC305-KGC composite solder joints was 74 kJ/mol and it exhibited lower diffusion coefficient as compared to non-reinforced SAC305.

(c) Meanwhile, the addition of KGC particles into SAC305 could suppress the growth of Cu_3Sn IMC layer by 24% as compared to non-reinforced SAC305 solder. In addition, the growth of Cu_3Sn IMC layer in this study was controlled by grain-boundary diffusion. The activation energy of Cu_3Sn IMC layer for SAC305-KGC composite solder joints was 104 kJ/mol and it exhibited lower diffusion coefficient even at high temperature as compared to non-reinforced SAC305.

(d) The average shear strength for all the solder joints decreased with the increase ageing time and temperature. However, the decrement of the strength was lower in SAC305-KGC composite solder joints than non-reinforced SAC305 solder. The average shear strength of SAC305-KGC composite solder was in the range of ca. 8–13 MPa. Meanwhile for non-reinforced SAC305, the average shear strength was in the range of ca. 5–9 MPa. In addition, SAC305-KGC solder joint possessed a combination of ductile and brittle fracture mode at higher temperature of 150 °C and 1000 h of ageing time.

Author Contributions: Conceptualization, methodology, and writing: N.S.M.Z.; supervision and resources: M.A.A.M.S.; visualisation and validation: A.V.S.; supervision and methodology: M.M.A.B.A.; investigation, formal analysis and data curation: N.S., S.Z.A.R. and P.V.; formal analysis and writing article—review and editing: R.M.S. and M.I.I.R. All authors have read and agreed to the published version of the manuscript.

Funding: This work was supported by Ministry of Higher Education, Malaysia and Newton fund under ISIS Neutron and Muon Source, Ministry of Education Malaysia (MOE), under reference no: JPT.S (BPKI)2000/016/018/019(29).

Institutional Review Board Statement: Not applicable.

Informed Consent Statement: Not applicable.

Data Availability Statement: The data presented in this study are available on request from the corresponding author.

Acknowledgments: The authors gratefully acknowledged the Ministry of Higher Education, Malaysia and Newton fund under ISIS Neutron and Muon Source, Ministry of Education Malaysia (MOE), under reference no: JPT.S (BPKI)2000/016/018/019(29) for the financial support. The authors would also like to acknowledge Centre of Excellent Geopolymer and Green Technology (CeGeoGTech), University Malaysia Perlis (UniMAP), Faculty of Chemical Engineering Technology, UniMAP and Faculty of Mechanical Engineering Technology, UniMAP for the support. The authors would also like to acknowledge the support from Nihon Superior.

Conflicts of Interest: The authors declare no conflict of interest.

References

1. Somidin, F.; Maeno, H.; Mohd Salleh, M.A.A.; Tran, X.Q.; McDonald, S.D.; Matsumura, S.; Nogita, K. Characterising the polymorphic phase transformation at a localised point on a Cu_6Sn_5 grain. *Mater. Charact.* **2018**, *138*, 113–119. [CrossRef]
2. Wang, F.; Zhou, L.; Wang, X.; He, P. Microstructural evolution and joint strength of Sn-58Bi/Cu joints through minor Zn alloying substrate during isothermal aging. *J. Alloys Compd.* **2016**, *688*, 639–648. [CrossRef]
3. Lee, C.-J.; Min, K.D.; Park, H.J.; Jung, S.-B. Mechanical properties of Sn-58 wt%Bi solder containing Ag-decorated MWCNT with thermal aging tests. *J. Alloys Compd.* **2020**, *820*, 153077. [CrossRef]
4. Tao, Q.B.; Benabou, L.; Nguyen Van, T.A.; Nguyen-Xuan, H. Isothermal aging and shear creep behavior of a novel lead-free solder joint with small additions of Bi, Sb and Ni. *J. Alloys Compd.* **2019**, *789*, 183–192. [CrossRef]
5. Wen, Y.; Zhao, X.; Chen, Z.; Gu, Y.; Wang, Y.; Chen, Z.; Wang, X. Reliability enhancement of Sn-1.0Ag-0.5Cu nano-composite solders by adding multiple sizes of TiO_2 nanoparticles. *J. Alloys Compd.* **2017**, *696*, 799–807. [CrossRef]
6. Zhang, L.; Fan, X.; Guo, Y.; He, C. Properties enhancement of SnAgCu solders containing rare earth Yb. *Mater. Des.* **2014**, *57*, 646–651. [CrossRef]
7. El-Daly, A.A.; El-Taher, A.M.; Gouda, S. Novel Bi-containing Sn–1.5Ag–0.7Cu lead-free solder alloy with further enhanced thermal property and strength for mobile products. *Mater. Des.* **2015**, *65*, 796–805. [CrossRef]
8. Chen, G.; Peng, H.; Silberschmidt, V.V.; Chan, Y.C.; Liu, C.; Wu, F. Performance of Sn–3.0Ag–0.5Cu composite solder with TiC reinforcement: Physical properties, solderability and microstructural evolution under isothermal ageing. *J. Alloys Compd.* **2016**, *685*, 680–689. [CrossRef]
9. Fix, A.R.; López, G.A.; Brauer, I.; Nüchter, W.; Mittemeijer, E.J. Microstructural development of Sn-Ag-Cu solder joints. *J. Electron. Mater.* **2005**, *34*, 137–142. [CrossRef]
10. Liu, P.; Yao, P.; Liu, J. Evolutions of the interface and shear strength between SnAgCu–xNi solder and Cu substrate during isothermal aging at 150 °C. *J. Alloys Compd.* **2009**, *486*, 474–479. [CrossRef]
11. Tang, Y.; Luo, S.M.; Wang, K.Q.; Li, G.Y. Effect of Nano-TiO_2 particles on growth of interfacial Cu_6Sn_5 and Cu_3Sn layers in Sn–3.0Ag–0.5Cu–xTiO_2 solder joints. *J. Alloys Compd.* **2016**, *684*, 299–309. [CrossRef]
12. Mohd Nasir, S.S.; Yahaya, M.Z.; Erer, A.M.; Illés, B.; Mohamad, A.A. Effect of TiO_2 nanoparticles on the horizontal hardness properties of Sn-3.0Ag-0.5Cu-1.0TiO_2 composite solder. *Ceram. Int.* **2019**, *45*, 18563–18571. [CrossRef]
13. Liu, Z.; Ma, H.; Shang, S.; Wang, Y.; Li, X.; Ma, H. Effects of TiO_2 nanoparticles addition on physical and soldering properties of Sn–xTiO_2 composite solder. *J. Mater. Sci. Mater.* **2019**, *30*, 18828–18837. [CrossRef]
14. Wu, J.; Xue, S.; Wang, J.; Wu, M.; Wang, J. Effects of α-Al_2O_3 nanoparticles-doped on microstructure and properties of Sn–0.3Ag–0.7Cu low-Ag solder. *J. Mater. Sci. Mater.* **2018**, *29*, 7372–7387. [CrossRef]
15. Mahim, Z.; Mohd Salleh, M.A.A.; Saud, N. Effect on microstructural and physical properties of Sn-3.0Ag-0.5Cu lead-free solder with the addition of SiC particles. *Eur. J. Mater. Sci.* **2019**, *4*, 37–43. [CrossRef]
16. Li, Z.H.; Tang, Y.; Guo, Q.W.; Li, G.Y. A diffusion model and growth kinetics of interfacial intermetallic compounds in Sn-0.3Ag-0.7Cu and Sn-0.3Ag-0.7Cu-0.5CeO_2 solder joints. *J. Alloys Compd.* **2019**, *818*, 152893. [CrossRef]
17. Mohamad Zaimi, N.S.; Mohd Salleh, M.A.A.; Abdullah, M.M.A.B.; Ahmad, R.; Mostapha, M.; Yoriya, S.; Chaiprapa, J.; Zhang, G.; Harvey, D.M. Effect of kaolin geopolymer ceramic addition on the properties of Sn-3.0Ag-0.5Cu solder joint. *Mater. Today Commun.* **2020**, *25*, 101469. [CrossRef]
18. Li, Q.; Chan, Y.C. Growth kinetics of the Cu_3Sn phase and void formation of sub-micrometre solder layers in Sn–Cu binary and Cu–Sn–Cu sandwich structures. *J. Alloys Compd.* **2013**, *567*, 47–53. [CrossRef]
19. Mookam, N.; Kanlayasiri, K. Evolution of Intermetallic Compounds between Sn-0.3Ag-0.7Cu Low-silver Lead-free Solder and Cu Substrate during Thermal Aging. *J. Mater. Sci. Technol.* **2012**, *28*, 53–59. [CrossRef]
20. Burduhos Nergis, D.D.; Vizureanu, P.; Corbu, O. Synthesis and Characteristics of Local Fly Ash Based Geopolymers Mixed with Natural Aggregates. *Rev. Chim.* **2019**, *70*, 1262–1267. [CrossRef]
21. Nergis, D.D.B.; Vizureanu, P.; Ardelean, I.; Sandu, A.V.; Corbu, O.C.; Matei, E. Revealing the Influence of Microparticles on Geopolymers' Synthesis and Porosity. *Materials* **2020**, *13*, 3211. [CrossRef]
22. Rabiatul Adawiyah, M.A.; Saliza Azlina, O. Comparative study on the isothermal aging of bare Cu and ENImAg surface finish for Sn-Ag-Cu solder joints. *J. Alloys Compd.* **2018**, *740*, 958–966. [CrossRef]
23. Sobhy, M.; El-Refai, A.M.; Mousa, M.M.; Saad, G. Effect of ageing time on the tensile behavior of Sn–3.5wt% Ag–0.5wt% Cu (SAC355) solder alloy with and without adding ZnO nanoparticles. *Mater. Sci. Eng. A* **2015**, *646*, 82–89. [CrossRef]
24. Gain, A.K.; Zhang, L. Nanosized samarium oxide (Sm_2O_3) particles suppressed the IMC phases and enhanced the shear strength of environmental-friendly Sn-Ag-Cu material. *Mater. Res. Express* **2019**, *6*, 066526. [CrossRef]
25. Feng, J.; Hang, C.; Tian, Y.; Liu, B.; Wang, C. Growth kinetics of Cu_6Sn_5 intermetallic compound in Cu-liquid Sn interfacial reaction enhanced by electric current. *Sci. Rep.* **2018**, *8*, 1775. [CrossRef]
26. Zhang, L.; Song, B.X.; Zeng, G.; Gao, L.L.; Huan, Y. Interface reaction between SnAgCu/SnAgCuCe solders and Cu substrate subjected to thermal cycling and isothermal aging. *J. Alloys Compd.* **2012**, *510*, 38–45. [CrossRef]
27. Chandra Rao, B.S.S.; Weng, J.; Shen, L.; Lee, T.K.; Zeng, K.Y. Morphology and mechanical properties of intermetallic compounds in SnAgCu solder joints. *Microelectron. Eng.* **2010**, *87*, 2416–2422. [CrossRef]

28. Wang, K.-K.; Gan, D.; Hsieh, K.-C. The orientation relationships of the Cu_3Sn/Cu interfaces and a discussion of the formation sequence of Cu_3Sn and Cu_6Sn_5. *Thin Solid Film.* **2014**, *562*, 398–404. [CrossRef]
29. Mohd Salleh, M.A.A.; McDonald, S.D.; Nogita, K. Effects of Ni and TiO_2 additions in as-reflowed and annealed $Sn_{0.7}$Cu solders on Cu substrates. *J. Mater. Process. Technol.* **2017**, *242*, 235–245. [CrossRef]
30. Li, C.; Hu, X.; Jiang, X.; Li, Y. Interfacial reaction and microstructure between the $Sn_3Ag_{0.5}$Cu solder and Cu-Co dual-phase substrate. *Appl. Phys. A* **2018**, *124*, 484. [CrossRef]
31. Dariavach, N.; Callahan, P.; Liang, J.; Fournelle, R. Intermetallic growth kinetics for Sn-Ag, Sn-Cu, and Sn-Ag-Cu lead-free solders on Cu, Ni, and Fe-42Ni substrates. *J. Electron. Mater.* **2006**, *35*, 1581–1592. [CrossRef]
32. Tian, R.; Hang, C.; Tian, Y.; Zhao, L. Growth behavior of intermetallic compounds and early formation of cracks in Sn-3Ag-0.5Cu solder joints under extreme temperature thermal shock. *Mater. Sci. Eng.* **2018**, *709*, 125–133. [CrossRef]
33. Yin, Z.; Lin, M.; Huang, Y.; Chen, Y.; Li, Q.; Wu, Z. Effect of Doped Nano-Ni on Microstructure Evolution and Mechanical Behavior of Sn-3.0Ag-0.5Cu (SAC305)/Cu-2.0Be Solder Joint during Isothermal Aging. *J. Mater. Eng. Perform.* **2020**, *29*, 3315–3323. [CrossRef]
34. Tay, S.L.; Haseeb, A.S.M.A.; Johan, M.R.; Munroe, P.R.; Quadir, M.Z. Influence of Ni nanoparticle on the morphology and growth of interfacial intermetallic compounds between Sn–3.8Ag–0.7Cu lead-free solder and copper substrate. *Intermetallics* **2013**, *33*, 8–15. [CrossRef]
35. Yoon, J.-W.; Noh, B.-I.; Kim, B.-K.; Shur, C.-C.; Jung, S.-B. Wettability and interfacial reactions of Sn–Ag–Cu/Cu and Sn–Ag–Ni/Cu solder joints. *J. Alloys Compd.* **2009**, *486*, 142–147. [CrossRef]
36. Jeon, S.-J.; Kim, J.-W.; Lee, B.; Lee, H.-J.; Jung, S.-B.; Hyun, S.; Lee, H.-J. Evaluation of drop reliability of Sn–37Pb solder/Cu joints using a high speed lap-shear test. *Microelectron. Eng.* **2012**, *91*, 147–153. [CrossRef]
37. Dele-Afolabi, T.T.; Azmah Hanim, M.A.; Norkhairunnisa, M.; Yusoff, H.M.; Suraya, M.T. Investigating the effect of isothermal aging on the morphology and shear strength of Sn-5Sb solder reinforced with carbon nanotubes. *J. Alloys Compd.* **2015**, *649*, 368–374. [CrossRef]
38. Zhao, J.; Cheng, C.-Q.; Qi, L.; Chi, C.-Y. Kinetics of intermetallic compound layers and shear strength in Bi-bearing SnAgCu/Cu soldering couples. *J. Alloys Compd.* **2009**, *473*, 382–388. [CrossRef]

Article

Experimental Research on the Cutting of Metal Materials by Electrical Discharge Machining with Contact Breaking with Metal Band as Transfer Object

Aurel Mihail Țîțu [1,2,*], Petrică Vizureanu [3,*], Ștefan Țîțu [4], Andrei Victor Sandu [3,5,*], Alina Bianca Pop [6], Viorel Bucur [1], Costel Ceocea [7] and Alexandru Boroiu [8]

1. Industrial Engineering and Management Department, Faculty of Engineering, "Lucian Blaga" University of Sibiu, 10 Victoriei Street, 550024 Sibiu, Romania; viorel.bucur@ulbsibiu.ro
2. The Academy of Romanian Scientists, 54 Splaiul Independenței, Sector 5, 050085 Bucharest, Romania
3. Faculty of Materials Science and Engineering, Gheorghe Asachi Technical University, Blvd. D. Mangeron 71, 700050 Iasi, Romania
4. The Oncology Institute "Prof. Dr. Ion Chiricuță" Cluj-Napoca, 34-36 Republicii Street, 400015 Cluj Napoca, Romania; stefan.titu@ymail.com
5. Romanian Inventors Forum, Str. Sf. P. Movila 3, 700089 Iasi, Romania
6. SC TechnoCAD SA, 72 Vasile Alecsandri Street, 430351 Baia Mare, Romania; bianca.bontiu@gmail.com
7. Department of Marketing and Management, The Faculty of Economic Sciences, "Vasile Alecsandri" University of Bacău, 157 Mărăști Street, 600115 Bacău, Romania; costelceocea@gmail.com
8. Automotive and Transport Department, Faculty of Mechanics and Technology, University of Pitești, 1 Târgul din Vale Street, 110040 Pitești, Romania; alexandru.boroiu@upit.ro

* Correspondence: mihail.titu@ulbsibiu.ro (A.M.Ț.); peviz@tuiasi.ro (P.V.); sav@tuiasi.ro (A.V.S.)

Received: 5 October 2020; Accepted: 17 November 2020; Published: 20 November 2020

Abstract: The scientific paper presents practical research carried out by a mixed team of Romanian researchers from universities and the business environment. The research consists in applying the process of cutting metallic materials through electrical discharge machining with contact breaking using a metal band as a transfer object. The research was implemented with the help of a specially designed installation in the laboratory and subsequently all the necessary steps were taken to obtain the patent for it. Various metallic materials were cut using this process, but first of all, high alloy steels. In the global research conducted by the authors, active experimental programs and classic experimental programs were used. The composite central factorial experiment was the method that led to the most effective results in terms of interpretations and conclusions. The research as a whole includes unique elements from an engineering point of view and here we can highlight the use of a metal band as a transfer object for this type of process as well as the designed, realized, and subsequently patented installation.

Keywords: electrical discharge machining with contact breaking; metal band; process modeling; objective functions; cutting; central composite design

---

1. Introduction

Dimensional processing through electrical discharge machining is one of the most widespread nonconventional processing processes in the world. Contact breaking electrical discharge machining is a process widely used today for cutting metal materials using mainly a disk-type transfer object. The authors of this research came to complete the range of possibilities available today using a metal band as a transfer object for processing through electrical discharge machining with contact breaking. This type of process and implicitly cutting metallic materials is mentioned in the literature by the Russian researchers Boris and Natalia Lazarenco, respectively, by the English researcher Priesley (1770).

We are witnessing a continuous evolution in the use of new types of metallic materials and increasing new modern technologies in fields such as aeronautics, automotive, car construction, etc., using the so-called nonconventional technologies in which material processing is done by using and directing energies in various forms [1,2].

The literature details an important range of researchers' concerns in order to improve the efficiency and effectiveness of processing processes in the field of nonconventional technologies [3,4]: the study of dielectrics and fluids, in particular, in the case of erosion processes, regardless of the type of process chosen [5–7]; complex studies and research on the processing and cutting of metallic materials and the issue of the effect of the electrode material on residual voltage [8,9]; studies on the rigidity of materials and cracks that occur in the material [10–12].

There are relevant analyses that have been taken into account regarding the rigidity of the processed surfaces correlated with the wear of the transfer object in order to determine the efficiency of the material sampling process depending on certain parameters such as energy source, electrical impulse duration, but also other process parameters [13–15].

It was found the extension of erosion processing technologies in the bio-medical field where the transfer object composition was modified by adding chromium in its component regardless of the form and type of processing [16–18].

It was determined that most of the contributions mentioned in the literature were found in terms of processing through electrical discharge machining with form copying [19–21], as well as the study of aspects worthy of consideration regarding issues of objective functions, processing productivity, volume wear, and relative wear [22–24]. Regarding the transfer object, specific properties of a carbon fiber reinforced polymer with a specific strength and extremely high rigidity, with a relatively low thermal expansion coefficient and a special property, namely the possibility to be able to easily model its surface, can be discussed [25–27]. The properties of this carbon fiber reinforced polymer are superior to other common materials.

Specific researches are presented in the literature where a redistribution of specific energies is observed in the processing of different types of materials that include aluminum alloys and sandwich panels [28–30]. Specific studies and research have been considered, which include issues of reliability and maintenance of material sampling processes using modern numerical simulations [31–33].

Through this research, it is desired to transmit a special approach regarding the modeling of technological parameters and objective functions for processing through electrical discharge machining with contact breaking using a transfer object made of a metal band, using a specially designed and subsequently patented installation and used in an industrial organization. Process parameters taken into account, together with the objective functions chosen in the context of the current process, lead to a modeling and subsequently to an optimization of the material sampling process as a whole.

2. General Consideration

The cutting of high steel alloys by electrical discharge machining with contact breaking (EDMCB) with electrode-tool—the metal band—represents one of the modern technological procedures of the non-conventional processing of some high steels alloy (hard and extra hard) categories, under the economic conditions of optimal efficiency [32,33]. The experimental research regarding the cutting process has emphasized, in particular, the technological aspects specific to the cutting (semi-fabricated of high steels alloy) through EDMCB, with Transfer Object (TO)—metallic band. Additionally, the existence of different values of the working parameters is highlighted, which is determined by the metal band use as a transfer object. The installations of the cutting steel process through EDMCB, has in its structure a tool—TO—in a metal disk form (this is the most commonly applied constructive solution), a solution subject to specific restrictions determined by the dimensions and a large size of semi-finished products. The replacement of the metal disk with the metal band, as a tool—TO—to carrying out the cutting operations, fundamentally changes the constructive solutions applied so far, determining completely new construction forms for the new installations. This replacement influences the range of phenomenological, constructive, and technological constraints, which limit the using possibilities of the metal disc, as a tool (determined by the range of the semi-finished products) in the specific processes of the metallurgical industry in particular. The multiple research options applied, led to the identification of optimized values of the construction elements of the metal band (length, width, thickness), as well as for the ends of the connection, to form a closed contour, strong and durable. At the same time, the correlation of the diameter of the flywheels was considered, with the distance between their axes and related to the elements in the structure of the experimental installation/experimental stand. The variables measurement of the working parameters used and presented in the paper is measured with standardized measurement and control gages used on a large scale. For example *Ra*—which is the arithmetical mean deviation of the assessed profile—was determined with the Surtronic S-116 Series Surface Roughness Tester (Taylor-Hobson, Warrenville, IL, USA). The measurement capability of this gauge has the range of 200 [μm] and the resolution of 100 [nm]. The stylus tip radius is 5 μm. It is able to calibrate to ISO 4287 roughness standard. The filter cut-off is 0.25 mm and the filter type is 2CR/Gaussian. The evaluation length is suitable for 0.25–17.5 mm, with a measuring speed of 1 mm/sec. During the experimental research, the cutting time was measured with the chronometer. The cutting was measured with the feller gauge and the thermal influenced zone was analyzed with Micro-Vu VERTEX 261 (Micro-Vu, Conde Lane Winsdor, CA, USA).

3. The Experimental Research Development

For the experimental research development under optimal conditions, the following aspects have been considered:

(a) ensuring the material conditions regarding the establishment of the processing objects of high steel alloys category, under the form of rolled steel:

- 34 MoCrNi15–ϕ 40 mm;
- RUL–1–ϕ 26 mm.

(b) the work parameters observed during the cutting process:

- mechanical;
- the metal band thickness (s) [mm];
- the relative speed (v_r) [m/s];
- the feed rate (v_a) [m/min].

(c) electrical:

- the working current (I), [A];
- discharge voltage (U), [V];
- polarity (+,−).

(d) technological:

- the cutting time (t_t), [s];
- the cutting width (l_t), [mm];
- the arithmetical mean deviation of the assessed profile (R_a), [μm];
- the thermal influenced zone (TIZ) [mm].

(e) validation conditions.

The conditions that need to be fulfilled by the whole parameters considered as state variables have been complied with the following:

- the direct measuring possibility;
- the lowest dispersion of the values during the experiments.

The analysis of the main categories of parameters (mechanical, electrical, technological) was performed using a mathematical modeling program within the research program [34–36].

3.1. Main Parameters Considered as State Variables

The research done on the experimental installation specifically aimed to some technological aspects related directly to the cutting by electrical discharge machining with contact breaking with electrode-tool—the metal band.

The article presents some of the obtained results within the "Research Program" after having analyzed the main parameters of the categories considered stating the following variables:

- Cutting time (t_t), [s];
- Cutting width (l_t), [mm];
- Arithmetical mean deviation of the assessed profile (R_a), [µm];
- Thermal influenced zone (TIZ), [mm].

As independent functions, considered as input measures, the following were taken into consideration:

- The working current (I), [A];
- The feed rate (v_a) of the Processing Object (PO), [m/min];
- The relative speed of the electrode-tool (v_r) of the TO, [m/s];
- The metal band thickness (s), [mm].

The influence of each work parameter (or factor) grouped was observed, while the others were maintained constant.

3.2. Drawing Dependency Diagrams for State Variables

Performing the experimental research in the processing/cutting by electrical discharge machining with contact breaking (EDMCB) with TO—metal band—according to the Research Program in Figure 1, ensured the succession of the experimental determination of the concrete values of some parameter categories (electrical, mechanical, and technological).

The dependency diagrams of the state variables are presented below, considering the values mentioned in Tables 1 and 2, for the two categories of steel (34 MoCrNi15-φ 40 mm and RUL-1-φ 26 mm).

Table 1. Material 34 MoCrNi15-φ 40 mm.

No.	Sample Mark	Input Parameters				Average Values			
		I [A]	v_r [m/s]	v_a [m/min]	s [mm]	t_t [s]	l_t [mm]	R_a [µm]	TIZ [mm]
0	1	2	3	4	5	6	7	8	9
1	1	800	20	0.021	0.5	70	2.5	7.7	5.0
2	2′	700	25	0.042	0.5	55	2.2	7.3	5.3
3	3′	950	30	0.063	0.5	44	2.0	8.9	5.5
4	4′	1050	35	0.084	0.5	36	2.0	11.8	5.5
5	5″	700	41	0.105	0.5	30	2.0	19.1	5.5
6	*	1000	20	0.021	1.0	54	2.5	13.1	5.0
7	**	1000	20	0.042	1.0	40	2.5	11.0	5.5
8	***	1050	30	0.063	1.0	36	2.5	9.2	6.0
9	2*	1050	35	0.084	1.0	33	2.0	17.5	6.5
10	3*	1050	41	0.105	1.0	29	2.5	18.2	6.5

Table 2. Material RUL-1-Φ 26 MM.

No.	Sample Mark	Input Parameters				Average Values			
		I [A]	v_r [m/s]	v_a [m/min]	s [mm]	t_t [s]	l_t [mm]	R_a [μm]	TIZ [mm]
0	1	2	3	4	5	6	7	8	9
1	1″	320	41	0.021	0.5	44	1.5	15.9	5.0
2	2″	320	41	0.042	0.5	34	1.8	15.1	5.0
3	3″	330	41	0.063	0.5	24	2.1	12.3	4.0
4	3‴	360	41	0.084	0.5	20	2.0	12.7	4.0
5	4″	380	41	0.105	0.5	16	2.0	8.9	3.6
6	1	600	20	0.021	1.0	40	2.3	7.3	5.0
7	5	610	41	0.042	1.0	33	2.5	5.9	5.6
8	2	650	20	0.063	1.0	26	2.3	8.4	6.2
9	3	670	20	0.084	1.0	26	2.5	7.0	7.0
10	7	650	20	0.105	1.0	23	2.5	5.6	7.5

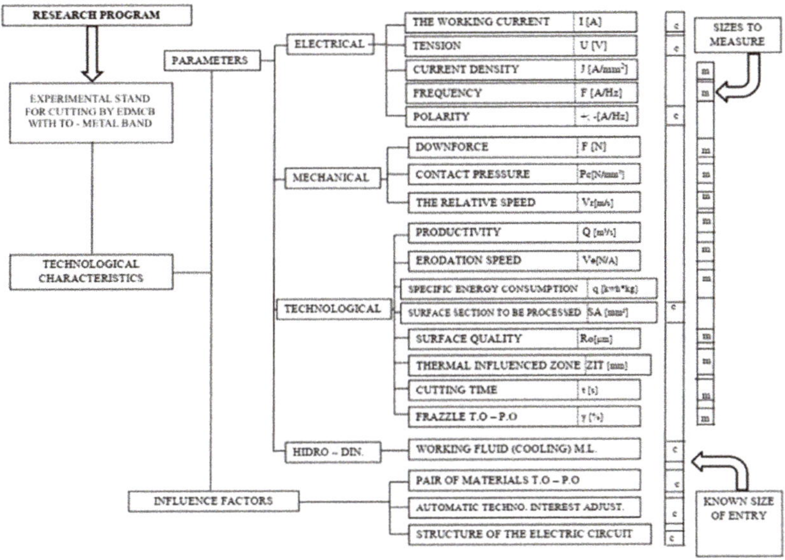

Figure 1. The experimental research programming through processing by electrical discharge machining with contact breaking (EDMCB) with Transfer Object (TO)—metal band.

3.2.1. For 34 MoCrNi15-φ 40 mm

(a) The dependence of the cutting time (t_t), the feed rate (v_a) of the PO, and the thickness of the metal band (s), is shown in Figure 2, based on the data presented in Table 3.

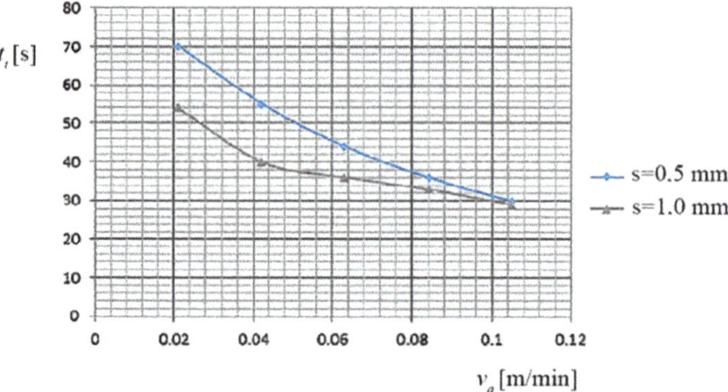

Figure 2. Dependence of the t_t, on v_a of the Processing Object (PO) and on the thickness (s) of the metal band.

Table 3. The dependence of the cutting time, the feed rate of the Processing Object (PO), and the thickness of the metal band.

No.	s [mm]	v_a [m/min]	t_t [s]
0	1	2	3
1	0.5	0.021	70
2	0.5	0.042	55
3	0.5	0.063	44
4	0.5	0.084	36
5	0.5	0.105	30
6	1.0	0.021	54
7	1.0	0.042	40
8	1.0	0.063	36
9	1.0	0.084	33
10	1.0	0.105	29

(b) The dependence of the cutting width (l_t), on the relative speed (v_r) of the TO and on the thickness of the metal band (s), is shown in Figure 3, based on the data in Table 4.

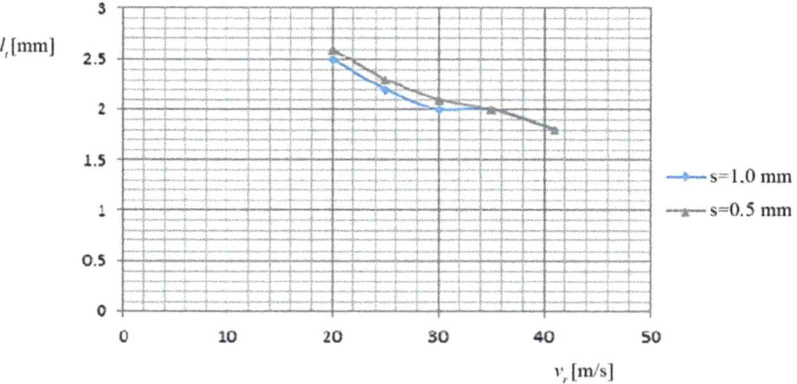

Figure 3. Dependence l_t, on the v_r of TO and on the thickness (s) of the metal band.

Table 4. The dependence of the width of the cut, on the relative speed of the Transfer Object (TO), and on the thickness of the metal band.

No.	s [mm]	v_r [m/s]	l_t [mm]
0	1	2	3
1	0.5	20	2.5
2	0.5	25	2.2
3	0.5	30	2.0
4	0.5	35	2.0
5	0.5	41	1.8
6	1.0	20	2.6
7	1.0	20	2.3
8	1.0	30	2.1
9	1.0	35	2.0
10	1.0	41	1.8

(c) The dependence of the R_a, on the working current (I), on the thickness of the metal band (s), and on the v_a of PO, is shown in Figure 4, based on the data presented in Table 5.

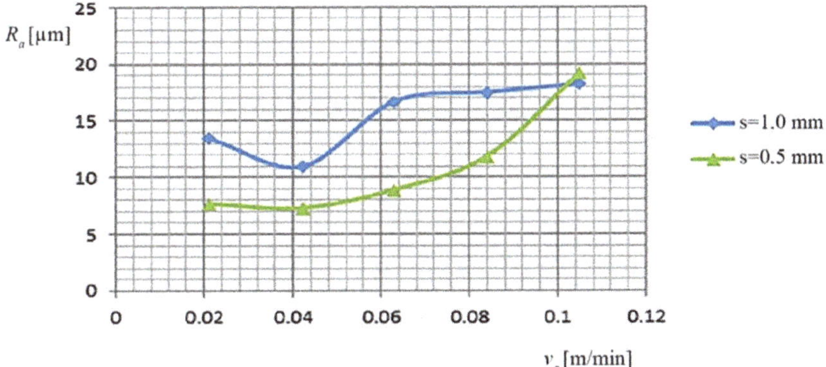

Figure 4. Dependence on the R_a by v_a of the PO and the thickness of the metal band (s).

Table 5. The dependence of the R_a, on the working current, on the thickness of the metal band, and on the v_a of PO.

No.	I [A]	R_a [μm]	v_a [m/min]	s [mm]
0	1	2	3	4
1	1000	13.5	0.021	1.0
2	950	11.0	0.042	1.0
3	1050	16.7	0.063	1.0
4	1050	17.5	0.084	1.0
5	1050	18.2	0.105	1.0
6	800	7.7	0.021	0.5
7	700	7.3	0.042	0.5
8	950	8.9	0.063	0.5
9	1050	11.8	0.084	0.5
10	700	19.1	0.105	0.5

(d) Dependence of the thermal influenced zone (TIZ), on the working current (I), on the feed rate v_a of PO, and by the thickness of the metal band (s), is shown in Figure 5, based on the data presented in Table 6.

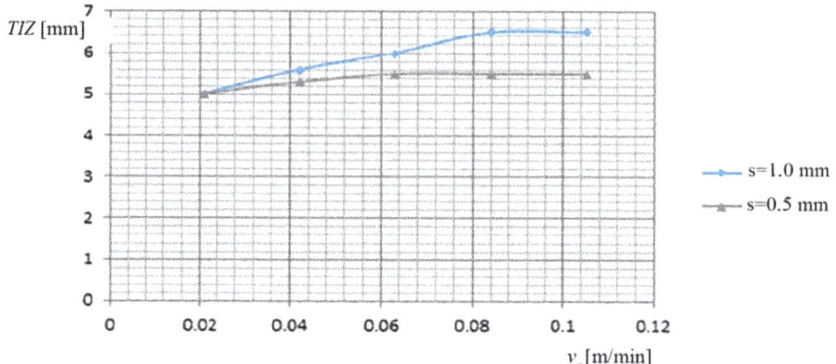

Figure 5. Dependence Thermal Influenced Zone (*TIZ*) on PO and thickness (*s*).

Table 6. Dependence of the thermal influenced area on the working current, on the feed rate of PO, and by the thickness of the metal band.

No.	v_a [m/min]	*I* [A]	*TIZ* [mm]	*s* [mm]
0	1	2	3	4
1	0.021	1000	5.0	1.0
2	0.042	1020	5.6	1.0
3	0.063	1050	6.0	1.0
4	0.084	1050	6.5	1.0
5	0.105	1070	6.5	1.0
6	0.021	500	5.0	0.5
7	0.042	550	5.3	0.5
8	0.063	550	5.5	0.5
9	0.084	650	5.5	0.5
10	0.105	700	5.5	0.5

3.2.2. For RUL-1-φ 26 mm

(a) The dependence of the cutting time (t_t), the feed rate (v_a) of the PO, and the thickness of the metal band (*s*), is shown in Figure 6, based on the data presented in Table 7.

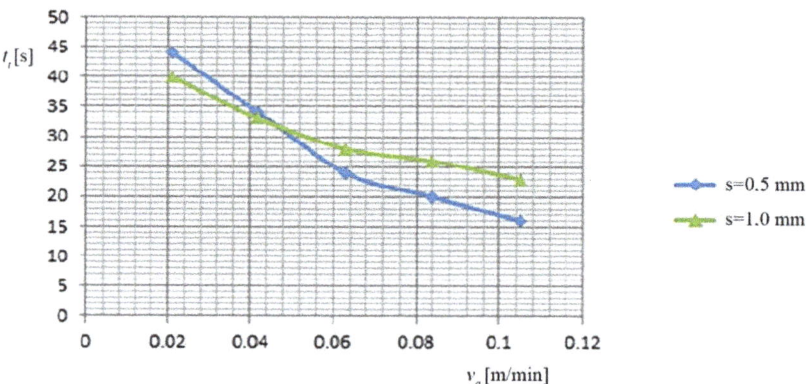

Figure 6. Dependence on the cutting time (t_t), the feed rate (v_a) of the PO, and the thickness of the metal band (*s*).

Table 7. The dependence of the cutting time, the feed rate of the PO, and the thickness of the metal band.

No.	s [mm]	v_a [m/min]	t_t [s]
0	1	2	3
1	0.5	0.021	44
2	0.5	0.042	34
3	0.5	0.063	24
4	0.5	0.084	20
5	0.5	0.105	16
6	1.0	0.021	40
7	1.0	0.042	33
8	1.0	0.063	28
9	1.0	0.084	26
10	1.0	0.105	23

(b) The dependence of the cutting width (l_t) on the relative speed (v_r) of the TO and on the thickness of the metal band (s), is shown in Figure 7, based on the data of Table 8.

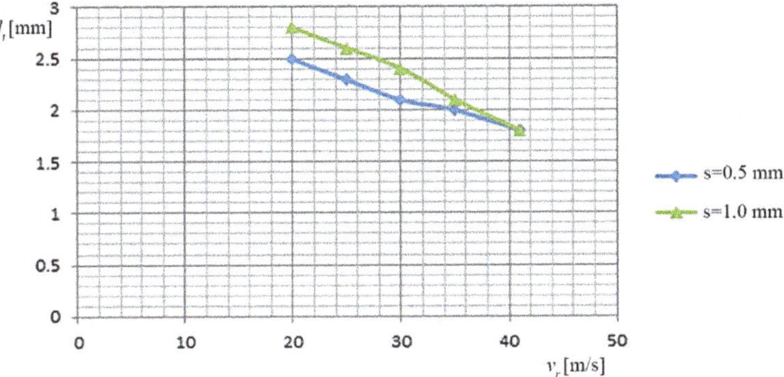

Figure 7. Dependence of the cut width (l_t) of the relative speed (v_r) and the thickness of the metal band TO (s).

Table 8. The dependence of the width of the cut, on the relative speed of the TO and on the thickness of the metal band.

No.	s [m]	v_r [m/s]	l_t [mm]
0	1	2	3
1	0.5	20	2.5
2	0.5	25	2.3
3	0.5	30	2.1
4	0.5	35	2.0
5	0.5	41	1.8
6	1.0	20	2.8
7	1.0	25	2.6
8	1.0	30	2.4
9	1.0	35	2.1
10	1.0	41	1.8

(c) The dependence of the R_a, on the feed rate v_a of PO and on the thickness of the metal band (s), is shown in Figure 8, based on the data of Table 9.

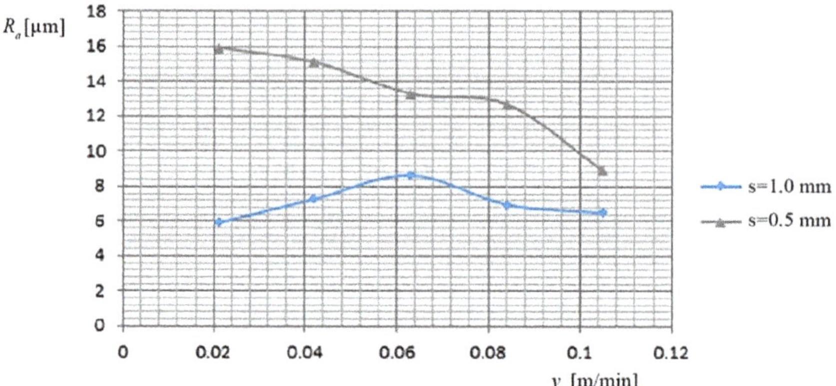

Figure 8. Dependence on the R_a, the feed rate of the PO and the thickness of the metal band.

Table 9. The dependence of the R_a, on the feed rate of *PO* and on the thickness of the metal band.

No.	v_a [m/min]	*I* [A]	R_a [µm]	s [mm]
0	1	2	3	4
1	0.021	600	5.9	1.0
2	0.042	600	7.3	1.0
3	0.063	600	8.7	1.0
4	0.084	610	7.0	1.0
5	0.105	475	6.5	1.0
6	0.021	320	15.9	0.5
7	0.042	320	15.1	0.5
8	0.063	330	13.3	0.5
9	0.084	360	12.7	0.5
10	0.105	380	8.9	0.5

(d) The dependence of the thermal influenced zone (*TIZ*), on the feed rate v_a of PO and on the thickness of the metal band (*s*), is shown in Figure 9, based on the data presented in Table 10.

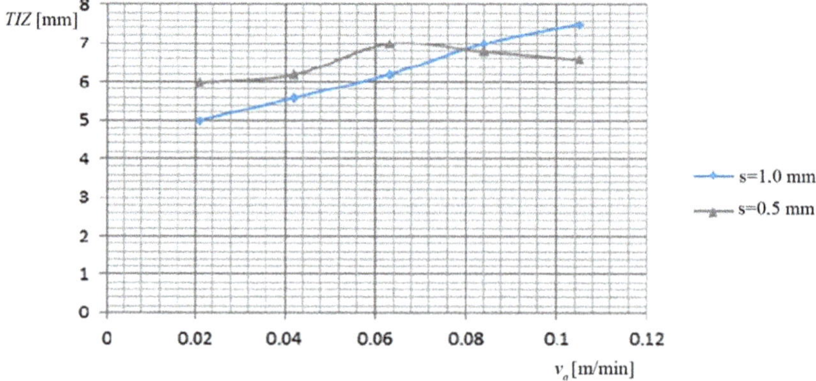

Figure 9. Dependence *TIZ* on the feed rate v_a of PO and on the thickness of the metal band (*s*).

Table 10. The dependence of the thermal influenced zone, on the feed rate of PO and on the thickness of the metal band.

No.	v_a [m/min]	*I* [A]	*TIZ* [mm]	*s* [mm]
0	1	2	3	4
1	0.021	600	5.0	1.0
2	0.042	610	5.6	1.0
3	0.063	650	6.2	1.0
4	0.084	670	7.0	1.0
5	0.105	650	7.5	1.0
6	0.021	450	6.0	0.5
7	0.042	500	6.2	0.5
8	0.063	700	7.0	0.5
9	0.084	650	6.8	0.5
10	0.105	475	6.6	0.5

3.3. Interpretation of the Dependency Graphs Charts

(a) The research highlights the following aspects:

- with the increasing speed of the PO, the cutting time decreases;
- when increasing the feed rate of PO by 0.021 [m/min] to 0.084 [m/min], the cut-off time is reduced to 55%, reflecting a significant processing productivity increase (as a result of the increase v_a of PO);
- for the metal band thicknesses (s = 1.0 mm), the range of the PO feed rate was limited to 5 levels (avoiding the breaking of the metal band at higher speeds);
- for increasing the cutting productivity, it is recommended to use a cutting time (t_t) correlated to the pressure between TO and PO, metal band thicknesses of s = 0.5 mm, and values of (v_a) of PO greater than 0.063 m/min.

(b) In this case, minimum deviations were observed for the 2 categories of materials cut:

- at the relative speeds of 20–25 [m/s] of the TO, due to the "extension" (during operation) of the metal band, its "sweeps and motions" appear, which cause an intensification of the "wrong springs" (lateral), between the TO and the PO, which causes an increase in the cutting width;
- the cutting width (l_t) has the optimum values (≈2 mm) at high relative speeds (41 m/s) and average values of the working current (7500 A), when the metal band greatly reduces its lateral play (sweeping tendency).

(c) The variation mode of the R_a according to the working regime (I, v_a, v_r, s) was followed.

- low values of R_a were observed at the v_a of PO, in the range of 0.042–0.084 m/min, and to the working current values between 600–1000 A;
- the thickness of the metal band (s) causes high, different roughness for small values of the working current, between 300 and 600A.

(d) The cutting of the two categories of materials (34 MoCrNi15-φ 40 mm and RUL-1-φ 26 mm), under different regimes and with the variation of the cooling agent, showed different *TIZ*, these being influenced by the other parameters (the working current, the feed rate of the PO, and the thickness (s) of the metal band).

- the cooling agent (technological water) was "ordered" to ensure the cooling of the work area;
- *TIZ* has maximum values, in the exit zone of the metal band from the material, due to the eroded metal, reached the melting state and due to the insufficient cooling, the main cause being both the working current and the advance speed of the PO. These frequently encountered situations were observed primarily visually, and then following the repeated experiments,

> it can be concluded the important main cause which "is due to the working current speed but also the feed rate of the object to be processed".

The thermally influenced zone has the maximum value in the output area of the metal band from the material, due to eroded metal so that it reaches the molten state and then cooled but insufficiently. Based on the authors experience in this field, it can be considered that the main cause is due to the working current but also to the feed rate of the processing object.

The different cutting regimes for the processed samples determined different influences on the main parameters pursued in the experimental research (t_t, l_t, R_a, and *TIZ*). In Figure 10, it is presented the macrostructural analysis of 4 samples which were cut with metal band by EDMCB with different regimes.

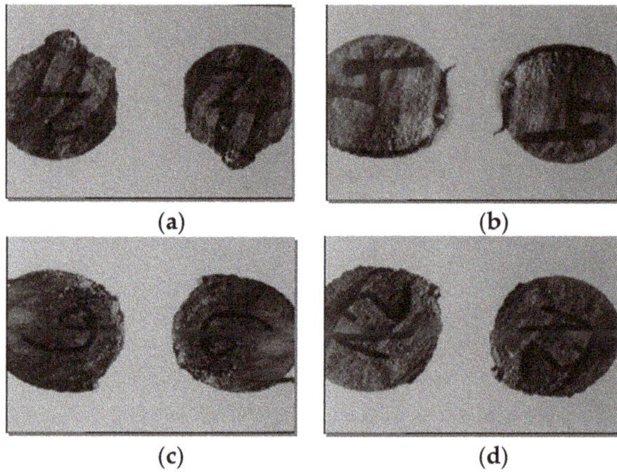

Figure 10. Dependence *TIZ* on the feed rate v_a of PO and on the thickness of the metal band (s): (**a**) Sample 1, (**b**) Sample 2, (**c**) Sample 3, (**d**) Sample 4.

The cut surfaces show specific cutting traces of EDMCB:

- different structures and sizes of *TIZ*, due to uneven cooling with coolant during the surface cutting process;
- areas of molten and re-solidified material (characteristic of the Joule–Lenz effect), especially in the exit part of the metal band from the PO (Figure 10a,c);
- the quality of the cut is not the same on the whole cut area;
- to the pressure forces higher than 400 N, the number of contacts between TO-PO increases, the metal band tends to "sweep" changing the perpendicularity and causing an accentuation of "erroneous" arc, between it and PO and a decrease of the quality of the cut surface.
- the extension of the metal band determines the amplification of the discharges in non-stationary arc, between it and the PO having as an effect the deterioration of the respective surfaces (Figure 10b,d).

4. Experimental Installation/Stand, for the Study of the Processing by EDMCB, With TO—Metallic Band, Of the High Steels Alloy (Hard and Extruded)

In order to establish the technological characteristics, the correlation between the factors and the parameters, an experimental installation/stand was designed and executed for the cutting of high steels alloy, considering the functions and the structure resulting by the dimensional analysis and the research results presented in the specific literature.

The experimental installation/stand allowed a wide limit of variation of some technological parameters and of their measurement, (this being designed and structured based on the idea diagrams and the morphological matrices).

NOTE: The analysis and realization of the structure of the experimental installation took into consideration the influence of some parameters and factors specific to the processing by EDMCB (some of them are also found in the machines that use as TO—the metal disk).

4.1. The Main Components of the Experimental Installation's Structure

The experimental installation is made out of two main subassemblies:

- The electricity supply source;
- The experimental cutting stand.

(a) The electricity supply source:

- Electric supply source 1000A, [A].

It has the role of generating the erosive agent, bearing the technical characteristics, mentioned in the "Power source book", being connected to a 380 V industrial network, complying with the norms of protection against electrocution.

- Electric command board;
- The control of the feeding systems (of the experimental stand), made with the help of a thyristor bridge that includes:
    - ○ electric motor;
    - ○ reducer;
    - ○ screw.

(b) The experimental cutting stand:

It is made out of a metal frame (frame—welded construction) which gives it an optimal behavior with dynamic shocks and vibrations.

The experimental stand scheme is presented in Figure 11.

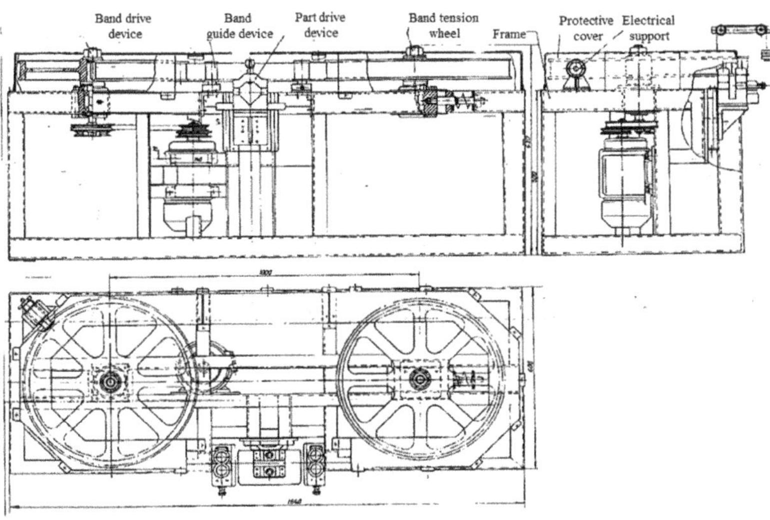

Figure 11. The experimental stand assembly.

4.2. The Main Elements Ensuring the Main Functions

The systems which ensure the realization of the main functions (engaging, lengthening of the metal band, guiding, etc.) are mounted on the frame. Additionally, mounted are:

- The mechanism of the processing objects;
- The orientation guides of the electrode-tool;
- The positioning and advance subassembly of the processing objects;
- The receiver subassembly of the working current;
- The cooling and evacuation subassembly of the eroded material and many others.

The parts of the experimental installation in the working area are isolated and covered by protection elements for the benefit of the operating workers which observe it.

The present installation is a part of the field of non-conventional processing and has as an object a cutting installation by electrical discharge machining with contact breaking, with metal band transfer object. The installation can be applied in the machine building industry, in the metallurgical industry, in foundries, etc., being indicated, in particular, for cutting parts of materials difficult to process by cutting, which have a high and very high hardness, or which are resistant to corrosion, and cavitation, refractory materials, brittle materials, etc. The installation can also be used to cut the supply masses and casting networks of some castings. This plant, although clearly superior to conventional cutting machines, still has the main drawback of the high cost of processing due to high electricity consumption. This high consumption of electricity is mainly generated by energy losses due to the lateral electrical discharges, produced between the front faces of the electrode disc used as an object of electricity transfer and the cutting surfaces already made in the processed object, discharges which no longer contribute to the actual cutting process. As the cutting progresses, the areas of the contact surfaces between which these parasitic side discharges occur also increase, having as a consequence a certain instability of the electrical discharge machining process as well as a relatively low productivity.

The technical problem solved by the installation consists of the reducing of the contact surfaces areas between the electricity transfer object and the processed object and thus the accentuated decrease of the energy losses by the lateral electric discharges produced between the two objects. The contact erosion cutting plant with contact breaking, with metal band transfer object, solves the mentioned technical problem and eliminates the above disadvantages. That consists of an endless metal band with the role of electricity transfer object, positioned vertically and performing a rectilinear, forward, and interacting motion, with a processing object arranged tangentially under the metal band. This is wound on two flywheels with vertical axes, of which a drive flywheel set in rotation by an electric motor and a belt drive. The second flywheel is a driven flywheel, supported by a movable bearing coupled to a metal belt tensioner which is also held firmly in a vertical plane, by means of at least two identical guiding devices, placed on either side of the workpiece caught in a fastening and advancing device which, moving on a guide plate, under the action of suspended counterweights at the end of a steel cable, performs the advance movement of the object to be processed, perpendicular to the trajectory of the metal band and the contact pressure necessary to carry out the process of electrical discharge machining, with contact breaking, supply electrical energy of the processing process being made by direct connection of the object to be processed to one pole of the power supply, the other pole being connected to the metal band, by means of several current receiving devices, arranged on both sides of the metal band, both on one side and on the other side of the object to be processed.

The endless metal band, with the role of transfer object, is made of narrow carbon steel band, cold rolled, having 30–50 mm width and 0.3–1.00 mm thickness. The drive flywheel and the driven flywheel are provided at their periphery with a groove for laying and guiding the metal band.

The tensioning device of the metal band consists of a stretching axis supported and guided axially by a plate-bearing integral with the frame and provided with a threaded area screwed into a fixed plate, stiffened with the bearing plate by some connecting columns, the tensioning shaft being coupled by an elastic ring to a bearing plate which supports the movable bearing of the driven flywheel. A metal band guide device consists of two guide rollers, with parallel axes, through which passes the metal band, the guide rollers rotating freely on a fixed axis and, respectively, on a movable axis, the fixed axis being mounted on a guide body in which a sledge can be supported which supports the movable shaft of the guide roller provided at the ends with sills and which, together with the sledge, can approach or move away from the guide roller with fixed shaft by manually driving. The device for fixing and advancing the object to be machined consists of an upper body, fixed to the end of the steel cable with counterweights and provided with screws for fixing the object to be machined on a prism fixed on a bracket that can move on a plate with guides, fixed on the frame, under the action of counterweights suspended on the steel cable. A current receiving device consists of several collecting brushes pressed on the side of the metal band by the force of compression springs, mounted together with the collecting brushes in a brush holder box made of insulating material, fixed on a support provided with some oval holes through which some fixing screws pass on an element of the frame.

5. Using the Central Composite Factorial Experiment When Processing the Results

The experimental results obtained have been processed with the help of a Program Package which made it possible to analyze, model, optimize, and statistically interpret the data. Thus, the data obtained (namely the equations which express the mathematical form of the objective functions, as well as the values of the regression coefficients) have been statistically modeled, thus making the modeling and optimization of the processing through electrical discharge machining with contact breaking with electrode-tool—metal band.

5.1. Interpreting the Graphs of the State Variables According to the Estimated Values and the Measured Ones

For the four state variables (R_a, *TIZ*), the graphs of the estimated values and the values measured during the experimental research are presented in Figures 12 and 13.

The R_a distribution of the estimated and measured values is shown in Figure 12.

The *TIZ* distribution of the estimated and measured values is shown in Figure 13.

The plan representation of the results interpretation was done using the correlation diagrams. It was seen that the ratio between the values measured and the estimated ones within the conducted research, is found within the K E [0.85, 1] limits; which lead to the conclusion that the values obtained experimentally are viable and reasonable. This aspect was also highlighted by the spreading manner of the experimental results obtained.

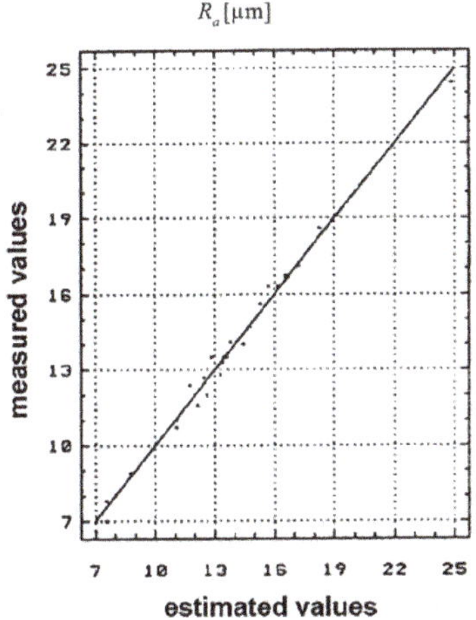

Figure 12. R_a measured values [μm].

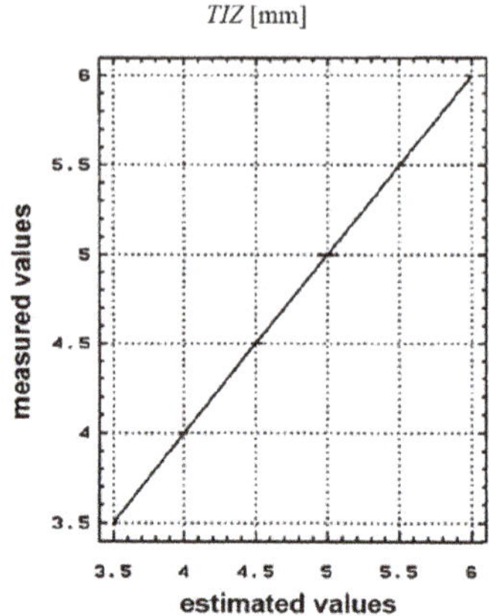

Figure 13. Thermal influenced zone values measured [mm].

5.2. *Interpreting the PARETO Diagrams*

For the state variables (t_t, R_a), in Figures 14 and 15, the standard effects are presented according to the influence of the input factors (v_a, v_r, s and I) and the chosen performance criteria.

It is found that the influence of the independent factors in the process depends on the performance criterion chosen.

In the influence values for the objective functions case–will be highlighted which of the input parameters (I, s, v_r and v_a) have the biggest influence on these functions.

(a) For the cutting time (t_t) the biggest influence is the working current (I), followed by the thickness of the metal band (s) and the relative speed (v_r) of the electrode-tool;

(b) For the cutting width (l_t) the biggest influence is the feed rate (v_a) of the processing objects followed by the working current (I) and thickness of the metal band (s);

(c) For the R_a the most important parameter is the metal band thickness (s), followed by the feed rate (v_a) of the processing objects and the relative speed (v_r) of the working current (I);

(d) For the thermal influenced zone (TIZ) the biggest influence is the processing objects feed rate (v_a), followed by the working current (I) and the TO relative speed (to a lower degree).

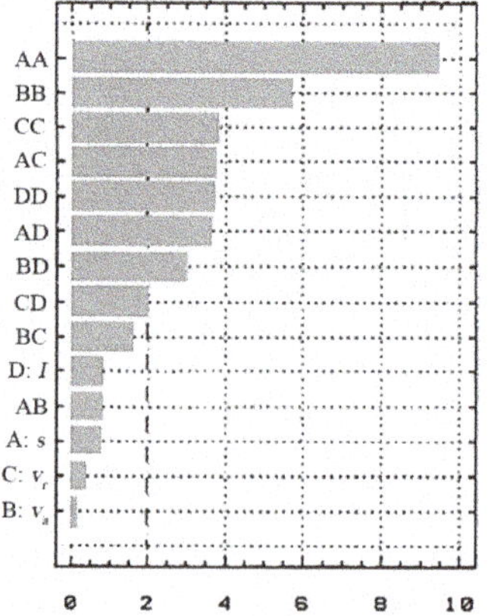

Figure 14. The input factors influence on t_t [s].

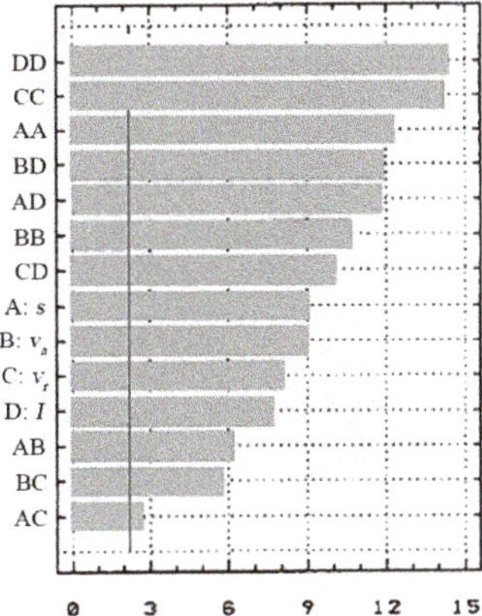

Figure 15. The input factors influence on R_a [μm].

The standard effects (the PARETO diagrams) give an overall view of the evolution and variation of the process's independent variable influences, according to the chosen and analyzed criteria.

In these PARETO diagrams, the terms coefficients which constitute the objective functions (t_t-A, l_t-B, R_a-C, *TIZ*-D) are represented by the markings: AA, BB, CC, DD, AB, AC, AD, BD, BC and CD.

5.3. Interpreting the Sections Through the Variation Graphs of the Objective Functions for Each State Variable According to the Determined Input Parameters

The analysis of the diagrams and sections of the state variables related to the dependent ones (considered as input measures) will be presented next for each of them.

5.3.1. For the Cutting Time (t_t)

- The dependency of the cutting time (t_t) to the feeding rate (v_a) of the processing object and the working current (I) is presented in Figure 16.
- Thus, the cutting time (t_t) reaches the optimal values (30–32) [s] when the input parameters within the working cycle have been set for the following values:

  v_a—between 0.042–0.084 [m/min];
  I—between 450–550 [A];
  s—between 0.5–0.8 [mm].

- The dependency of the cutting time (t_t) to the relative speed (v_r) of the electrode-tool and the working current (I).

The cutting time (t_t) reaches the optimal values (30–32) [s] when the input parameters are set for the following values:

v_r—between 30–35 [m/s];
I—between 450–550 [A];
s—between 0.5–0.8 [mm].

Each of the input parameters determines a certain influence on the cutting time (t_t).

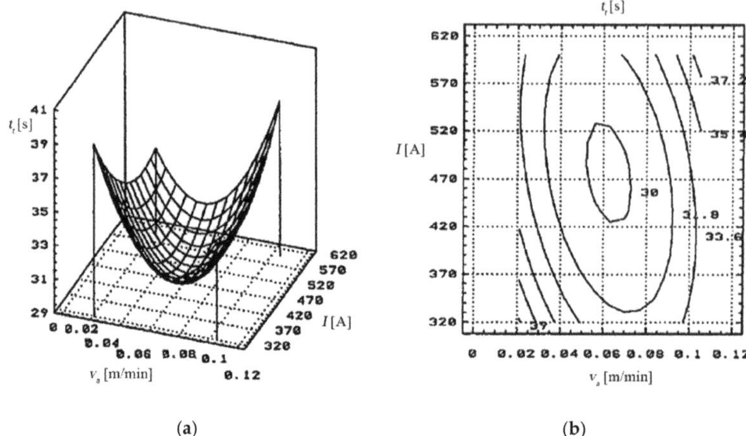

(a) (b)

Figure 16. The dependency of the cutting time (t_t) to the feeding rate (v_a) and the working current: (**a**) 3D graph, (**b**) 2D graph.

5.3.2. For the cutting width (l_t)

The dependency of the cutting width (l_t) to the relative speed of the electrode-tool and the working current (I).

The length of the cutting found at values between 1.8–2.2 mm was obtained with the following values of the input measures:

- the working current: I between 550–600 [A];
- the relative speed of the electrode-tool $v_r > 40$ [m/s];
- the metal band thickness: s between 0.6–0.8 [mm].

During the experiments the cutting's width variation (l_t) according to the work parameters and the work cycle but not necessarily in a pre-established dependency, was noted.

The dependency of the cutting's width (l_t) to the metal band thickness (s) and the feeding rate (v_a) of the processing object is presented in Figure 17.

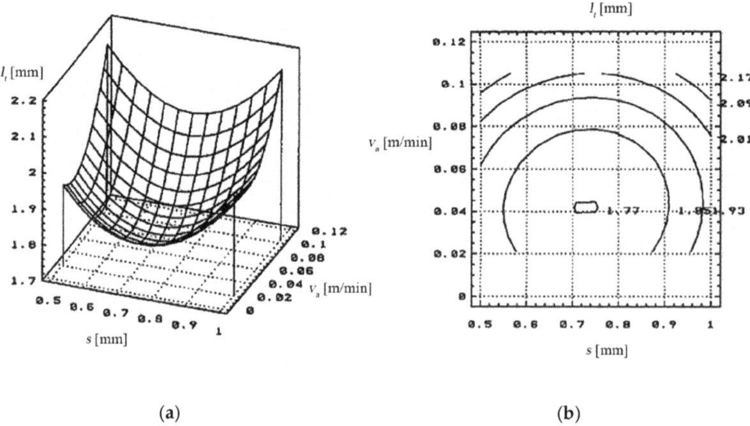

(a) (b)

Figure 17. The dependency of the cutting width (l_t) to the thickness (s) of the metal band and the feeding rate (v_a) of the processing object: (**a**) 3D graph; (**b**) 2D graph.

Variations of the cutting width between 1.8–2.2 mm were obtained with the following values of the input variables:

- the working current: I between 400–500 [A];
- the feed rate of the processing objects: v_a between 0.021–0.042 [m/min];
- the metal band thickness: s between 0.5–0.7 [mm].

5.3.3. For the Arithmetical Mean Deviation of the Assessed Profile (R_a)

The surface quality according to the feed rate (v_a) of the processing object and the working current (I) is presented in Figure 18.

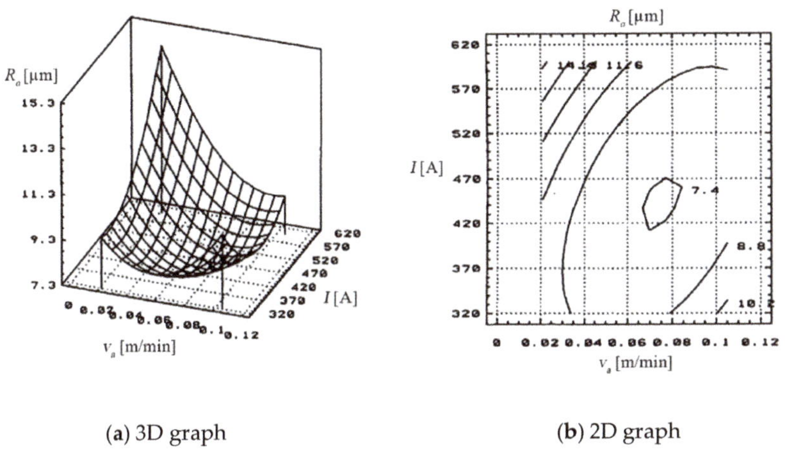

(**a**) 3D graph (**b**) 2D graph

Figure 18. The dependency of the arithmetical mean deviation of the assessed profile (R_a) to the feeding rate (v_a) of the processing object and the working current (I): (**a**) 3D graph, (**b**) 2D graph.

The optimal values of the rigidness $R_a = 9.5$ [µm] were obtained the following values of the input measures:

- the feed rate: v_a between 0.042–0.063 [m/min];
- the working current: I between 400–500 [A].

The quality of the cut surface was influenced by the working cycle and the agent for evacuating eroded products.

The dependency of the arithmetical mean deviation of the assessed profile (R_a) to the metal band thickness (s) and the relative speed (v_r) of the electrode-tool is presented in Figure 19.

The R_a obtained after cutting, comprises a wide range of values (5–20 [µm] for R_a, determined by the working regime (hard/soft) and by the cooling and evacuation agent of erosive products. The cooling agent of erosive products (technological water), is essential to obtain a quality of the surfaces cut as best as possible.

Overheating and slow evacuation of erosion products, caused by insufficient cooling agent and improper cooling, cause deterioration of the cut surface.

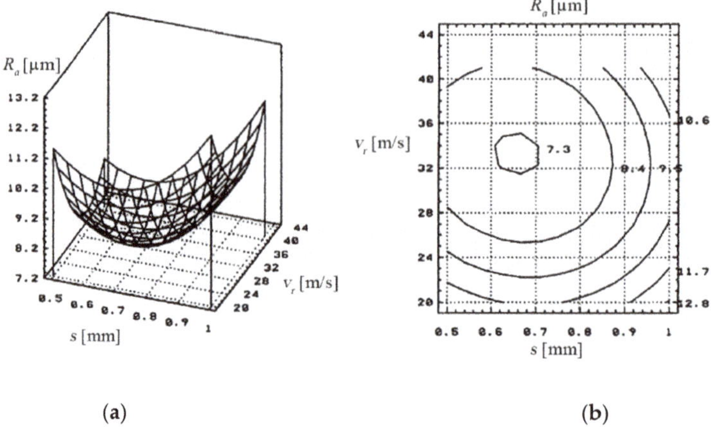

Figure 19. The dependency of the arithmetical mean deviation of the assessed profile (R_a) to the thickness (s) of the metal band and the relative speed (v_r) of the electrode-tool; (**a**) 3D graph, (**b**) 2D graph.

5.3.4. For the Thermal Influenced Zone

The dependency of the thermal influenced zone (*TIZ*) to the relative speed (v_r) of the electrode-tool and the working current (*I*) is presented in Figure 20.

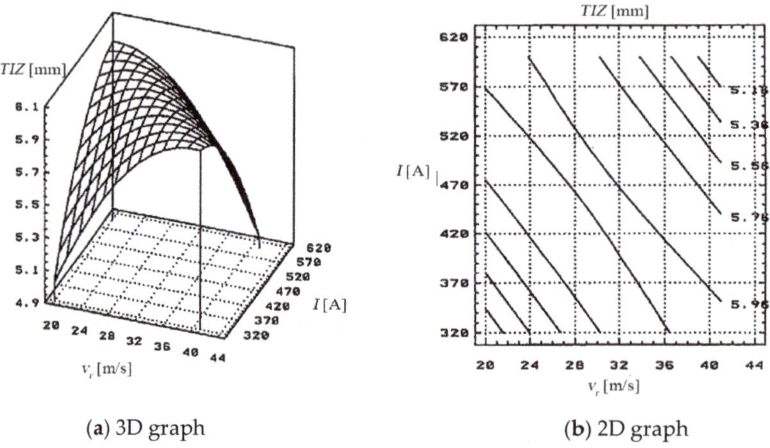

Figure 20. The dependency of the thermal influenced zone (*TIZ*) to the relative speed (v_r) of the transfer object and the working current (*I*): (**a**) 3D graph, (**b**) 2D graph.

The thermal influenced zone had a different manifestation after the experimental tests were done, due to the impact of the working cycle (hard/soft) and the cooling agent applied.

Figure 20 highlights the dependency of the thermal influenced zone, whose values surpass 6 [cm] at a "hard" working cycle, for values of the working current between 500–620 [A] and relative speeds of the electrode-tool between 24–36 [m/s].

From the thermal influenced zone, variation graphs analysis, and of the sections using graphs, it is shown that the working parameters have been set at these values:

- the feed rate of the processing objects: v_a between 0.020–0.042 [m/min];
- the relative speed: v_r between 20–30 [m/s];

- the working current: I between 400–500 [A];
- the metal band thickness: s between 0.5–1 [mm].

After using these working parameters, thermal influenced zone with values smaller than 5 [cm] were obtained.

6. Results and Discussions

The scientific research has a wide applicability in the industry. The used stand was patented and subsequently optimized structurally and functionally, which led to its manufacture in small series to be used in the industry. The use of this equipment can lead to the assurance of optimal process parameters when cutting through electrical discharge machining with contact breaking with metal band as the transfer object.

The research conducted on the obtained results with the Program Package (Factorial Programs help) has allowed emphasis to be placed on state variables (t_t, l_t, R_a and TIZ) of some of the values which, statistically interpreted, lead to substantial conclusions and motivations.

The experimental installation which considered the structure and the functions resulted from the dimensional analysis allowed the realization of the work process, ensuring multiple possibilities of technological research.

The experimental results obtained have been processed using a Statistic Program ("Statistic Data System 2000" Program Package) which made it possible to analyze, model, optimize and statistically interpret this data. Thus, the obtained data (namely the equations which express the mathematical form of the objective functions, as well as the values of the regression coefficients) have been statistically modeled, thus showing the modelling and optimization of the processing through electrical discharge machining with contact breaking with electrode-tool—metal band.

By interpreting the graphs of the state variables according to both the estimated values and the measured ones, it was seen that the ratio between the values measured and the ones estimated by the conducted research is found within the K E [0.85, 1] limits. This leads to the conclusion that the values obtained experimentally are viable and reasonable; this aspect also being highlighted by the spreading manner of the experimental results obtained.

By interpreting the Pareto diagram, it shows which of the input parameters (I, s, v_r and v_a) have the biggest influence on the objective functions:

- For the cutting time (t_t)—the biggest influence is the working current (I);
- For the cutting width (l_t)—the feed rate (v_a) of the processing objects;
- For the arithmetical mean deviation of the assessed profile (R_a)—the most important parameter is the metal band thickness (s);
- For the thermal influenced zone TIZ—the biggest influence is the feed rate (v_a) of the processing objects;

By interpreting the sections through the variation graphs of the objective functions for each state variable according to the determined input parameters, it is seen that:

- For the cutting time (t_t)—each of the input parameters determines a certain influence on the cutting time (t_t).
- During the experiments, it was noted that the variation of the cuttings width (l_t)—according to the work parameters and the work cycle, is not necessarily in a pre-established dependency.
- The quality of the cut surface was influenced by the working cycle and the agent for evacuating eroded products.
- The values of the rigidness have been influenced by overheating due to the inadequate cooling and insufficiency of the cooling agent, evacuation and slow evacuation of the eroded products.
- The thermal influenced zone had a different manifestation after the experimental tests were done, due to the influence of the working cycle (hard/soft) and the cooling agent applied.

For specialists in the field, understanding these interpretations may constitute a starting point in implementing this performing, efficiency cutting procedure.

7. Conclusions

Economic efficiency is an essential criterion that allows the establishment of cutting technology that provides the greatest economic effects, essentially maximizing the effect/effort ratio.

The experimental results highlight the economic advantages obtained when cutting semi-finished products of alloy steels, by EDMCB, using as TO (tool), the metal band.

Following the experiments performed and the results obtained, it became possible to apply a new technology for cutting alloy steels, which can successfully replace existing technologies, specific to conventional fields (by cutting, electric arc, thermal energy, etc.).

Characteristic for the new cutting technology by EDMCB developed are:

- increasing the productivity of social work, respectively saving social work;
- improving the quality of the cut areas;
- facilitating work and changing its character.

The present scientific paper through its wide approach of the problems related to the cutting by electrical discharge machining with contact breaking using as a transfer object the metal band, has proved that it is one of the modern technologies that can solve some problems related to the processing materials with special characteristics (hard, extra hard, etc.) in conditions of adequate technical and economic efficiency. The analysis and research carried out during the elaboration of this scientific paper in the field of non-conventional processing (of cutting materials by EDMCB) have highlighted the performances and advantages offered by this process. The research activity carried out led to the highlighting of some technological and managerial aspects related to the cutting of some alloy steel materials, using as a metal object of transfer, the metal band, and the completion of the established research directions allowed to make important original contributions.

The obtained results in the theoretical and experimental research, as well as the original contributions made during the practical experiments performed with the presented installation (State Patent Office for Invention and Trademarks no. 122346) allowed the elucidation of technological and managerial aspects considered by the authors. Even if in the approached field there could be other things left to research and elucidate, the authors express their conviction that the elaborated scientific paper includes elements of novelty and solutions, tangible results presented in qualitative and quantitative form, future trends and clearly explained reasons, which applied in industry, will directly lead to the elimination of problems of a technological nature that still persist today, as well as to outstanding decisions and results.

The arguments underlying the above statements take into account the calculation of the optimal manufacturing batch, the economic calculation justifying the adoption of the proposed variant of EDMCB cutting technology with TO—metal band—as well as a comparative analysis of labor productivity. All these aspects constitute further directions of research.

Author Contributions: Conceptualization, A.M.Ț., V.B.; methodology V.B., P.V.; software, Ș.Ț.; validation, A.M.Ț., A.V.S. and Ș.Ț.; formal analysis, A.B.P.; investigation, V.B., P.V.; resources, A.B.; data curation, Ș.Ț.; writing—original draft preparation, V.B., A.M.Ț.; writing—review and editing, A.M.Ț., A.V.S., Ș.Ț.; visualization, C.C.; supervision, A.M.Ț.; project administration, A.M.Ț., A.V.S. and Ș.Ț. All authors have read and agreed to the published version of the manuscript.

Funding: This research received no external funding.

Conflicts of Interest: The authors declare no conflict of interest.

References

1. Chakrabortya, S.; Deya, V.; Ghoshb, S.K. A review on the use of dielectric fluids and their effects in electrical discharge machining characteristics. *Precis. Eng.* **2015**, *40*, 1–6. [CrossRef]
2. Torres, A.; Puertas, I.; Luis, C.J. Mechanical, Modelling of surface finish, electrode wear and material removal rate in electrical discharge machining of hard-to-machine alloys. *Precis. Eng.* **2015**, *40*, 33–45. [CrossRef]
3. Yoo, H.K.; Kwon, W.T.; Kang, S. Development of a new electrode for micro-electrical discharge machining (EDM) using Ti(C,N)-based cermet. *Int. J. Precis. Eng. Manuf.* **2014**, *15*, 609–616. [CrossRef]
4. Somashekhar, K.P.; Mathew, J.; Ramachandran, N. A feasibility approach by simulated annealing on optimization of micro-wire electric discharge machining parameters. *Int. J. Adv. Manuf. Technol.* **2012**, *61*, 1209–1213. [CrossRef]
5. Steuer, P.; Weber, O.; Bähre, D. Structuring of wear-affected copper electrodes for electrical discharge machining using Pulse Electrochemical Machining. *Int. J. Refract. Met. Hard Mater.* **2015**, *52*, 85–89. [CrossRef]
6. Singh, P.; Chaudhary, A.K.; Singh, T.; Rana, A.K. Comparison of outputs for dry EDM and EDM with oil: A review. *Int. J. Res. Emerg. Sci. Technol.* **2015**, *2*, 45–49.
7. Khatri, B.C.; Rathod, P.; Valaki, J.B. Ultrasonic vibration-assisted electric discharge machining: A research review. *Proc. Inst. Mech. Eng. Part B J. Eng. Manuf.* **2015**, *230*, 319–330. [CrossRef]
8. Pandey, A.; Singh, S. Current research trends in variants of electrical discharge machining: A review. *Int. J. Eng. Sci. Technol.* **2010**, *2*, 2172–2191.
9. Huanga, T.S.; Hsieha, S.F.; Chenb, S.L.; Linb, M.H.; Oua, S.F.; Changa, W.T. Surface modification of TiNi-based shape memory alloys by dryelectrical discharge machining. *J. Mater. Process. Technol.* **2015**, *221*, 279–284. [CrossRef]
10. Dave, H.K.; Desai, K.P.; Raval, H.K. Modelling and analysis of material removal rate during electro discharge machining of Inconel 718 under orbital tool movement. *Int. J. Manuf. Syst.* **2012**, *2*, 12–20.
11. Azhiri, R.B.; Teimouri, R.; Baboly, M.G.; Laseman, Z. Application of Taguchi, ANFIS and grey relational analysis for studying, modelling and optimization of wire EDM process while using gaseous media. *Int. J. Adv. Manuf. Technol.* **2014**, *71*, 279–295. [CrossRef]
12. Chen, Z.; Huang, Y.; Huang, H.; Zhang, Z.; Zhang, G. Three-dimensional characteristics analysis of the wire-tool vibration considering spatial temperature field and electromagnetic field in WEDM. *Int. J. Mach. Tools Manuf.* **2015**, *92*, 85–96. [CrossRef]
13. Zhao, W.; Gu, L.; Xu, H.; Li, L.; Xiang, X.A. Novel High Efficiency Electrical Erosion Process—Blasting Erosion Arc Machining. *Procedia CIRP* **2013**, *6*, 621–625. [CrossRef]
14. Naumoski, H.; Riedmüller, B.; Minkowb, A.; Herr, U. Investigation of the influence of different cutting procedures on the global and local magnetic properties of non-oriented electrical steel. *J. Magn. Magn. Mater.* **2015**, *392*, 126–133. [CrossRef]
15. Sahu, R.K.; Hiremath, S.S.; Manivannan, P.V.; Singaperumal, M. An innovative approach for generation of aluminium nanoparticles using micro electrical discharge machining. *Procedia Mater. Sci.* **2014**, *5*, 1205–1213. [CrossRef]
16. Sahu, R.; Somashekhar, S.; Manivannan, P. Investigation on Copper Nanofluid obtained through Micro Electrical Discharge Machining for Dispersion Stability and Thermal Conductivity. *J. Proc. Eng.* **2013**, *64*, 946–955. [CrossRef]
17. Ghodsiyeh, D.; Golshan, A.; Shirvanehdeh, J.A. Review on current research trends in wire electrical discharge machining (WEDM). *Indian J. Sci. Technol.* **2013**, *6*, 4128–4140. [CrossRef]
18. Banu, A.; Ali, M.Y.; Rahman, M.A. Micro-electro discharge machining of non-conductive zirconia ceramic: Investigation of MRR and recast layer hardness. *Int. J. Adv. Manuf. Technol.* **2014**, *75*, 257–267. [CrossRef]
19. Yang, R.T.; Tzeng, C.J.; Yang, Y.K.; Hsieh, M.H. Optimization of wire electrical discharge machining process parameters for cutting tungsten. *Int. J. Adv. Manuf. Technol.* **2012**, *60*, 135–147. [CrossRef]
20. Chaudhari, R.; Vora, J.J.; Patel, V.; López de Lacalle, L.N.; Parikh, D.M. Surface Analysis of Wire-Electrical-Discharge-Machining-Processed Shape-Memory Alloys. *Materials* **2020**, *13*, 530. [CrossRef]
21. Ezeddini, S.; Boujelbene, M.; Bayraktar, E.; Ben Salem, S. Optimization of the Surface Roughness Parameters of Ti–Al Intermetallic Based Composite Machined by Wire Electrical Discharge Machining. *Coatings* **2020**, *10*, 900. [CrossRef]

22. Abu Qudeiri, J.E.; Saleh, A.; Ziout, A.; Mourad, A.-H.I.; Abidi, M.H.; Elkaseer, A. Advanced Electric Discharge Machining of Stainless Steels: Assessment of the State of the Art, Gaps and Future Prospect. *Materials* **2019**, *12*, 907. [CrossRef] [PubMed]
23. Koshy, P.; Boroumand, M.; Ziada, Y. Breakout detection in fast hole electrical discharge machining. *Int. J. Mach. Tools Manuf.* **2015**, *50*, 922–925. [CrossRef]
24. Wang, F.; Liu, Y.; Tang, Z.; Ji, R.; Zhang, Y.; Shen, Y. Ultra-high-speed combined machining of electrical discharge machining and arc machining. *J. Eng. Manuf.* **2013**, *228*, 663–672. [CrossRef]
25. Țîțu, M.; Oprean, C.; Boroiu, A. *Experimental Research Applied in Increasing the Quality of Products and Services*; AGIR: Bucharest, Romania, 2011; ISBN 978-973-720-362-5.
26. Ahmad Mufti, N.; Rafaqat, M.; Ahmed, N.; Qaiser Saleem, M.; Hussain, A.; Al-Ahamri, A.M. Improving the Performance of EDM through Relief-Angled Tool Designs. *Appl. Sci.* **2020**, *10*, 2432. [CrossRef]
27. Rafaqat, M.; Mufti, N.A.; Ahmed, N.; Alahmari, A.M.; Hussain, A. EDM of D2 Steel: Performance Comparison of EDM Die Sinking Electrode Designs. *Appl. Sci.* **2020**, *10*, 7411. [CrossRef]
28. Ablyaz, T.R.; Shlykov, E.S.; Muratov, K.R.; Mahajan, A.; Singh, G.; Devgan, S.; Sidhu, S.S. Surface Characterization and Tribological Performance Analysis of Electric Discharge Machined Duplex Stainless Steel. *Micromachines* **2020**, *11*, 926. [CrossRef]
29. Niamat, M.; Sarfraz, S.; Ahmad, W.; Shehab, E.; Salonitis, K. Parametric Modelling and Multi-Objective Optimization of Electro Discharge Machining Process Parameters for Sustainable Production. *Energies* **2020**, *13*, 38. [CrossRef]
30. Świercz, R.; Oniszczuk-Świercz, D.; Chmielewski, T. Multi-Response Optimization of Electrical Discharge Machining Using the Desirability Function. *Micromachines* **2019**, *10*, 72. [CrossRef]
31. Markopoulos, A.P.; Papazoglou, E.-L.; Karmiris-Obratański, P. Experimental Study on the Influence of Machining Conditions on the Quality of Electrical Discharge Machined Surfaces of aluminum alloy Al5052. *Machines* **2020**, *8*, 12. [CrossRef]
32. Chen, J.; Gu, L.; He, G. A review on conventional and nonconventional machining of SiC particle-reinforced aluminium matrix composites. *Adv. Manuf.* **2020**, *8*, 279–315. [CrossRef]
33. Sabyrov, N.; Jahan, M.P.; Bilal, A.; Perveen, A. Ultrasonic Vibration Assisted Electro-Discharge Machining (EDM)—An Overview. *Materials* **2019**, *12*, 522. [CrossRef] [PubMed]
34. Marrocco, V.; Modica, F.; Bellantone, V.; Medri, V.; Fassi, I. Pulse-Type Influence on the Micro-EDM Milling Machinability of Si3N4–TiN Workpieces. *Micromachines* **2020**, *11*, 932. [CrossRef] [PubMed]
35. Sindhu, M.K.; Nandi, D.; Basak, I. Electric discharge phenomenon in dielectric and electrolyte medium. *Adv. Manuf.* **2018**, *6*, 457–464. [CrossRef]
36. Khanna, N.; Pusavec, F.; Agrawal, C.; Krolczyk, G.M. Measurement and evaluation of hole attributes for drilling CFRP composites using an Indigenously developed cryogenic machining facility. *Measurement* **2020**, *154*, 107504. [CrossRef]

Publisher's Note: MDPI stays neutral with regard to jurisdictional claims in published maps and institutional affiliations.

Article

The Comparative Study of the State of Conservation of Two Medieval Documents on Parchment from Different Historical Periods

Maria Boutiuc (Haulică) [1,2], Oana Florescu [1,3], Viorica Vasilache [4,*] and Ion Sandu [4,5,6,*]

1 Faculty of Geography and Geology, Doctoral School of Geosciences, "Alexandru Ioan Cuza" University of Iasi, 22 Carol I Blvd., 700506 Iasi, Romania; boutiucmariahaulica@gmail.com (M.B.); of.poni@gmail.com (O.F.)
2 County Service of the National Archives Iasi, Blvd. Carol I no. 26, 700505 Iasi, Romania
3 "Poni-Cernatescu" Museum of Iasi, 7B Kogalniceanu St., 700454 Iasi, Romania
4 ARHEOINVEST Centrum, Institute of Interdisciplinary Research, "Alexandru Ioan Cuza" University of Iasi, 11 Carol I Blvd., 700506 Iasi, Romania
5 Romanian Inventors Forum, Str. Sf. P. Movila 3, 700089 Iasi, Romania
6 Academy of Romanian Scientists (AOSR), 54 Splaiul Independentei St., Sect. 5, 050094 Bucharest, Romania
* Correspondence: viorica_18v@yahoo.com (V.V.); ion.sandu@uaic.ro (I.S.)

Received: 23 September 2020; Accepted: 20 October 2020; Published: 26 October 2020

Abstract: The paper explores the potentiality of an experimental multianalytic protocol with appropriate methodology for determining the chemical and morphostructural characteristics of two old documents on parchment support. Such a protocol can authenticate and assess the state of conservation under the influence of environmental factors during storage and archival documentation, thus advancing preventive and prophylactic measures in "treasure" deposits such as the National Archives of Romania, where these documents are kept. The work methodology consisted of three stages. The first stage consists of visual observation for identifying deteriorations and degradations, alongside the selection of representatives' areas from where micro-samples were collected. The second stage involves Scanning Electron Microscopy coupled by Energy Dispersive X-ray Spectrometry (SEM-EDX) analysis, for highlighting the morphology and determining the elemental composition; lastly, the Fourier-transform infrared (FTIR) analysis and correlation of results establish the chemical and morphostructural changes. The use of this gradual system of analyses allowed determining the differences between these two documents in terms of the materials used for producing them, their manufacturing technologies, the writing and ornamentation, and the overall state of conservation. The results provided the first accurate picture of the chemical nature and manufacturing of the two parchment documents by determining the main characteristics of the collagen and of the finishing, writing, and decoration materials, in view of the natural aging through the oxidative and gelatinization processes of the collagen. The SEM-EDX results revealed the morphological changes of the parchment that occurred at various levels in the collagen fibrous mesh and established the state of conservation of the support, writing, and decorations.

Keywords: parchment; collagen; state of conservation; OM; SEM-EDX; ATR-FTIR

---

1. Introduction

Parchment is obtained from untanned or slightly tanned animal skins (lamb, sheep, goat, deer, calf, oxen, camel, crocodile, seal, etc.) by removing the epidermis and hypodermis, involving various physical–chemical and mechanical treatments. The main component of the dermal tissue is collagen, which is a fibrillary protein with a framing and supporting role. The fibrous nature of the collagen is due to long sequences of amino acids (approximately 20) forming long chains, with a molecular mass of 90,000–95,000. The collagen fiber consists of three long chains, linked in α-helix [1], and each polypeptide chain has crystalline areas where the amino acids are nonpolar and amorphous areas where the amino acids are polar or polarizable. The polypeptide chains will undergo hydrolytic cleavage during the preliminary operations of removing the epidermis and hypodermis. These cleavages leave their mark on the evolution of the rate of aging, being an endogenous factor that interacts easily with exogenous ones (pollution, microorganisms, atmospheric humidity, temperature, illumination, etc.). A specific characteristic of collagen is thermal denaturation and that produced by strong acids and bases, and gelatinization in hot water; both groups of processes render parchment insoluble in water, diluted solutions of acids, bases and neutral salts, and organic solvents at normal temperatures [2,3].

Therefore, documents and other old writings on parchment, on account of their heritage value, have been at the center of various studies throughout time [4–7], aiming to identify the main components of the support, writing, and miniature materials, with regard to the evolution of the state of conservation in correlation with endogenous and exogenous factors, which affected their integrity, aesthetics, and historical message.

The use of high-resolution optical microscopy (OM), Scanning Electron Microscopy coupled by Energy Dispersive X-ray Spectrometry (SEM-EDX), and attenuated total reflection Fourier-transform infrared (ATR-FTIR) is well established for characterizing the natural degradation of old parchments [8], specifically for determining the main characteristics of the collagen and establishing their organic and inorganic composition, alongside assessments based on changes in crystallinity pertaining to natural degradation.

Similarly, there are other techniques used for assessing the state of conservation of ancient or medieval parchments [9,10], which are of particular importance for developing adequate/compatible procedures for the preventive and prophylactic preservation of valuable historical documents.

Assessing the authenticity of old documents on parchment involves a series of analysis methods that allow determining certain archaeometric characteristics or the chemometric characteristics of archaeometric value; these methods are generally used for identifying forged and counterfeit documents [11].

The present paper advances an experimental multianalytic protocol with a specific work methodology for determining the chemical and morphostructural characteristics of two old documents on parchment, in order to authenticate and establish the evolution of the state of conservation under the influence of environmental and anthropic factors during storage and archival documentation, as to warrant taking urgent preventive and prophylactic preservation measures in archival deposits.

The two medieval documents are of high historical and artistic value, were produced in different periods, and are kept by the Iasi County Service of the National Archives of Romania, where they are assigned to the "Treasure" collection. The two documents were analyzed by corroborating the OM, SEM-EDX, and ATR-FTIR techniques, in order to determine the state of conservation for establishing the evolution of structural and aesthetic–artistic integrity across time. We analyzed the chemical nature and a series of characteristics of the support, the materials used for writing, ornaments, and miniatures, which allow us to identify the evolutionary effects of deterioration and degradation.

2. Experimental Part

2.1. Description of Documents and Their State of Conservation

For this study, we selected two old parchments from the collection of medieval documents of high heritage value (classed in the treasure fund), which were created 154 years ago using very different artistic techniques and execution technologies. This study aims to obtain information regarding the nature of the component materials and their state of conservation, alongside a series of other archaeometric and morphostructural characteristics, in order to trace their integrity diachronically.

These documents were indexed with the initial S followed by the order number 1 and, respectively, 2. These documents were sorted according to their heritage value.

Figure 1 presents the image of document S1, which is a charter issued on 15 October, 7272/‹1763› by the princely chancellery of Moldavia, during the rule of Voivode Grigore Ioan Callimachi, which cancelled different taxes (on vineyards and winemaking).

Figure 1. Parchment S1 with the index codes of the areas under scrutiny.

The document is rectangular in shape and was kept rolled or folded in three; it measures 140.5 × 65.5 cm and was manufactured from three pieces of parchment by binding their overlapping edges (a 5.0 × 65.6 cm stripe at the top, a 81.5 × 65.6 cm stripe at the middle, and a 54.0 × 65.6 cm at the bottom). After hydric stabilization and cutting out, the ornaments and writing were inserted. The substrate of the document is yellow in color because the stage of tanning was skipped. Moreover, it has a silky texture, particularly the back. It is embellished with a sumptuous decoration in the form of a carefully executed frame filled with ornaments and miniatures and the coat of arms of

Moldavia (an official sign acknowledged and perpetuated for centuries), an auroch head seen from the front, with a star in-between the horns. The auroch head is surrounded by Christian iconographic elements, which are surmounted by a crown. The drawing and ornaments are made in black, red, green, and golden ink. In the lower part of the document, there is the seal impression made in red ink and surrounded by decorative rosette, which in modern terms is assimilated with a stamp. The writing, in black ink, is aligned and symmetrical, while the auroch head, crown, title, symbolic invocation, and capitals are in golden ink. In addition to its documentary importance, this document is an exceptional work of calligraphic and miniature art. It is part of the richly decorated documents of Moldavia, which are much rarer than those of Wallachia. This is due to the limited attention given by the princely chancellery of Moldavia to coloristic art, even though this was a flourishing era for miniature art overall. Even though the throne of Moldavia was occupied by three princes of Wallachian origin (Radu Mihnea, Simion Movila, and Alexandru Ilias), who tried to impose the richly decorated style, calligraphic writing, and miniature decoration of documents, but without consequences in Moldavia. Such a tradition existed in this principality only in monastery workshops before the age of Stephen the Great.

When collected from the archives, the document presented degradations and denaturation in the form of cornification and crumbling. In addition, the overlapping of the components of the support and margins were fragilized, and in some areas, it was fringed, with microfibers failing to bind together. Other defects were a series of restricted areas with fouling halos and spots of tegumentary lipids contaminated with atmospheric dust, particularly the superior part of the backside of the document.

The second document, indexed as S2 (Figure 2), is an original Slavonic parchment, with a hanging seal, containing the charter of Constantin Moghilă, the Prince of Moldavia, reconfirming the rights and privileges of the Secu Monastery. The document was issued in the year 7117 (1609), on 4 February, in Iasi. The manuscript is rectangular in shape, measuring 63.4 × 57.5 cm, of which 8 cm represent the width of the sleeve (the lower cuff of the document folded at the front, fitted with four rhombic perforations placed at the center, through which the string of the seal runs).

Figure 2. Parchment S2 with the index numbers of the analyzed areas.

The document was kept folded in three, in a sleeve. The seal is the element that authenticates the document; it is large, round, and impressed once (on the front) on warm red wax with the coat of arms of Moldavia.

With respect to ornamentation, compared to the first document, this one is simpler, with fewer colors: black and golden. The golden ink was used for writing the title, the invocation with "the cross ahead", the initials, the first two rows, certain expressions, and the capital letters in the body of the text. The initial and the invocation are much larger in size than the rest of the writing, of 16 and 12 rows.

This document is better preserved than the first document, but it has traces of use, with loss of material along the folding lines. All these makes the document highly fragile. The support has the collagen fibers well individualized and presents a better flexibility during manipulation.

2.2. Samples and Collecting Areas of Documents

A series of samples was taken from the areas marked in Figures 1 and 2 and indexed according to the area, as follows:

- S1-SD, degraded collagenic support (samples S1-SD1 and S1-SD2);
- S1-CN, writing ink based on black pigment;
- S1-PR, ink used for the red-pigment miniature;
- S1-PA ink used for the golden/yellow pigment miniature;
- S1-PV ink used for the green pigment miniature;
- S2-CN, black ink;
- S2-CA, golden ink.

2.3. Methods and Analysis Techniques

In the first stage, the documents were the subject of direct analysis, by visual analysis, with the naked eye or with optical magnifying instruments: lens, binocular stereo lens, and stereomicroscope.

Optical microscopy was conducted using a Carl Zeiss Axio Imager A1m microscope (Berlin, Germany) at magnifications between ×50 and ×200, attached to an Axiocam camera and using specialized software.

For the SEM-EDX analysis, an SEM model VEGA II LSH produced by Tescan (Brno, Czech Republic) and coupled to an EDX detector type Quantax QX2 made by Bruker/Roentec (Berlin, Germany) has been used. The analysis of the samples was carried out at a magnification of 200 ... 2500×, with an acceleration tension of 30 kV, and the work pressure was lower than 1×10^{-2} Pa. The resulting image was constituted by secondary electrons (SE) and backscattered electrons (BSE). The SEM-EDX microscopy involved the collecting and processing of the samples.

FTIR spectra were obtained using a Vertex 70 FTIR (Bruker, Berlin, Germany) equipped with accessories: ATR mode and RAMAN II. The spectra were recorded in the range of 4000–700 cm^{-1} with a resolution of 4 cm^{-1}.

3. Results and Discussions

3.1. Visual Analysis

The direct visual analysis of these two documents point out that the parchment supports are in good quality, without major defects and with uniform thickness, light color, lusterless appearance, smooth, velvety surfaces, and the writing area well prepared. Both documents meet all the characteristics of a document on parchment issued by the princely chancellery of Moldavia. The deteriorations and degradations found were caused by both endogenous and exogenous factors, which left their marks on the evolution of the aging rate, respectively on the structural–functional and aesthetic–artistic integrity of the artefacts.

Document S1 (Figure 1) was analyzed by a handheld lens and stereomicroscope, and it has a series of deteriorations and degradations caused by careless handling and use, such as fine tears and frays caused by hydrolytic embrittlement and limited ungluing in the overlap area of the three constituting pieces of parchment caused by cornification processes through chemical or thermal denaturation, particularly in the top area. Regarding the writing and miniatures, document S1 presents fine erosions (thinning of the layers) and exfoliations of the writing and miniature inks caused by fluctuations of the microclimate parameters beyond the optimal values (18–22 °C for temperature, 50 lx for illumination, and 50–60% for atmospheric humidity). Pollution, illumination, and atmospheric humidity turned the originally white parchment to yellow-gray, as seen on both the front and back side of the document, particularly the top area of the document, since that was the most exposed part.

Document S2 (Figure 2) is better preserved; it is darker in color in the fold lines areas because of the dust collected on the surface. The document has small areas where the material was lost along the folding lines, particularly in the case of multiple folds, where the corners resulting from folding are highly fragile. Examination under a stereo lens highlights the substrate with well-defined fibers in terms of its morphology and entanglement system. The parchment is still sufficiently flexible for manipulation.

3.2. Analysis of the Texture

These two documents were analyzed by optical microscopy, without collecting samples, directly through reflection, in white light, dark field, using the selected areas by experimental protocol. Therefore, for document S1, we obtained microphotographs of the front of the parchment in the marginal area free of writing (S1-SD1, Figure 3a) between the written rows in the binding area (S1-SD2, Figure 3b,c) and also from the same areas of the back of the parchment (Figure 3d), while for the document S2, we analyzed from the front a degraded area (Figure 4a), one without conspicuous degradation (Figure 4b), and another one from the back of this document (Figure 4c).

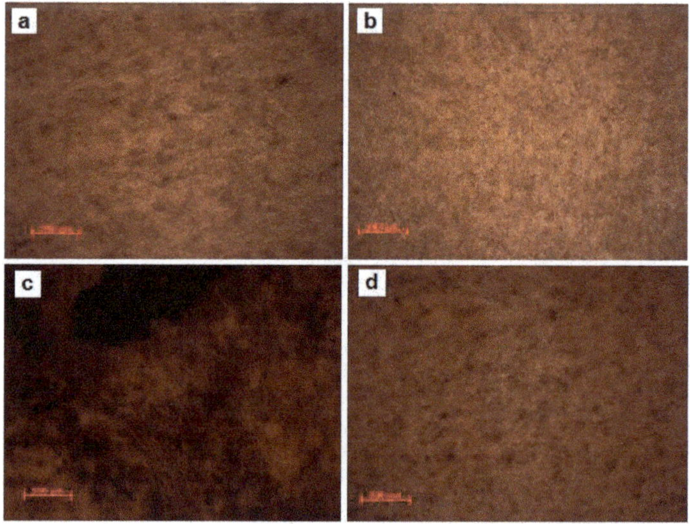

Figure 3. Optical microscopy (OM) images (×50) of document S1: (**a**) marginal area (S1-SD1); (**b**) between the rows of writing (S1-SD2); (**c**) the binding area of the support (S1-SD2); and (**d**) the back of the document.

Figure 4. OM images (×50) of document S2: (**a**) the degraded and fringed area on the front; (**b**) well preserved area on the front; and (**c**) well preserved area on the back.

Following the analysis of the granular configuration determined by the arrangement of the pilose follicles (removed by physical–chemical and mechanic processes from the top layer of the dermis), we found that both documents were obtained from veal leather, which has a denser and finer structure than mature bovine skin, and that the alveoli are smaller and the pilose follicles are distributed uniformly, in rows of one and in relatively linear lines [2,3,12].

Figures 5 and 6 present the OM (×50) microphotographs for the writing and miniature inks of document S1 from the following areas: S1-PR (red pigment), S1-PV (green pigment), S1-PA (golden pigment), and S1-CN (black pigment). For S2, microphotographs were obtained for areas S2-CA (golden ink) and S2-CN (black ink).

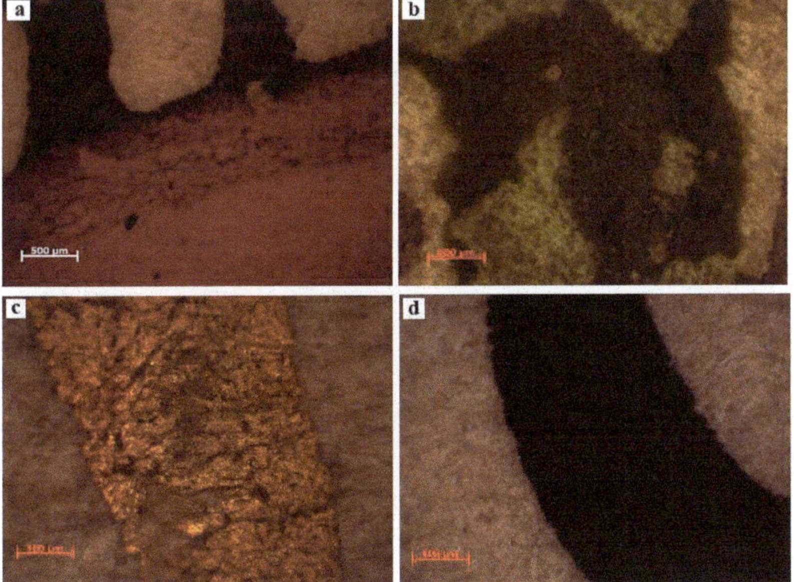

Figure 5. OM images (×50) for the writing and miniature inks of document S1: (**a**) S1-PR (red pigment); (**b**) S1-PV (green pigment); (**c**) S1-PA (golden pigment); and (**d**) S1-CN (black pigment).

The analysis of the images of the inks used for writing and miniatures reveals a series of aspects, as they are presented below.

The order of application of the inks on document S1 was first black ink and then red ink (Figure 5a).

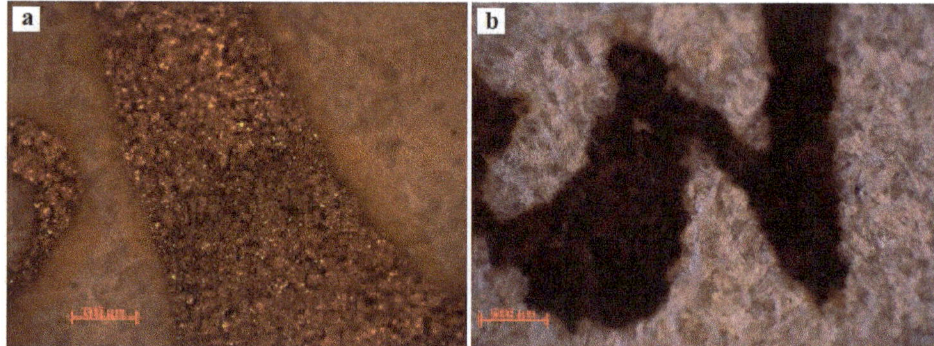

Figure 6. OM images (×50) for the writing and miniature inks of document S2: (**a**) S2-CA; (**b**) S2-CN.

Both red and green ink (Figure 5a,b) present reticular fissures (cracks) and micro-lacunae resulting from the detachment of the inks from the support. These forms of deterioration are due to the fragilization in time of the binder in the two inks, meaning that the process of hydrolytic alteration determines the destruction of the writing and miniature works. Low humidity alongside high temperature causes dehydration, which is followed by the contraction and breakup of the inks, producing age cracks.

More wear is seen in the case of the golden ink on document S1 (Figure 5c), where the detachments are more critical. In this case, besides the causes described for the red and green inks, the different hygroscopicities of the collagen support and of the two components of the golden ink (pigment and binder) are also relevant, which at fluctuant humidity produce different dilatations and contractions. On the other hand, the golden ink from document S2 (Figure 6a) has traces left by the elevated humidity of the air up to the point of saturating the fiber, causing the migration of the binder to the vicinity of its placement on the document along its contour, which causes a better attachment to the support by increasing the contact area through the resulting halo.

The microscopic analysis of the black ink from document S1 (Figure 5d), which was made from carbon pigment, showed signs of dividing and the detachment of microparticles, despite the fine macroscopic appearance. In the case of document S2, in Figure 6b, we noticed the ingress of the ink into the collagen support, which proves that the ink was ferogallic and was likewise confirmed by the SEM-EDX and ATR-FTIR analyses. However, document S2 also presents a detachment of microparticles that are darker in color, which proves that the ink was a mixture of iron gall and carbon: a recipe commonly used in the past in order to mitigate the shortcoming of illegibility of ferogallic once it is used. Since it is a solution, the iron gall ink was absorbed by the support, coloring it irreversibly, whereas carbon ink, notwithstanding its excellent stability in time, does not adhere well to the smooth surface of the parchment because the dispersed fine carbon does not enter the collagen fibers, but only affixed on the surface. It follows that the sole culprit for the degradation of the carbon ink is the binder (gum Arabic or animal glues), which is more sensitive to the action of exogenous factors.

3.3. The Analysis SEM-EDX

For obtaining structural details beyond those provided by optical microscopy, we employed SEM. The micro-samples analyzed by SEM-EDX were collected from the same areas as in the case of OM.

Figures 7 and 8 present the SEM microphotographs of the two parchments for the areas most affected by destructions and alterations.

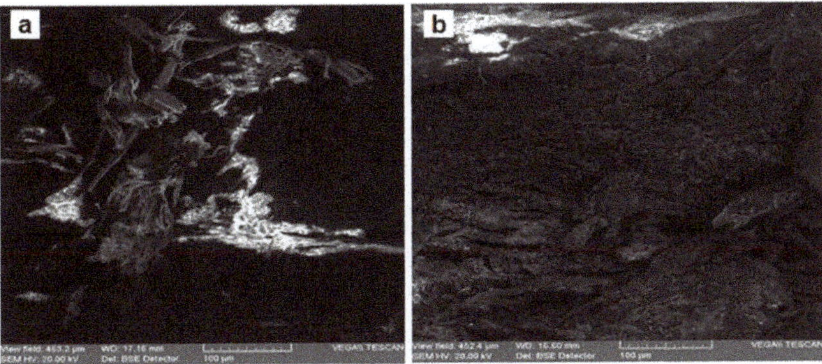

Figure 7. SEM microphotographs (×500, backscattered electrons (BSE)) for the support of parchment S1: (**a**) degraded support, area S1-SD1; (**b**) degraded support, area S1-SD2.

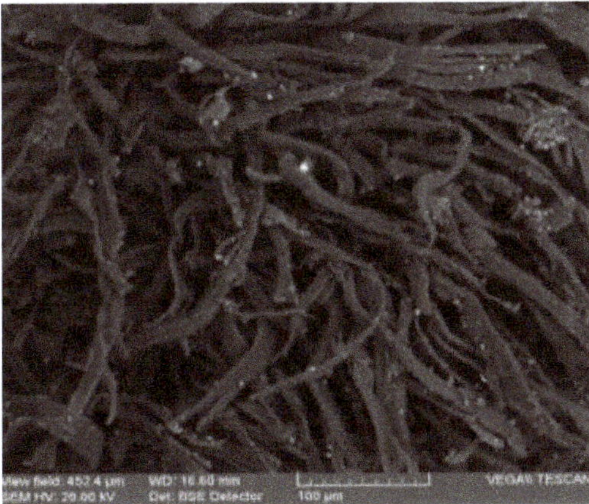

Figure 8. SEM microphotographs (×500, BSE) for the support of parchment S2, the degraded areas S2-S.

The analysis of the images from Figures 7 and 8 reveals the following:

- The collagen support of document S1 is more degraded by denaturation and gelatinization, presenting fibers compacted by monolithization and with a reticulation of three-dimensional fissures and micro-crevices caused by the fracturing occurring when the document was rolled up (Figure 7a,b), with the frayed fibers clearly distinct on the surface from the monolithized ones.
- The sources of these deteriorations were initially due to alteration processes, which led to evolving effects of degradation caused by higher temperatures of the environment in which the document was kept, and by air humidity above 70%, which are optimal conditions for activating microbiological attack; subsequently, damage had also resulted from constrictions at high temperatures and environmental humidity below 40%, which led to dehydration and the loss of hygroscopic water, resulting in contractions after the elimination of the hydrogen bonds from the teloproteins of the polypeptide chains of the collagen molecules [13].
- The advanced loss of hygroscopic water down to 5% rendered the parchment supports rigid, creasy, and shriveled; in case of accidental moistening, the supports will disintegrate in thin sheets.
- The visual analysis using the handheld lens or the stereomicroscope of the various forms of shredding and fringing shows multiple amorphous areas in the collagen fibers.

- In the case of document S2, the collagen fibers are very well individualized within a 3D meshing (Figure 8), with conspicuous microparticles of calcium carbonate and inorganic salt crystallites left as residues of the preliminary processing of the skins.
- Compared to S1, document S2 is more flexible, which is noticeable during manipulation.
- The SEM images of the inks of these two documents, as presented in Figures 9 and 10, confirm the results obtained by the OM analysis. Furthermore, the SEM detailing allowed a better highlighting of the cracks, exfoliations, micro-zonal lacunae, and scaly structures in the form of plates/lamellae for the golden pigment (Figures 9c and 10a) and non-uniform granules, with a reticulation of fissures and micro-crevices for the black carbon-based pigment (Figures 9d and 10b).

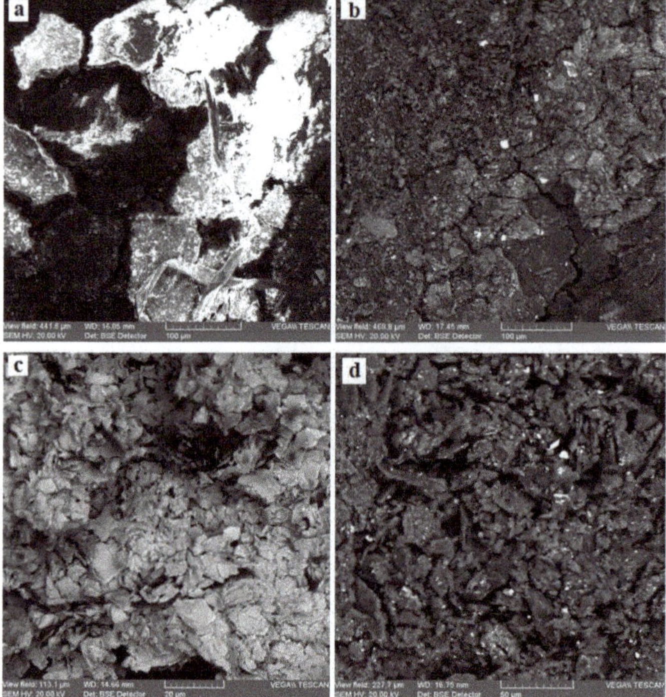

Figure 9. SEM microphotographs for the inks of document S1: (**a**) S1-PR (×500); (**b**) S1-PV (×500); (**c**) S1-PA (×2000); and (**d**) S1-CN (×1000).

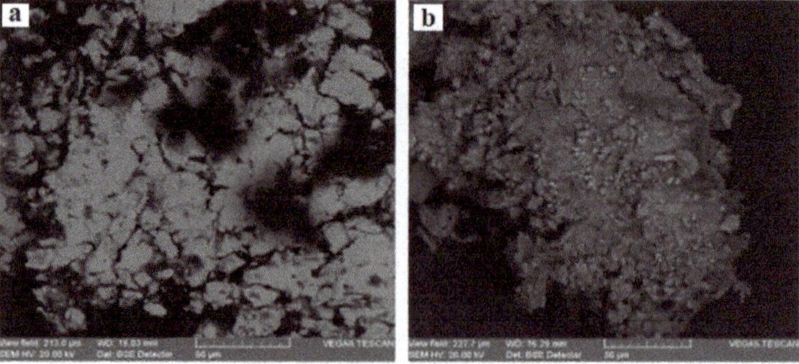

Figure 10. SEM microphotographs (×500, BSE) for the inks of document S2: (**a**) S2-CA; (**b**) S2-CN.

The EDX spectra provided the elemental composition of the structural components of the two documents, as presented in Table 1.

Table 1. The elemental composition in gravimetric percentages for the samples collected from the two documents under scrutiny and indexed according to the area and type of material.

Sample	**Elemental Composition (Gravimetric Percentages)**														
	Na	**K**	**Mg**	**Ca**	**Fe**	**Cu**	**Au**	**Hg**	**Al**	**C**	**Si**	**P**	**O**	**S**	**Cl**
S1-SD$_1$	0.807	1.065	0.254	3.779	-	-	-	-	0.169	40.733	0.460	-	51.498	0.543	0.691
S1-SD$_2$	1.261	1.772	0.508	7.080	-	-	-	-	0.431	32.634	1.011	-	53.553	0.901	0.848
S1-CN	0.524	1.526	0.065	5.159	-	-	-	-	0.129	35.745	0.491	-	56.084	0.278	-
S1-PR	1.505	0.756	0.799	2.974	-	-	-	8.235	0.627	36.826	0.587	-	46.075	1.256	0.358
S1-PA	-	-	-	-	-	-	70.690	-	-	15.609	-	-	13.702	-	-
S1-PV	1.477	1.853	1.966	4.582	0.655	7.846	-	-	1.372	23.151	3.569	0.620	51.066	1.195	0.648
S2-S	1.171	1.032	0.213	1.295	-	-	-	-	0.359	41.772	0.377	0.097	52.154	0.635	0.895
S2-CN	0.292	1.622	0.055	8.391	5.842	-	-	-	0.109	26.024	0.768	0.152	53.781	2.702	0.262
S2-CA	-	-	-	2.135	-	-	52.851	-	-	20.121	-	-	24.893	-	-

The elemental composition of the micro-samples allowed the following observations:

- The calcium in $CaCO_3$ left as a residue of the manufacturing technology renders the support alkaline, making it more resistant to acids and micro-organism attack [13–16]. In the case of the support of parchment S1, the concentration of Ca is much higher (approximately five times higher) than that of S2, which shows that the technologies for obtaining the two writing supports were different. The production of S1 involved the treatment with calcium hydroxide (slaked lime), while for S2, the treatment consisted of calcium hydroxide and sodium sulfite, which expedited the removal of the epidermis and hypodermis, including their components (hairs/wool, glandes, etc.). If S1 had been stored in the same microclimate conditions as S2, the former document would have been better preserved than the latter.
- Sulfur, in the case of document S1, originates from sulfuric amino acids, having a lower value in the area most degraded (where the amorphous areas of the collagen fibers predominate), and from the sulfur found in the ink pigments diffused through migration into the support. In the case of document S2, the sulfur concentration is greater, since it originates from the treatment with sodium sulfite added in the calcium hydroxide bath and from the migration into the support of a portion of the sulfur contained by the ferogallic ink.
- The presence of aluminum and silicon in these two parchments is due to the atmospheric dust that adhered to the surface of the documents in the form of aluminosilicates and to impurities found in the calcium hydroxide used in the preparation stage of the parchment; these two elements are found in greater quantities in S1 than S2.
- Chlorine and sodium originate from the sodium chloride (salt) used for treating the hides just after skinning.
- The presence of the chemical elements potassium, magnesium, sodium, and calcium in all inks except the golden one can be explained by the use as a binder of gum Arabic, but also from the impurities found in the hydrated lime (slaked lime) used for manufacturing the parchment.
- The composition of pigments, with traces of aluminum, sodium, potassium, magnesium, and silicon suggests that they were obtained from colored earths.
- The green pigment may be natural ultramarine—$3Na_2O{\cdot}3Al_2O_3{\cdot}6SiO_2{\cdot}3Na_2S$—combined with iron yellow ochre—Fe_2O_3–, green apatite—$Ca_5(PO_4)_3(Cl, OH)$—or malachite (basic copper carbonate—$[2CuCO_3Cu(OH)_2]$).
- The red pigment, with a mercury content of 8.325% and sulfur content of 1.256%, suggests that it originates from cinnabar (known in its natural form since antiquity and the artificial form since the 13th century). This pigment turns brown under the action of natural light radiation, but in

our case, the document's writing and ornamentation were protected from light by rolling and folding it.

- The golden pigment contains a large quantity of gold, specifically 70.69% in the case of parchment S1 and 52.85% in the case of S2; the golden ink of the latter document also contains traces of calcium and carbon originating from the gum Arabic and the calcium hydroxide used during the execution of the manuscript.
- The black ink used for writing and producing the ornaments of document S1 is carbon, and the one used in document S2 is a ferogallic ink combined with a small quantity of carbon.

3.4. Analysis by ATR–FTIR

In order to highlight the degradations of these supports and inks (pigments and binders), the spectrum of these two supports were overlapped in Figure 11 and respectively in Figure 12. Table 2 presents the functional groups and peaks corresponding to each component of these two documents.

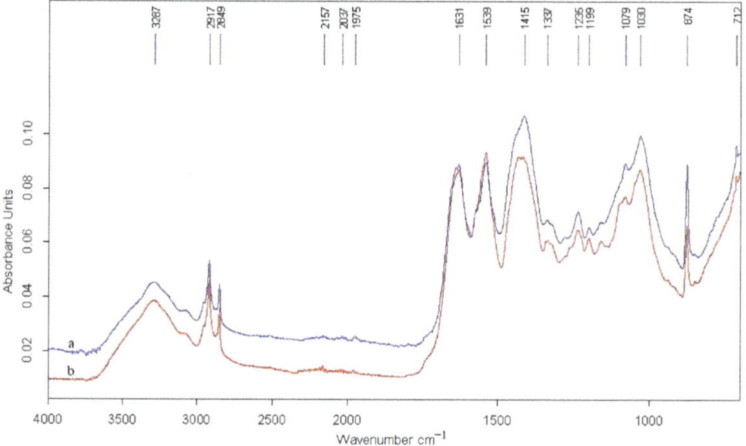

Figure 11. Fourier-transform infrared (FTIR) spectra on the support of the two documents: (a) S1-SD; (b) S2-S.

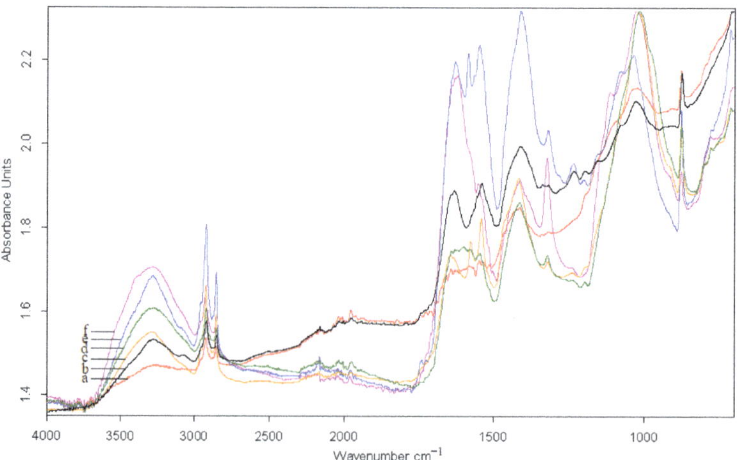

Figure 12. The FTIR spectra for the ink on the two documents: (a) S2-CA; (b) S1-PA; (c) S1-PR; (d) S1-CN; (e) S1-PV; and (f) S2-CN.

Table 2. Representative peaks and spectral bands of the ions identified in the samples analyzed by ATR-FTIR.

Type Ion	Theoretical Spectral Bands (cm^{-1})	Peak Present in Samples (cm^{-1})	Samples Analyzed
Amide A	3400–3100	3287 3277 3304 3281 3284	S1-PR S1-CN S1-PA S1-PV S2-CN
Amide I	1643–1631	1642 1633 1630 1630	S1-PR S1-PA S1-PV S2-CN
Amide II	1550–1532	1540 1545 1540 1546	S1-PR S1-CN S1-PA S1-PV
Amide III	1240–1232	1238 1233 1234	S1-PR S1-PA S1-PV
Aluminosilicates	1175–860	1027 1024 1029 1027 1023 1027	S1-PR S1-CN S1-PA S1-PV S2-CA S2-CN
Sulfate and sulfide	1035–960	1017 1014	S1-PR S1-CN
Carbonate	890–800; 1100–1040; 1530–1320	800, 873; 1416; 1320 874; 1414; 1321 874; 1414 828; 1411; 1319 875; 1413 875; 1412; 1322	S1-PR S1-CN S1-PA S1-PV S2-CA S2-CN
Phosphate	920–830; 1900–1600; 2500–2150; 2900–2750	1600; 2157 1607 2165	S2-CN S1-PR S1-PV
Chloride	1050–900	1034	S1-PV

We notice a perfect overlapping of the peaks of the supports, but their intensities differ. The peak from 3287 cm^{-1} shows that there was registered valence vibrations ν for the -OH and -NH groups. They belong to amide A with a band range of 3400–3100 cm^{-1} and form the first component of the ν(NH) frequency in the Fermi resonance of amide II.

Amide B has the band between 3080 and 3070 cm^{-1} and is attributed to the second Fermi resonance, which is between the ν(NH) valence vibration and the end of amide II.

Amide I, with the band 1643–1630 cm^{-1} is attributed to the ν(C = O) frequency (valence vibrations of the peptide bonds).

Amide II, with the band 1550–1532 cm^{-1}, is attributed to the ν(CN) valence vibration and the δ(NH) deformation vibration: respectively, the δ(CN) deformation vibrations.

Amide III, with the band 1240–1235 cm^{-1}, the absorbance of which comes from δ(NH) and the δ(CH_2) deformation vibration of glycine and the lateral chain of proline [17–20].

The peaks from 2917 cm^{-1} and 2849 cm^{-1} are specific to the valence bands of carboxylic groups.

The associated bands are representative in the IR spectral range of proteins (deformation vibrations), particularly the amides A_I and A_{II}.

For assessing the rate of denaturation of collagen, we calculated the difference of positioning between the two amides and the ratio of their intensities:

P1: I(AI)/I(AII) = I(A1633)/I(A1540) = 0.136/0.147 = 0.93

P2: I(AI)/I(AII) = I(A1631)/I(A1542) = 0.094/0.095 = 0.99.

The value of the ratio of intensities of the two amides, which is approximately equal to 1, demonstrates that the parchment was slightly degraded by hydrolysis (in the case of the investigated micro-zones).

In general, this ratio of intensities of the bands of amides A_I and A_{II} is the indicator of the degree of breaking of the peptide bonds [21].

P1: $\Delta\nu = \nu(AI) - \nu(AII) = 1633 - 1540 = 93\ cm^{-1}$
P2: $\Delta\nu = \nu(AI) - \nu(AII) = 1631 - 1542 = 89\ cm^{-1.}$

The difference between the positions of the two amides, with the value of ≈93 cm^{-1}, respectively 89 cm^{-1}, demonstrates that an even more advanced form of degradation of the collagen fibers has occurred in the form of gelatinization or cornification, which is easily seen in the SEM microphotographs. These phenomena occur in the case of a prolonged hydration and repeated desiccations at high temperatures. In such cases, a thermal disorganization occurs (caused by the thermal activation energy, which imparts sufficient energy to the water molecules to compete with the hydrogen and the van der Waals bonds, which maintain the triple-helix configuration/helicoid structure), alongside the breaking of certain peptide links or the partial loss of the constitution water of the hydroxyl groups, or those between abutting amine and carboxyl groups.

In terms of degradation dynamics, the process of denaturation of collagen can be caused by inter- or intra-molecular changes at the level of the hydrogen bonds, as well as at the level of the constitutive amino acids, which leads to the destabilization of the ordered structure [22,23].

The same conclusion follows from the significant change in the FTIR spectra, which consists of the shifting of the peaks of amide II from 1550 to 1539 cm^{-1}. The shift toward smaller numbers is due to the irreversible transformation of the native structure (triple helix) of the collagen into a disorganized structure (gelatinization) or a monolithic structure (cornification).

Moreover, the damage of the integrity of the triple helix of collagen is indicated by the ratio of the intensities of the bands from 1434 cm^{-1} (for P1) and 1415 cm^{-1} (for P2) and of amide III, which is

P1: $I(A_{1434})/I(A_{1236}) = 0.144/0.100 = 1.44$
P2: $I(A_{1415})/I(A_{1235}) = 0.117/0.072 = 1.63.$

The value is greater than 1, which demonstrates that the structure of the collagen was modified by cornification compared to the initial one. It is known that a collagen with an intact conformation of the triple helix (the integrity of the collage is unaffected) has a ratio between 1.0 and 0.5 [24,25].

Regarding the composition of the carbonate ion, peaks with values of 1435 cm^{-1} and 874 cm^{-1}, which are characteristic to the absorption bands of calcium carbonate ($CaCO_3$), confirm its presence in these two documents under scrutiny and originate from the treatments used during the technological process. The collagenic support S1 has all the intensities of the absorption bands smaller than the collagenic support S2, meaning that the first parchment is more affected than the second.

The deformation band found at 1079 cm^{-1} is attributed to the oxidation of the lateral chains of amino acids as well as to methionine, while the band at 1030 cm^{-1} can be attributed to δ(CN), which is found in amino acids such as glycine, proline, and hydroxyproline.

Another deformation band found at 712 cm^{-1} can be attributed to the O-C-O grouping from the CO_3^{2-} ion, which is the allotropic form vaterite [26].

The ATR-FTIR spectra of the inks used for writing and decorating the two documents have similar peaks as the supports, with certain lower signals or shifts caused by diffusion of the ink into the support (Figure 12 and Table 2).

For the red pigment, peaks were identified at 1607, 1027, 1017, and 800 cm^{-1}, which are attributed to cinnabar.

For green, the peaks are at 2165, 1411, 1319, 1034, 1027, and 828 cm^{-1}, which are attributed to malachite.

For identifying the charcoal black in the writing and miniature ink, we used the peaks specific to tar remains from 1014 cm^{-1}, where the most intense bands were registered. For the presence of calcium carbonate in the black ink, which can originate from the overlapping with the band of the calcium carbonate from the collagenic support, we identified the peaks at 1412 cm^{-1} and 875 cm^{-1}.

In the case of golden inks used for writing and decorating the two documents on collagenic support, given that gold is a chemical element that does not absorb in IR, the majority of the peaks from the two spectra presented in Figure 12a,b, respectively, have lower absorbance than the other pigments, which is explained by the shielding of the support by the golden film.

4. Conclusions

Regarding the results from the technical analyses, we determined the chemical nature and the state of conservation of the support and inks.

Optical microscopy provided information on the color, porosity, roughness, granulation, and morphology of the microstructural components. Therefore, the analysis of the granular configuration, which was determined from the arrangement of the hair follicles, showed that these two documents were made of calf leather. With respect to the state of conservation of the inks used for writing and decorating these two documents, inks were applied onto the collagen support (first the black ink, then the red one). The degradations (reticular fissures/cracks, detachments of writing material, and the appearance of micro-lacunae) were caused by the fragilization in time of the binder as a result of the physical parameters of the climate, which strayed from the optimal values for storage, and of the different hydrophily of the constitutive materials.

The SEM analysis for the support and inks confirms the conclusion drawn from the OM analysis. By magnifying the areas analyzed, it was possible to highlight the microparticles of calcium carbonate and inorganic salt crystallites left as residues of the processing of the skins from which the parchment was produced.

The EDX analysis assessed the elemental composition of the constitutive materials of these two documents and provided information concerning the technologies employed for producing the supports. Therefore, S1 has a higher quantity of calcium, which is approximately five times higher than S2, meaning that the manufacturing technologies were different: for S1, the process involved the treatment with calcium hydroxide (slake lime), while for S2, the process employed either calcium hydroxide and sodium sulfite, for increasing the speed of removal of the other skin sub-products (epidermis and hypodermis, hair/wool, glandes, etc.).

By corroborating the data from the EDX and the ATR-FTIR analyses, we identified the natural pigments used for producing inks with which these documents were written and decorated: the natural pigment malachite (basic copper carbonate [$2CuCO_3Cu(OH)_2$]) for the green pigment; cinnabar for the red pigment; gold for the golden pigments; the binder of organic origin (gum arabic/xantan and slaked lime as mineral binder); carbon for the writing ink of S1; and in the case of S2, only a small quantity of carbon was added to the ferogallic ink.

The ATR-FTIR spectroscopy provided information regarding chemical changes caused by oxidation, hydrolysis, and denaturation processes of the collagen, which is the base material of the parchment, as well as the chemical changes of inks (particularly the binders). Moreover, it has been found that degradation was enhanced by the denaturation and gelatinization of the collagen, which led to its monolithization by compacting fibers, which present a lattice of three-dimensional fissures and micro-crevices resulting from fracturing during roll-up. These supports also registered dehydrations with the loss of the hygroscopic water, leading to contractions assisted by the elimination of the hydrogen bonds from the structure of the collagen's polypeptide chains.

Author Contributions: Conceptualization, M.B. and I.S.; methodology, V.V. and O.F.; validation, M.B., I.S., V.V. and O.F.; investigation, V.V. All authors have read and agreed to the published version of the manuscript.

Funding: This research is funded by the Ministry of Research and Innovation within Program 1—Development of the national RD system, Subprogram 1.2—Institutional Performance—RDI excellence funding projects, Contract no. 34PFE/19 October 2018. Also this work was co-funded by the European Social Fund, through Operational Programme Human Capital 2014–2020, project number POCU/380/6/13/123623, project title ‹‹PhD Students and Postdoctoral Researchers Prepared for the Labour Marcket!››.

Conflicts of Interest: The authors declare no conflict of interest.

References

1. Mills, J.S.; White, R. *The Organic Chemistry of Museum Object*, 2nd ed.; Butterworth-Heinemann: Oxford, UK, 1994.
2. Oprea, F. *Etiopathogenesis of Artwork and Structural Materials*; OSIM: Bucureşti, Romania, 2010; p. 44.
3. Sandu, I. *Degradation and Deterioration of the Cultural Heritage*; "Al. I. Cuza" University Publishing House: Iaşi, Romania, 2008; Volume II, p. 125.
4. Cicero, C.; Mercuri, F.; Paoloni, S.; Orazi, N.; Zammit, U.; Glorieux, C.; Thoen, J. Integrated adiabatic scanning calorimetry, light transmission and imaging analysis of collagen deterioration in parchment. *Thermochim. Acta* **2019**, *676*, 263–270. [CrossRef]
5. Cucos, A.; Budrugeac, P.; Miu, L.; Mitrea, S.; Sbarce, G. Dynamic mechanical analysis (DMA) of new and historical parchments and leathers: Correlations with DSC and XRD. *Thermochim. Acta* **2011**, *516*, 19–28. [CrossRef]
6. Budrugeac, P.; Badea, E.; Della Gatta, G.; Miu, L.; Comănescu, A. A DSC study of deterioration caused by environmental chemical pollutants to parchment, a collagen-based material. *Thermochim. Acta* **2010**, *500*, 51–62. [CrossRef]
7. Popescu, C.; Budrugeac, P.; Wortmann, F.-J.; Miu, L.; Demco, D.E.; Baias, M. Assessment of collagen-based materials which are supports of cultural and historical objects. *Polym. Degrad. Stab.* **2008**, *93*, 976–982. [CrossRef]
8. Hajji, L.; Seghrouchni, G.I.; Lhassani, A.; Talbi, M.; El Kouali, M.; Bouamrani, M.L.; Yousfi, S.; Hajji, C.; Carvalho, M.L. Characterization of natural degradation of historical Moroccan Jewish parchments by complementary spectroscopic techniques. *Microchem. J.* **2018**, *139*, 250–259. [CrossRef]
9. Schuetz, R.; Maragh, J.M.; Weaver, J.C.; Rabin, I.; Masic, A. The Temple Scroll: Reconstructing an ancient manufacturing practice. *Sci. Adv.* **2019**, *5*, eaaw7494. [CrossRef] [PubMed]
10. Budrugeac, P.; Cucos, A.; Miu, L. The use of thermal analysis methods for authentication and conservation state determination of historical and/or cultural objects manufactured from leather. *J. Therm. Anal. Calorim.* **2011**, *104*, 439–450. [CrossRef]
11. Al-Bashaireh, K.; ElSerogy, A.; Hussein, E.; Shakhatreh, M. Genuine or forged? Assessing the authenticity of a confiscated manuscript using radiocarbon dating and archaeometric techniques. *Archaeol. Anthropol. Sci.* **2017**, *9*, 337–343. [CrossRef]
12. Sandu, I.; Sandu, I.C.A.; Van Saanen, A. *Scientific Expertize of the Art Works*; "Al. I. Cuza" University Publishing House: Iaşi, Romania, 1998; Volume I, p. 34.
13. Oprea, F. *Biodeterioration of Works of Art and Structural Materials*; OSIM: Bucureşti, Romania, 2012; p. 37.
14. Oprea, F. *Manual for the Restoration of Old Books and Graphic Documents*; MNLR: București, Romania, 2009; p. 76.
15. Oprea, F. *Biology for the Conservation and Restoration of Cultural Heritage*; Maiko: Bucureşti, Romania, 2006; p. 33.
16. Oprea, F.; Lungu, C.M. *Preservation and Restoration of Archive Documents*; Fundația România de Mâine: București, Romania, 2008; p. 24.
17. Della Gatta, G.; Badea, E.; Mašic, A.; Ceccarelli, R. *Improved Damage Assessment of Parchment(IDAP) Collection and Sharing of Knowledge*; EC Research Report No. 18; Larsen, R., Ed.; Dictus Publishing: Luxembourg, 2007; p. 68.
18. Boyatzis, S.C.; Velivasaki, G.; Malea, E. A study of the deterioration of aged parchment marked with laboratory iron gall inks using FTIR-ATR spectroscopy and micro hot table. *Herit. Sci.* **2016**, *4*, 13. [CrossRef]
19. Mallamace, F.; Baglioni, P.; Corsaro, C.; Chen, S.-H.; Mallamace, D.; Vasi, C.; Stanley, H.-E. The influence of water on protein properties. *J. Chem. Phys.* **2014**, *141*, 165104. [CrossRef] [PubMed]
20. Barth, A. Infrared spectroscopy of proteins. *Biochim. Biophys. Acta* **2007**, *1767*, 1073–1101. [CrossRef] [PubMed]

21. Derrick, M. *Evaluation of the State of Degradation of Dead Sea Scroll Samples Using FT-IR Spectroscopy;* The Book and Paper Group Annual, American Institite of Conservation: Washington, DC, USA, 1991; Volume 10, pp. 1–13.
22. Badea, E.; Miu, L.; Budrugeac, P.; Giurginca, M.; Mašić, A.; Badea, N.; Della Gatta, G. Study of deterioration of historical parchments by various thermal analysis techniques, complemented by SEM, FTIR, UV-VIS-NIR and unilateral NMR investigations. *J. Therm. Anal. Calorim.* **2008**, *91*, 17–27. [CrossRef]
23. Badea, E.; Poulsen Sommer, D.V.; Muhlen Axelsson, K.; Larsen, R.; Kurysheva, A.; Miu, L.; Della Gatta, G. Damage ranking of historic pachments: From microscopic studies of fibre structure to collagen denaturation assessment by micro DSC. *e-Preserv. Sci.* **2012**, *9*, 97–109.
24. Botta, S.B.; Ana, P.A.; Santos, M.O.; Zezell, D.M.; Matos, A.B. Effect of dental tissue conditioners and matrix metalloproteinase inhibitors on type I collagen microstructure analyzed by Fourier transform infrared spectroscopy. *J. Biomed. Mater. Res. B* **2012**, *100*, 1009–1016. [CrossRef] [PubMed]
25. George, A.; Veis, A. FTIRS in water demonstrates that collagen monomers undergo a conformational transition prior to thermal self-assembly in vitro. *Biochemistry* **1991**, *30*, 2372–2377. [CrossRef] [PubMed]
26. Hajji, L.; Boukir, A.; Assouik, J.; Lakhiari, H.; Kerbal, A.; Doumenq, P.; Mille, G.; De Carvalhoc, M.L. Conservation of Moroccan manuscript papers aged 150, 200 and 800 years. Analysis by infrared spectroscopy (ATR-FTIR), X-ray diffraction (XRD), and scanning electron microscopy energy dispersive spectrometry (SEM–EDS). *Spectrochim. Acta A* **2015**, *136 Pt B*, 1038–1046. [CrossRef]

Publisher's Note: MDPI stays neutral with regard to jurisdictional claims in published maps and institutional affiliations.

MDPI
St. Alban-Anlage 66
4052 Basel
Switzerland
Tel. +41 61 683 77 34
Fax +41 61 302 89 18
www.mdpi.com

Materials Editorial Office
E-mail: materials@mdpi.com
www.mdpi.com/journal/materials

www.ingramcontent.com/pod-product-compliance
Lightning Source LLC
LaVergne TN
LVHW070143170726
843515LV00007B/1855

* 9 7 8 3 0 3 6 5 3 0 9 2 5 *